APPLIED MOLECULAR GENETICS

APPLIED MOLECULAR GENETICS

Roger L. Miesfeld

Professor of Biochemistry
The University of Arizona
Tucson, Arizona

A JOHN WILEY & SONS, INC., PUBLICATION
New York • Chichester • Weinheim • Brisbane • Singapore • Toronto

This book is printed on acid-free paper. ∞

Published simultaneously in Canada.

The cover image is a protein structure of the Y66H blue emission variant of the green fluorescent protein (GFP) from the Pacific Northwest jellyfish. This graphical image was derived with atomic coordinate data deposited in the Brookhaven Protein Data Bank (PDB) by S. James Remington and colleagues at the University of Oregon. The PDB data file, and freeware rendering algorithm used to generate the protein structure were obtained over the Internet using a World Wide Web browser program. Ectopic expression of the jellyfish GFP gene in transfected human cells, in vitro mutagenesis of the GFP chromophore tripeptide coding sequence, and retrieval of GFP protein data through the Internet, are representative examples of molecular genetic applications.

Library of Congress Cataloging-in-Publication Data:

Miesfeld, Roger L.

Applied molecular genetics. / Roger L. Miesfeld.

p. cm.

Includes bibliographical references and index.

ISBN 0-471-15676-0 (pbk. : alk. paper)

1. Molecular genetics—Methodology. 2. Genetic engineering —Methodology. I. Title

QH442.M5425 1999

572.8—dc21 98-29973

Printed in the United States of America.

10 9 8 7 6 5 4 3 2 1

Dedicated to Nicole, Noelle, and R. Barrett

CONTENTS

PREFACE

The fields of biochemistry, genetics, and cell biology have been dramatically altered in the past two decades by the development of an experimental approach that has been called recombinant DNA technology. In the past five years, this laboratory methodology has expanded into a much wider domain of scientific investigation I have chosen to call *applied* molecular genetics. One of the difficulties students have with applied molecular genetic concepts is understanding *how* various molecular genetic techniques are combined to accomplish a research goal. For many students, a conceptual gap exists between the methodological details found in laboratory protocol books and the factual statements contained in biological textbooks. Undergraduate life science students often begin their first hands-on research experience with only a vague idea of what "gene cloning" is all about, and they usually lack a clear understanding of how to design an experimental strategy. *Applied Molecular Genetics* attempts to fill this knowledge gap by presenting key biochemical and cell biological principles using simple descriptive terms and illustrated flow schemes.

The organization of the book was chosen to facilitate its use as a resource for life science students and researchers who need to see the forest among the trees. Toward this end, *Applied Molecular Genetics* is divided into three sections to reflect the level of expertise required by the reader. Section 1 is called Laboratory Techniques and it includes Chapters 1–3. The topics presented in these chapters provide a "nuts and bolts" overview of fundamental concepts in recombinant DNA technology. The material in Section 1 is best suited for readers who have little or no experience performing nucleic acid biochemistry or gene cloning. Section 2, Core Methods, includes Chapters 4–6 and it represents the heart of applied molecular genetic concepts. This material systematically describes the primary approaches and research objectives that characterize three major areas of applied molecular genetics: manipulation of genomic DNA, isolation and characterization of gene coding sequences, and the polymerase chain reaction. Section 2 is a good starting point for readers who have a basic understanding of molecular genetic methods but lack a clear idea of how various experimental tools are best applied to the study of complex biological problems. Section 3, entitled Specialized Applications, comprises Chapters 7–9. Though it would be expected that most life science laboratory researchers would be interested in the core methodologies described in Sections 1 and 2, classroom discussions may benefit more from material in

Section 3 which describes cell culture models, animal transgenesis and medical molecular genetics.

Each chapter concludes with a *laboratory practicum* describing the sequence of steps most often used in the laboratory to implement a specific experimental objective. These laboratory practicums are similar in style to the presentation of medical case studies in that the problem is first described, and then with text and illustrations, a hypothetical flow scheme is presented. An extensive bibliography is also contained in every chapter to guide readers both to landmark publications and to the most current literature in the field of applied molecular genetics. In addition there are seven appendices (standard abbreviations, properties of nucleic acids, amino acids, restriction enzymes, description of genetic markers, gel electrophoresis information, and URL addresses of useful Internet sites) in Section 4 to provide laboratory researchers with essential biochemical information.

The descriptive style of the text makes it an ideal companion book for upper division undergraduate, graduate, and medical school courses utilizing one of the larger molecular biology, biochemistry, or cell biology texts, in which applied molecular genetic concepts are relegated to "shaded boxes" that contain only cursory descriptions of essential techniques. *Applied Molecular Genetics* could also be used as a stand-alone text for elective/special topics courses typically offered to advanced life science students with some prior course work in biochemistry, molecular biology, or cell biology. In this context, the many examples of molecular genetic strategies, extensive primary references, and carefully chosen appendix materials provide the instructor with materials to teach students how to design complex molecular genetic strategies using lab exercises and Internet resources.

ACKNOWLEDGMENTS

The illustrations in this book would not have been possible without the artistic talent and altruistic dedication of my wife, Elizabeth J. Miesfeld. Her interpretation of my hand-drawn sketches, along with the aid of intuitive computer graphics tools from Adobe Systems Inc. and Apple Computer Inc., led to the production of her 123 inviting illustrations. Her goal was to bring a human, nontechnical look to what she often thought was a collection of not very exciting engineering schematics. In my opinion, she succeeded brilliantly. Much credit also goes to the editors and staff at John Wiley & Sons who made this book a reality. I also want to thank my graduate students for the many good ideas they contributed to the laboratory practicums, for their generosity in supplying examples of real data, and for their patience with me. These students include Mark Chapman, Dave Askew, David Whitacre, Unsal Kuscuoglu, Tom Leptich, and Mark DeBoer. Just as importantly, Sharon Pascoe, Susan Kunz, and Mistie Quigley, three excellent lab technicians, deserve special thanks for not only keeping the lab running smoothly, but also for reminding me that difficult concepts can best be understood using real-life examples. I also want to thank my faculty colleagues in the Departments of Biochemistry, Molecular & Cellular Biology, and the College of Medicine at the University of Arizona for giving expert technical advice; Hans Bohnert, Lynn Manseau, John Little, David Mount, Bill Montfort, Carol Dieckmann, Rick Hallick, John Bloom, Fernando Martinez, Don Nelson, John Anderson, and Sue Roberts. Finally, my view of molecular genetics and love of science would not have fully developed without the help of three outstanding mentors, Professors David C. Shepard, Norman Arnheim, and Keith R. Yamamoto.

Roger L. Miesfeld
www.biochem.arizona.edu/miesfeld

SECTION

1

LABORATORY TECHNIQUES

CHAPTER

1

BIOCHEMICAL BASIS OF APPLIED MOLECULAR GENETICS

Molecular genetics employs known principles of DNA structure and function to investigate the molecular basis for genotype directed phenotypes under normal and pathological conditions. The DNA segments most often studied by molecular geneticists are those that encode genes, the smallest unit of genetic heredity. The term *applied molecular genetics* is used here to describe a rapidly growing set of laboratory-based research tools that exploit the information potential of organismal DNA. As an introduction to the field of applied molecular genetics, we begin with a brief review of information transfer from DNA to RNA to protein, highlighting the most relevant concepts. This is followed by a discussion of basic nucleic acid biochemistry and a short description of the most common molecular genetic laboratory reagents and techniques. Laboratory practicum 1 illustrates how to identify the transcriptional start site of a newly isolated gene.

FLOW OF GENETIC INFORMATION: DNA → RNA → PROTEIN

Deoxyribonucleic acid (DNA) and ribonucleic acid (RNA) polymers consist of repeating units of deoxynucleotides and ribonucleotides, respectively. With the exception of some viruses, almost all organisms on this planet store their cellular blueprints for life in double-strand DNA molecules called chromosomes. In eukaryotic cells, chromosomes are copied during cell division, recombined and shuffled as a result of sexual reproduction, and transcribed into complementary RNA molecules through a process called gene expression. Figure 1.1 shows a schematic representation of how chemical information stored in the DNA coding sequences of a gene is transmitted to the protein synthesis machinery in the cell by mRNA "transcripts." This relationship between the DNA, RNA, and protein sequence information of a gene is sometimes referred to as the biochemical flow of genetic information. Although there are a few examples in nature where these simple hierarchical relationships do not hold true, the core principles of applied molecular genetics follow, for the most part, the classic paradigm of DNA → RNA → protein.

To understand fundamental molecular genetic principles, it is useful to think about how chemical information, stored in a simple nucleic acid polymer, could direct the development and maintenance of a living complex organism. First, remember that each

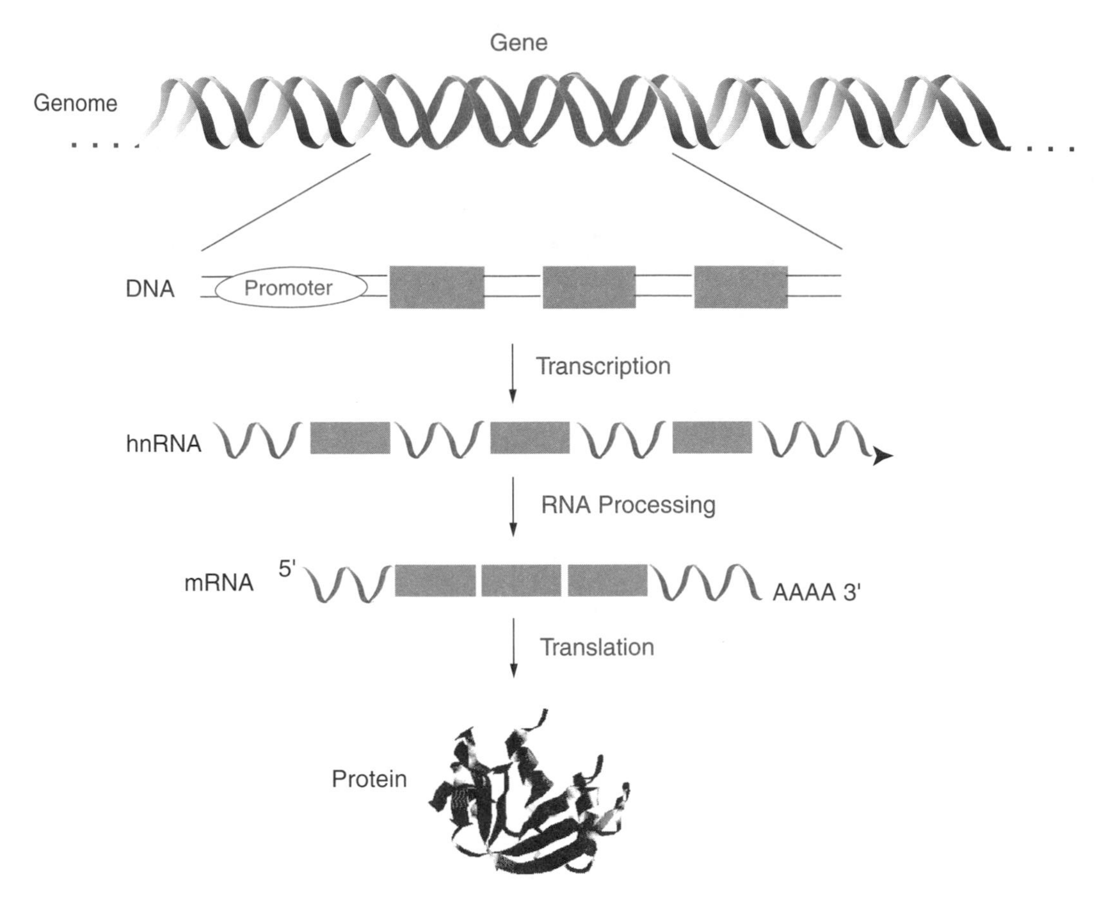

Figure 1.1 Schematic representation of genetic information processing in the cell. A gene is the fundamental unit of information storage and includes both coding sequence and transcriptional regulatory sequences. In eukaryotes, most gene coding sequences are contained within segments of DNA called exons (gray boxes) that are separated by noncoding sequences called introns. Transcriptional initiation begins at the 5′ end of the gene in the promoter region. Following elongation and transcriptional termination, post-transcriptional RNA processing mechanisms fuse exonic coding sequences by RNA splicing and add a 3′ polyadenylate tail (AAAA). Protein synthesis (mRNA translation) occurs in the cytoplasm of eukaryotes, whereas in prokaryotes, transcription and translation are coupled processes.

deoxyribonucleotide unit in DNA contains one of the four bases, guanine (G), adenine (A), cytosine (C), or thymine (T). Second, calculate the total number of sequence combinations that are possible for an oligonucleotide that is just 10 nucleotides long. Because each position in the oligonucleotide could be a G, A, C, or T, there would be 4^{10}, or 1,048,576, different sequence possibilities. Because the amount of DNA in a human cell contains more than 10^9 nucleotides, it is easy to imagine that there is more than enough storage space in the chromosomal DNA "hard drive" of a cell (4^{billion} sequence combinations) to encode someone even as unique as yourself! Nature's ability to encrypt the essence of life in long strings of DNA sequence is therefore not a problem of storage space, but rather how accurately the cell machinery is able to retrieve and interpret this vast amount of information. Over the past three decades, biochemists, geneticists, and cell biologists have been able to decipher the basic components of this information processing in the cell, which together have laid the groundwork for modern molecular genetics. Indeed, applied molecular genetics is the exploitation of this knowledge to investigate and utilize the processes of DNA synthesis (replication), RNA synthesis (transcription), and protein synthesis (translation) not only to access, but also to manipulate the information potential of organismal DNA. Let's briefly review the important characteristics of these three information processing "algorithms" to understand better the key principles of applied molecular genetics.

DNA Synthesis

Cell division requires that DNA be duplicated to produce an exact chromosomal copy. The two major concepts to remember about DNA replication are the following:

1. DNA is a double-strand molecule that is stabilized by hydrogen bonding between complementary base pairs in two antiparallel strands: guanine bonds with cytosine (G-C base pair) and adenine bonds with thymine (A-T base pair).
2. Initiation of DNA synthesis requires a template primer with a free 3′ hydroxyl group and polymerization always proceeds in the 5′ to 3′ direction.

RNA Synthesis

RNA synthesis, also known as DNA transcription, requires RNA polymerase enzymes that use single-strand DNA as a template to make complementary copies of information stored in the DNA sequence. RNA polymerases synthesize RNA in the 5′ to 3′ direction just as DNA polymerases do. However, there are two important differences between RNA and DNA synthesis. First, uridine is the ribonucleotide base that pairs with adenine (rather than thymine), and second, RNA polymerases do not require a template primer.

It is important to keep in mind the following points about RNA synthesis:

1. RNA synthesis is required for the transcription of DNA information contained within the genetic unit called a gene. RNA synthesis begins at a specific initiation site on the 5′ end of a gene (upstream) and terminates at the 3′ end (downstream).
2. Multiple copies of short-lived RNA molecules are synthesized from a single DNA template; the number of RNA transcripts synthesized per unit time depends on the rate of transcriptional initiation by RNA polymerase.

Protein Synthesis

The primary cellular machine involved in protein synthesis, also called translation, is the ribosome. This very abundant macromolecule contains a well-characterized arrangement of large ribosomal RNA molecules and numerous ribosomal proteins. Three main concepts in protein synthesis must be emphasized:

1. The DNA sequence of a gene, as faithfully copied into mRNA, contains information for protein synthesis in the form of triplet codons. There are 64 possible

triplet codons, of which 61 can specify the 20 amino acids (with redundancy), and three correspond to termination codons. This is called the genetic code, a copy of which is printed on the inside cover of the book.

2. The 5′ end of mRNA directs ribosome binding and the subsequent initiation of protein synthesis. The first amino acid, usually the methionine codon AUG, corresponds to the amino terminus (*N*-terminus) of the encoded protein. The ribosome "reads" the mRNA in the 5′ to 3′ direction until reaching the penultimate codon in the mRNA, which specifies the carboxy terminal amino acid. The ribosome disengages at the subsequent termination codon.
3. Because the genetic code is based on triplets, and the 5′ end of mRNA contains nucleotides upstream of the initiator methionine codon to accommodate ribosome binding, the ribosome can theoretically begin translation in any of three possible protein coding registers called "reading frames."

NUCLEIC ACID BIOCHEMISTRY

To understand many of the principles underlying applied molecular genetics, it is important to be familiar with two chemical properties that affect the behavior of DNA and RNA in solution: (1) the molecular forces that affect the structure of nucleic acid polymers and (2) the kinetic parameters that determine rates of denaturation and renaturation of complementary heteroduplexes. In addition, because many of the starting points for molecular genetic methods require the design of custom oligonucleotides, we examine the basic steps required to produce nucleic acid polymers synthetically using solid support chemistry.

Structure of Nucleic Acid Polymers

The chemical structure of a DNA–RNA heteroduplex is shown in Figure 1.2. The key features to note are (1) the phosphodiester linkages between repeating nucleotide units; (2) the antiparallel polarity of the DNA–RNA heteroduplex, such that the 5′ to 3′ DNA strand is base paired with the 3′ to 5′ RNA strand; and (3) the complementary DNA and RNA strands joined by hydrogen bonding between T-A, G-C, and A-U base pairs. The chemical structure of the bases present in DNA (G, C, A, T) and RNA (G, C, A, U), allow for the formation of hydrogen bonds between opposing purine (G and A) and pyrimidine (C, T, and U) bases. Importantly, the number of hydrogen bonds formed between G-C base pairs is three, whereas only two hydrogen bonds are formed between A-T and A-U base pairs. This difference in hydrogen bonding capacity contributes directly to the thermal stability of double-strand DNA molecules.

Noncovalent interactions are responsible for the three-dimensional structure of the DNA double helix. The two primary sources of noncovalent interactions are the base to base hydrogen bonds formed between antiparallel strands, and the hydrophobic interactions that occur between adjacent bases on the same strand (van der Waals interactions) (Fig. 1.3). Although double-strand DNA appears rigid in molecular models, it is actually quite flexible in the presence of DNA binding proteins, which can bend DNA helices to angles greater than 100 degrees. The ability of DNA strands to dissociate and reassociate locally is critical to its function as a molecular database. DNA replication, recombination, and transcription require unwinding of the DNA double helix, which is

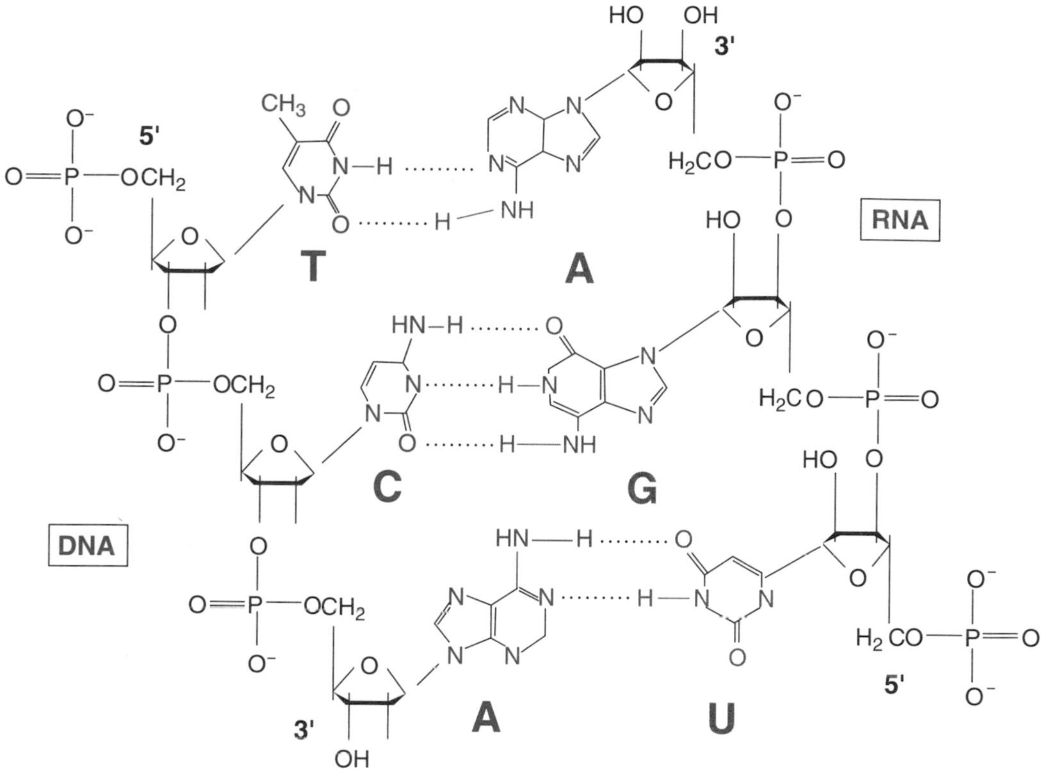

Figure 1.2 A DNA–RNA heteroduplex is formed during RNA synthesis and in the priming step of DNA replication. The bases have been rotated relative to the phosphate backbone to illustrate the hydrogen bond formation between base pairs. Note that the RNA strand has 2′-OHs on the ribose as well as uridine in place of thymine, and that purines (G, A) hydrogen bond with pyrimidines (C, T, U). Noncovalent interactions between G and C residues are more stable than A-T or A-U base pairs because of the extra hydrogen bond that is formed.

accomplished in vivo by specialized helix-destabilizing proteins. Pure double-strand DNA can be unwound in vitro using elevated temperatures or chemical denaturants as described in the next section.

Denaturation and Renaturation of Nucleic Acid Duplexes

The relative amount of single- or double-strand DNA in solution can be experimentally determined using spectrophotometery to measure ultraviolet light absorbance at a wavelength of 260 nanometers (optical density, OD_{260}). The aromatic bases in DNA are less accessible to ultraviolet light in the double-strand, compared to single-strand form, which creates a measurable difference in the observed OD_{260}. Using this empirical difference in absorbance, it is possible to observe the effect of temperature on DNA structure by monitoring OD_{260} over a temperature range of 55–90°C. Figure 1.4 shows a melting curve of double-strand DNA that demonstrates that the amount of denatured DNA rises sharply over a narrow temperature range, indicating that denaturation is a

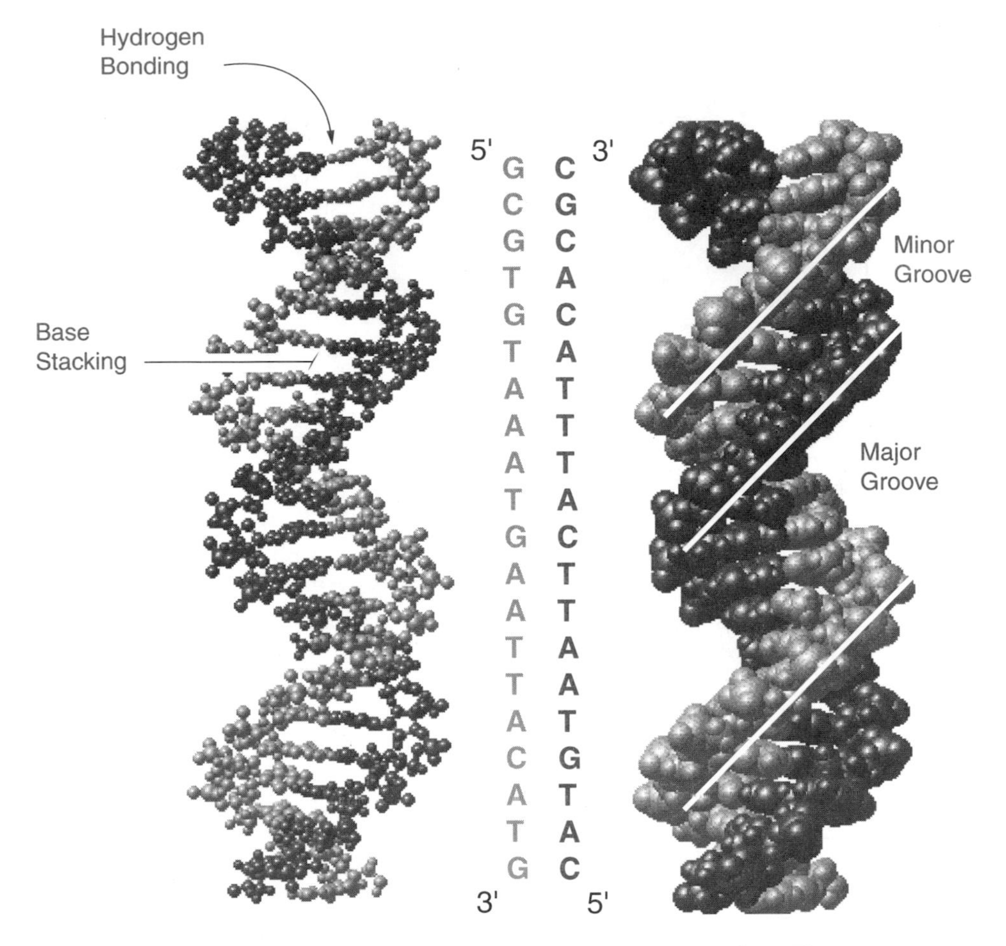

Figure 1.3 The molecular structure of a DNA double helix in the B form. The phosphodiester backbone is antiparallel and the bases from each strand are noncovalently bound through hydrogen bonding. Additional hydrophobic interactions between adjacent bases on the same strand also contribute to helix stability. DNA binding proteins associate with the double helix through noncovalent interactions with bases in portions of the DNA structure called the major and minor grooves.

cooperative process. This cooperativity indicates that once the DNA hybrid has been locally disrupted, it requires only a small amount of additional energy to separate the two strands completely. The temperature at which 50% of the DNA is denatured is called the T_m or melting temperature.

The T_m of a nucleic acid duplex is strongly affected by three factors: (1) base composition, (2) duplex length, and (3) ionic strength of the solution. Base composition is an important determinant of the T_m because G-C base pairs contain one more hydrogen bond than A-T base pairs. Therefore, duplex molecules that contain a high G-C content are more stable and have a higher T_m than do A-T-rich molecules. Duplex length of the hybrid affects the T_m because the overall stability of a double-strand molecule is directly proportional to the number of base pairs. This is especially evident for duplex molecules containing less than 150 consecutive base pairs. Two molecular genetic applications

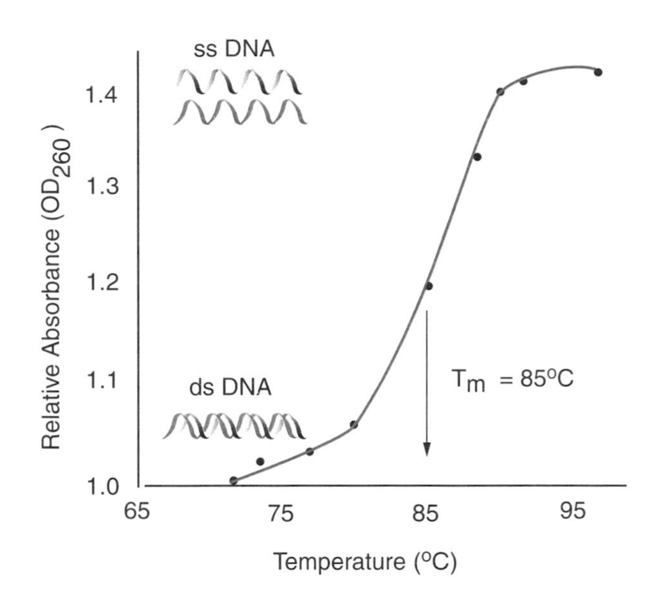

Figure 1.4 Helix denaturation can be monitored by recording the OD_{260} of a DNA solution over a range of temperatures. The T_m is the temperature at which 50% of the DNA is denatured. Duplexes are mostly double-strand (ds) below the T_m and single-strand (ss) above the T_m. Note that the denaturation of DNA is a cooperative reaction as seen by the large increase in absorbance over a narrow temperature range.

where extent of duplex formation is an important consideration are the use of short oligonucleotides in hybridization reactions and heteroduplex formations between molecules that are less than 100% complementary. The use of homologous, but not identical, DNA molecules in hybridization reactions is common when sequence divergence exists between two genes, for example, across species or among members of a related gene family. The T_m for heteroduplexes decreases by approximately 1°C for every 1% sequence mismatch. The third factor influencing the T_m of a given heteroduplex is the ionic strength of the solution. In high Na^+ concentrations (1*M*) the T_m is increased owing to electrostatic shielding of the negative phosphate charges in the DNA backbone, whereas in low Na^+ concentrations (0.1*M*) the T_m is decreased. It is possible to approximate the T_m of a short complementary oligonucleotide (10–20 bases) in a solution containing 1*M* Na^+ using the empirical formula:

$$T_m\,(^\circ C) = 2(\text{number of A} + \text{T}) + 4(\text{number of G} + \text{C})$$

The reverse of nucleic acid denaturation is renaturation, also referred to as reassociation or hybridization. This bimolecular process is most affected by temperature, ionic strength, molar concentration of the two complementary strands, and reaction time. Two other factors that can be introduced experimentally are the effect of denaturing agents such as formamide or urea, both of which lower the T_m, and the inclusion of dextran sulfate in hybridization reactions, which increases the rate of reassociation by as much as 10-fold. Under optimal conditions of temperature and ionic strength, which is usually 10–15°C below the T_m in a solution containing 0.2*M* Na^+, the concentration of nucleic acid becomes the rate-limiting step in the hybridization reaction. The term C_0t is used to describe the kinetics of hybridization between two nucleic strands in solution and is defined by the product of [nucleic acid] × (time). Put simply, when the concentration of two complementary strands in a solution is high, it takes a shorter time for hybridization to occur than it does when one or both of the strands are present at a low concentration.

C_0t curves plot percent reassociation versus C_0t (mole-seconds/liter) and are used to measure the sequence complexity of DNA samples (Fig 1.5). DNA from organisms with

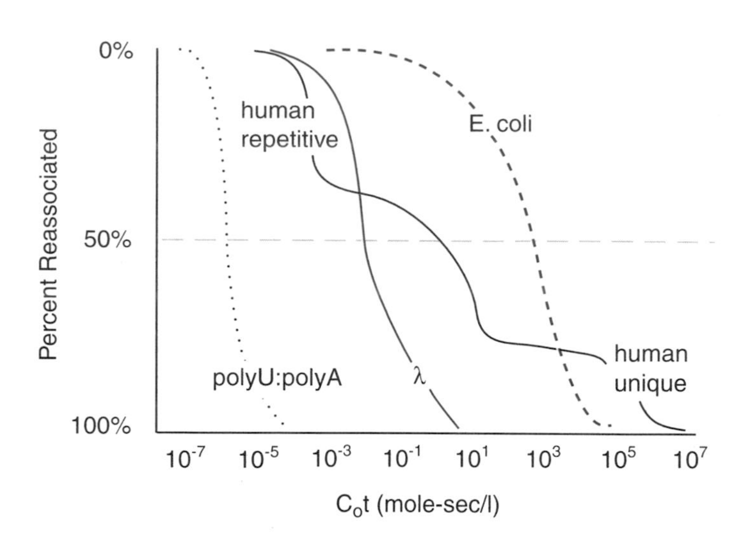

Figure 1.5 C_0t curves show the relationship between sequence complexity and hybridization kinetics. In this curve, a duplex molecule of polyU:polyA has the lowest sequence complexity and reassociates at the lowest C_0t value. Note that human genomic DNA contains repetitive sequence elements that reassociate at low C_0t values and unique single-copy gene sequences that reassociate at very high C_0t values. Lambda (λ) is a bacteriophage that has a genome size of ~50 kb.

small genomes have low sequence complexity and reassociate at much lower C_0t values than do denatured DNA samples from more complex organisms.

One way to think about how C_0t values affect hybridization kinetics, and how this relates to sequence complexity, is to imagine that at the same total DNA concentration, fragments of λ bacteriophage DNA require significantly less time to sort through all possible complementary strands than do similar sized fragments of *E. coli* DNA that are derived from a genome that is 1000 times larger. The C_0t curve obtained from the analysis of human DNA is a mixture of curves (Fig. 1.5). This irregularity is due to the abundance of repetitive DNA sequences that reassociate at low C_0t values, compared to the unique DNA sequences representing single-copy human genes that require much longer times to reassociate owing to their very low concentrations.

Table 1.1 summarizes factors that affect denaturation and renaturation of nucleic acid duplexes.

Chemical Synthesis of DNA and RNA

The ability to synthesize single strands of DNA or RNA using solid support chemistry has had a dramatic impact on the development of applied molecular genetic methods.

TABLE 1.1 Factors that affect the denaturation and renaturation of nucleic acid duplexes

Parameter	Effect on T_m	Effect on rate of renaturation
Base composition	↑ T_m with ↑% G-C	No effect
Hybrid length	↑T_m with ↑length >500 bp; no effect on T_m	↑Rate with ↑length
Ionic strength	↑T_m with ↑[Na^+]	Optimal at 1.5 M Na^+
% bp mismatch	↓T_m with ↑% mismatch	↓Rate with ↑% mismatch
DNA concentration	No effect	↑Rate with ↑[DNA]
Denaturing agents	↓T_m with ↑[formamide], [urea]	Optimal at 50% formamide
Temperature	Not applicable	Optimal at 20°C below T_m

(a)

Dimethoxytrityl

Deoxyadenosine

Phosphoramidite

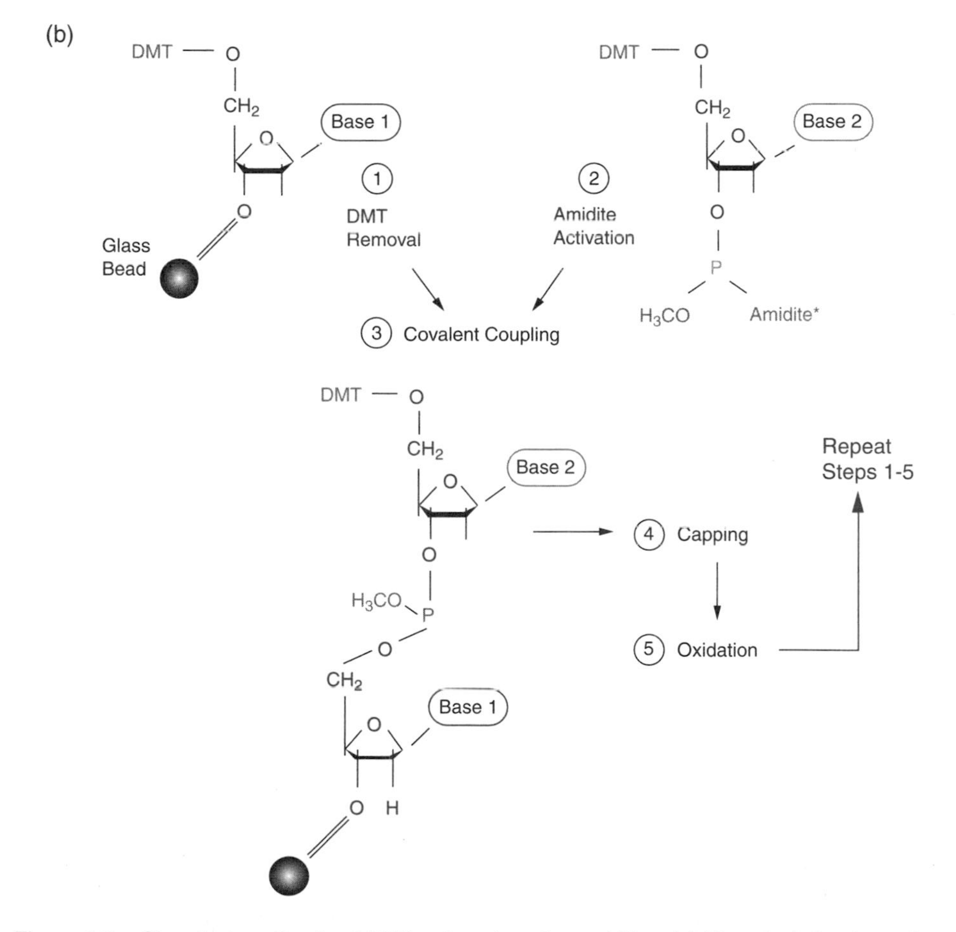

Figure 1.6 Chemical synthesis of DNA using phosphoramidites. (a) Chemical structure of a phosphoramidite showing the dimethoxytrityl (DMT) blocking group on the 5′ carbon and the methylated 3′-phosphite and diisopropylamine groups attached to the 3′ carbon. (*b*) Five sequential steps are required for each cycle of nucleotide extension.

Custom-designed oligonucleotides are available commercially and are used routinely in numerous experimental procedures. For example, oligonucleotides are used as template primers in DNA sequencing and PCR reactions (Chapter 6) and for the incorporation of site-specific mutations in cloned genes (Chapter 3). In addition, chemically modified ribonucleotides can be used to synthesize large quantities of RNA for use as "antisense" inhibitors of RNA function (Chapter 7). Figure 1.6 outlines the basic steps required for in vitro DNA synthesis using the phosphoramidite method.

In vitro DNA synthesis reactions take place inside sealed columns that contain glass beads that serve as the solid support for the sequential chemical reactions. Single phosphoramidites for each of the four bases are added to a growing chain that is initiated at the 3′ end. The five chemical steps required for each nucleotide addition are (1) deblocking the 5′ end by DMT removal, (2) amidite activation of the incoming phosphoramidite, (3) coupling of the nucleotides through a 5′–3′ linkage, (4) capping of unreacted nucleosides to prevent extension of incomplete products, and (5) oxidizing the phosphate triester to stabilize the 5′-3′ linkage. After the required number of linkages are formed through repeated cycling, the oligonucleotide products are demethylated, released from the column, and chemically treated to produce a population of 5'-hydroxylated molecules. The efficiency of each coupling step in the reaction is critical and must be >98% to produce significant yields of a full-length product.

DNA METABOLIZING ENZYMES

Enzymes that modify and metabolize nucleic acids are essential tools for many applied molecular genetic methods. The commercial availability of these enzymes has led to the development of molecular genetic "enzyme kits," which can often be useful components of molecular genetic research strategies. However, as many researchers who use these molecular genetic kits will attest to, it can sometimes be difficult to troubleshoot a failed experiment when the protocol reads "combine equal volumes of solution A (red cap) with buffer B (yellow cap) and incubate for 30 minutes at room temperature in 1/10 volume of enzyme reaction mix C (blue cap)." Therefore, when using DNA metabolizing enzymes, it is important to have an understanding of the function (and limitation) of each enzyme and to follow standard biochemical laboratory practices to optimize enzyme activity. Three classes of enzymes are described here that represent the primary biological reagents for the most common molecular genetic applications: sequence-specific DNA restriction enzymes, ligases and kinases, and DNA and RNA polymerases. Nucleases are another important class of nucleic acid metabolizing enzymes that are described at the end of this chapter as reagents in laboratory practicum 1.

Sequence-Specific DNA Restriction Enzymes

Biochemists discovered more than 30 years ago that most bacteria contain endonucleases that degrade the DNA genomes of infectious bacteriophages. They found that the host bacteria are "immune" to these nucleolytic enzymes because of site-specific DNA methylations that prevent endonuclease attack on the bacterial genome. This form of bacterial immunity requires two distinct enzymatic activities: (1) restriction endonucleases that cleave double-strand DNA at specific sites, and (2) DNA methylases that modify bases at these same sites in the host cell genome (Fig. 1.7). Different species of

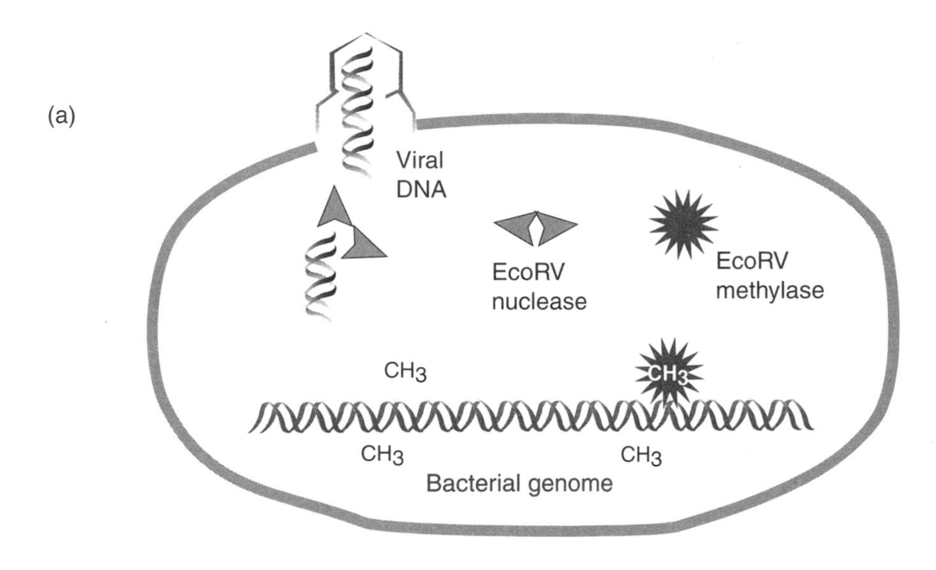

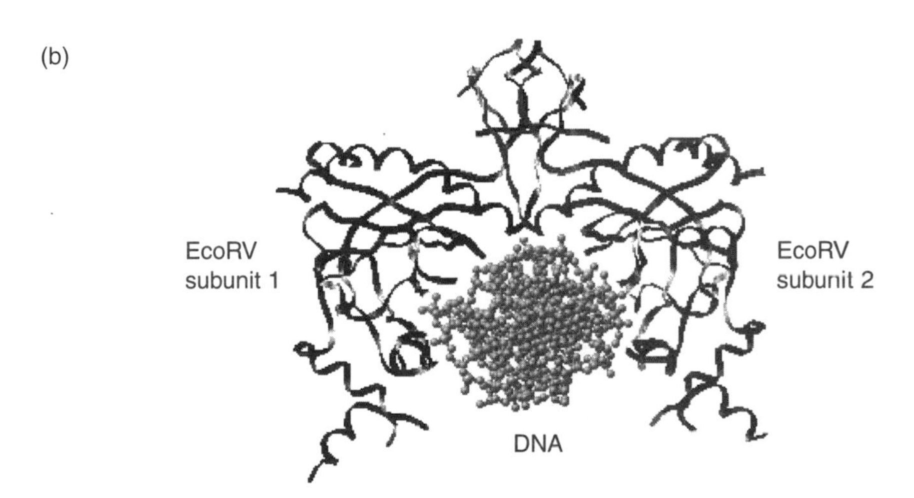

Figure 1.7 Restriction enzymes are site-specific endonucleases that cleave double-strand DNA. (*a*) Restriction enzymes and their corresponding methylases function in bacteria to protect against bacteriophage infection. (*b*) Type II restriction enzymes bind to specific DNA sequences as homodimers and produce a double-strand break in the phosphodiester backbone.

bacteria contain their own sets of endonucleases and corresponding methylases. The term "restriction" refers to the function of these enzymes in restricting the host range of bacteriophage infection.

Based on mechanistic differences between several types of bacterial restriction systems, three classes of restriction enzymes have been described. Type II restriction enzymes are used in molecular genetic applications because they can be used in vitro to recognize and cleave within DNA sequences typically consisting of four to eight nucleotides. By using DNA lacking the specific methylations for a given restriction enzyme (e.g., eukaryotic DNA or plasmid DNA obtained from a methylase-deficient bacterial strain), it is possible to cleave any DNA molecule that contains the recognition sequence for a particular Type II restriction enzyme.

A large number of restriction enzymes have been characterized and shown to bind as dimers with high affinity to specific DNA sequences (Fig. 1.7). The DNA recognition sites for most all restriction enzymes are palindromes and double strand cleavage produces a 5′ phosphate and 3′ hydroxyl at the DNA termini (Fig. 1.8). There are three types of restriction enzyme cleavage reactions, two of which result in the formation of staggered or "sticky" ends having a 5′ or 3′ extended terminus, and a third that results in flush or "blunt" DNA termini on both strands. Applications of restriction enzymes to recombinant DNA cloning methods are discussed in Chapter 2 and biochemical properties of the most common restriction enzymes are listed in Appendix D.

Ligases and Kinases

Ligases and kinases are another class of enzymes that play an important role in recombinant DNA methodologies. DNA ligases catalyze the formation of 5′–3′ phosphodiester bonds in double-strand DNA molecules. In vivo, ligases function in DNA replication to repair single-strand nicks resulting from DNA repair processes, and to join adjacent Okasaki DNA fragments produced in the lagging strand of a DNA replication fork. The T4 bacteriophage DNA ligase is an ATP-dependent ligase that is com-

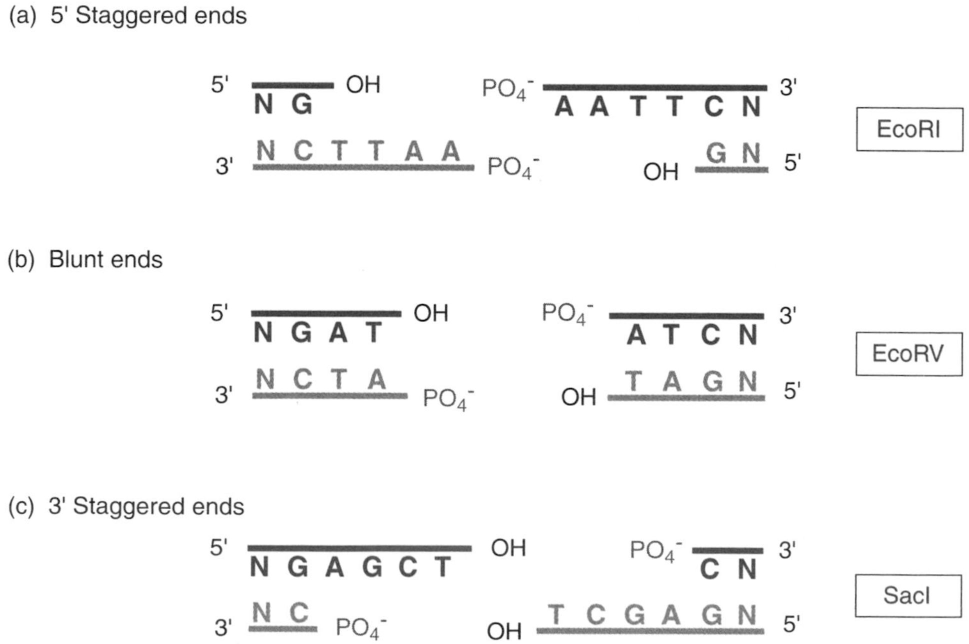

Figure 1.8 Restriction enzymes cleave palindromic DNA sequences to produce double-strand breaks. Restriction enzyme cleavage results in the formation of 5′ PO_4^- and 3′ OH termini with (*a*) 5′ staggered ends, (*b*) blunt ends, or (*c*) 3′ staggered ends.

monly used in DNA cloning strategies to "ligate" two DNA fragments (Fig. 1.9). *E. coli* DNA ligase is also used for some applications, however, it differs from T4 DNA ligase in that it uses NAD for a co-factor and cannot efficiently ligate blunt-end fragments.

Kinases are enzymes that phosphorylate specific substrates by covalently attaching the γ phosphate from ATP to a reactive group on the target molecule. T4 polynucleotide kinase is an enzyme that phosphorylates 5′ hydroxyl termini on DNA and RNA (Fig. 1.10). T4 polynucleotide kinase is often used to label DNA radioactively using [γ-^{32}P]ATP for the purpose of making high specific activity radioactive probes. This

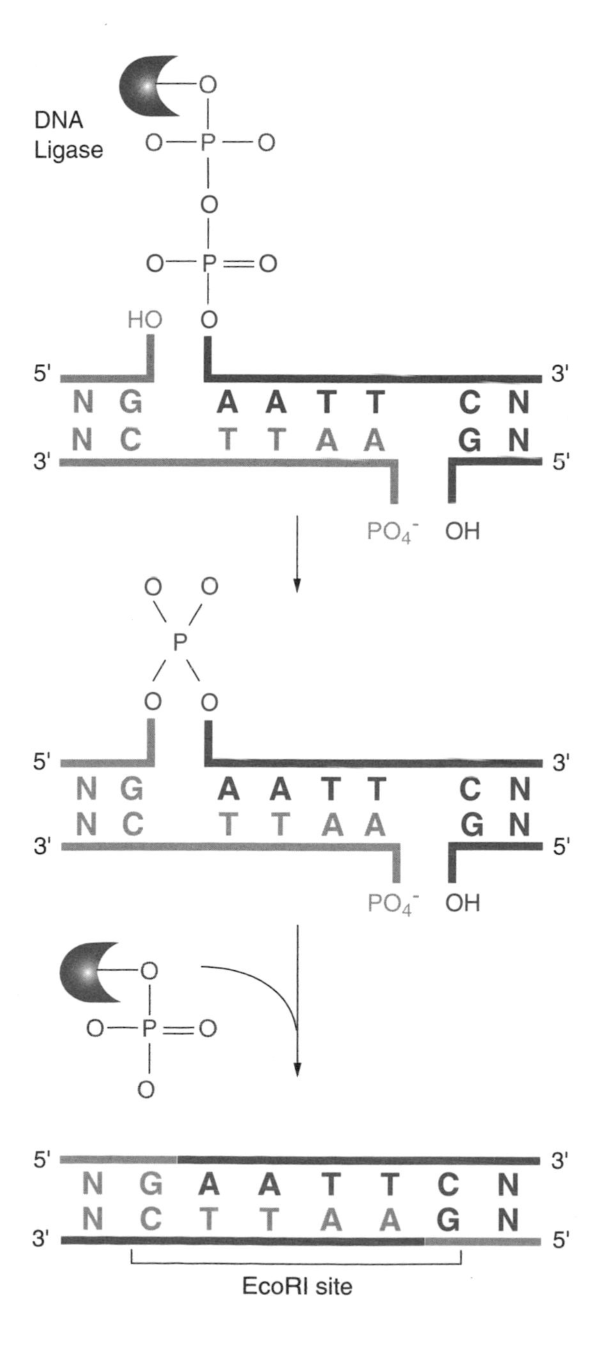

Figure 1.9 DNA ligases catalyze the formation of a phosphodiester bond in nicked double-strand DNA. A ligase–AMP complex forms a transient intermediate with the 5′ phosphate initiating a nucleophilic attack on the 3′ hydroxyl group. The reaction shown here illustrates the two-step ligation of heterologous EcoRI DNA fragments producing covalently closed double-strand DNA.

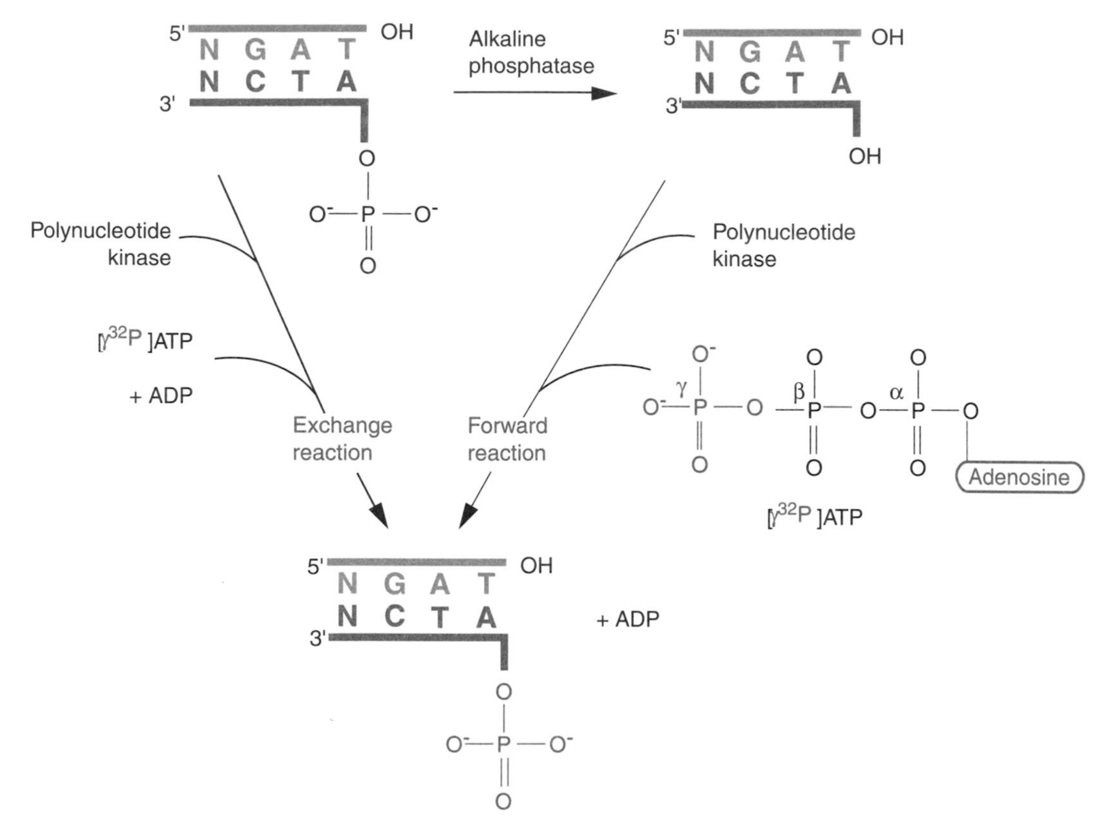

Figure 1.10 T4 polynucleotide kinase can be used to label radioactively the 5′ ends of DNA in a reaction using [γ-^{32}P]ATP. DNA termini with 5′ phosphates can be dephosphorylated by the enzyme bacterial alkaline phosphatase to produce a 5′ hydroxyl that is the optimal substrate for polynucleotide kinase. In addition to the forward reaction, phosphorylated DNA can also be labeled with T4 DNA kinase in a two-step exchange reaction utilizing unlabeled ADP as an intermediate.

enzyme can also be used to phosphorylate synthetic DNA for certain DNA cloning strategies that utilize oligonucleotides. Bacterial alkaline phosphatase is an enzyme that removes the 5' phosphate from DNA termini. Treatment of DNA with alkaline phosphatase is used to dephosphorylate DNA termini prior to labeling with T4 polynucleotide kinase and [γ-^{32}P]ATP or as a strategy to prevent re-ligation of vector DNA (Chapter 2). Polynucleotide kinase can also be used in a two-step exchange reaction to label 5′ phosphorylated DNA with [γ-^{32}P]ATP in the presence of excess ADP (Fig. 1.10).

DNA and RNA Polymerases

DNA and RNA polymerases direct the synthesis of complementary nucleic acids using single-strand DNA as the template. DNA synthesis requires a preexisting DNA or RNA primer with a 3′ hydroxyl, whereas RNA synthesis can initiate synthesis de novo. The in vitro enzymatic synthesis of DNA and RNA has become a central component in a variety of molecular genetic applications, for example, the amplification of DNA sequences using the polymerase chain reaction (PCR). The polymerase chain reaction requires the use of a thermostable DNA polymerase called *Taq* DNA polymerase, which has optimal activity at 75°C (Chapter 6). One very important use of DNA polymerases in molecular genetic applications is for determining the nucleotide sequence of cloned DNA. This is done using a modified bacteriophage T7 DNA polymerase, called Sequenase™, in an in vitro reaction that contains dideoxynucleotides that serve as chain terminators. Other uses of DNA polymerases are for the production of complementary DNA (cDNA) using reverse transcriptase (Chapter 5) and for radioactively labeling DNA fragments with the Klenow fragment of *E. coli* DNA polymerase I.

RNA polymerases are used to synthesize single strand radioactive RNA probes using cloned DNA as a template and to produce mRNA suitable for in vitro protein synthesis or microinjection into cells. As with the DNA polymerases, commercially available bacteriophage RNA polymerases are especially useful in molecular genetic methods because of their high specific activity.

BIOCHEMICAL METHODS TO STUDY DNA AND RNA

Purification of Nucleic Acids

The most common step in any nucleic acid procedure is precipitation of DNA and RNA with ethanol in the presence of monovalent cations. If contaminating proteins must be removed prior to nucleic acid precipitation, then the sample is first extracted with an organic solvent such as phenol. Ethanol precipitation of nucleic acids from aqueous solutions yields a white or clear pellet that can easily be dissolved in an appropriate buffered solution such as a Tris-EDTA (pH 7). EDTA chelates Mg^{2+}, which functions as a co-factor for DNA nucleases (DNases) that may be present at very low levels in the sample. DNA and RNA can be separated using cesium chloride buoyant density gradient centrifugation in the presence of the DNA intercalating dye ethidium bromide. This type of density gradient is used routinely to separate different forms of DNA molecules based on the amount of ethidium bromide absorbed. Ethidium bromide intercalation causes local unwinding of the DNA helix, which reduces the molecular density of DNA. As shown in Figure 1.11, large linear genomic DNA molecules can bind more ethidium bromide than supercoiled plasmid DNA owing to the difference in topology of these two molecules. As described in Chapter 2, when plasmids are propagated in bacteria, topoisomerases introduce supercoils into circular DNA molecules. Because supercoiled plasmid DNA cannot be easily unwound, much less ethidium bromide can intercalate. In contrast, linear and relaxed circular DNA molecules are not as topologically constrained and can therefore absorb more ethidium bromide molecules, resulting in an overall decrease in molecular density.

It is also possible to separate DNA and RNA of different molecular weights using direct physical methods, such as size exclusion column chromatography and gel electrophoresis. Affinity matrices have also been developed using silica gels or anion exchange resins that preferentially bind nucleic acids under appropriate conditions and allow for the removal of proteins and polysaccharides from DNA preparations using crude cell lysates.

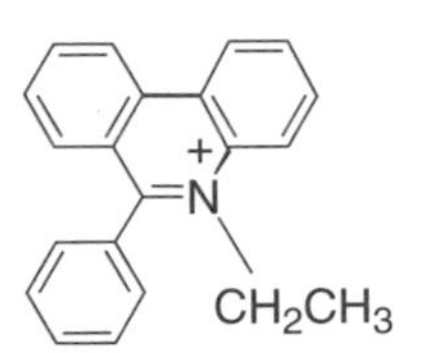

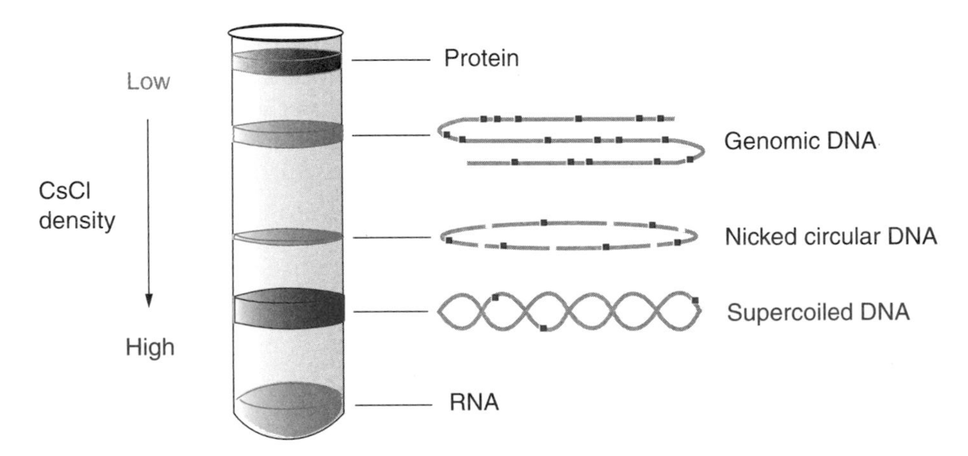

Figure 1.11 Cesium chloride density gradients can be used to separate genomic DNA and plasmid DNA based on differential buoyant densities resulting from the amount of ethidium bromide absorbed. The chemical structure of ethidium bromide is shown.

Gel Electrophoresis

The separation of nucleic acids by electrophoretic mobility is used for both analytical and preparative methods. DNA and RNA molecules are negatively charged owing to the phosphate backbone and the polymer structure results in a constant charge to mass ratio. Therefore, in a uniform electric field, nucleic acids migrate through a solid support matrix toward the positively charged anode at a rate that is inversely proportional to the $\log_{10}$ of the molecular weight. Physical measurements indicate that nucleic acids move through gel matrices as rods and that electrophoretic mobility of circular double-strand DNA can be affected by molecular topology and by the amount of superhelicity (supercoiled molecules migrate faster than relaxed circles). Two types of matrices, agarose and acrylamide, are used to separate DNA and RNA by gel electrophoresis. Because of the differences in pore size of these two gel matrices, standard agarose gels are used to separate nucleic acid molecules of 0.2–10 kb, and acrylamide gels are best suited for resolving nucleic acids less than 500 nucleotides long (see Appendix F).

Agarose is a purified linear polysaccharide polymer derived from a red seaweed that is commonly harvested for commercial applications. Liquid agarose gel is made by adding powdered agarose to a solution of electrophoresis buffer and heating to the boiling point to produce a homogeneous mixture. This solution is then poured into a Plexiglas gel support system and allowed to cool, resulting in the formation of a horizontal slab gel containing slots at one end for sample loading (Fig. 1.12). Platinum electrodes are connected to a power supply and an electric field is established (~5 V/cm) in the presence of a Tris-acetate (or Tris-borate) electrophoresis buffer. Tracking dyes such as bromophenol blue or xylene cyanol are loaded with the samples to monitor electrophoretic mobility. Agarose concentration (mass/volume) determines the pore size and thus differentially affects the migration of small and large molecules. Applied voltage across the gel determines the current (field strength), which directly affects the velocity of electrophoretic mobility against a constant resistance. Nucleic acids are visualized by staining the gel with ethidium bromide, which fluoresces when exposed to ultraviolet light.

The other commonly used gel matrix for separating nucleic acids is polyacrylamide. Polyacrylamide, $[CH_2 = CHCONH_2]_n$, is formed by cross-linking chains of acrylamide with methylenebis-acrylamide $(CH_2 = CHCONH_2)_2CH_2$ in the presence of ammonium persulfate to generate free radicals, and TEMED (*N,N,N′,N′*-tetramethylethylenediamine), which stabilizes the free radicals and sustains the chemical reaction. An acrylamide matrix is formed with a porosity that is determined by both the concentration of acrylamide in the gel and the ratio of acrylamide to cross-linking agent (see Appendix F).

DNA Sequencing

A fundamental component of most applied molecular genetic strategies is a working knowledge of the DNA sequences being utilized in the approach. The nucleotide sequences of all cloned DNA fragments that have ever been published are archived in a large database managed by the U.S. government called GenBank. This database can be readily accessed through the Internet using the World Wide Web (Chapter 9). Up until

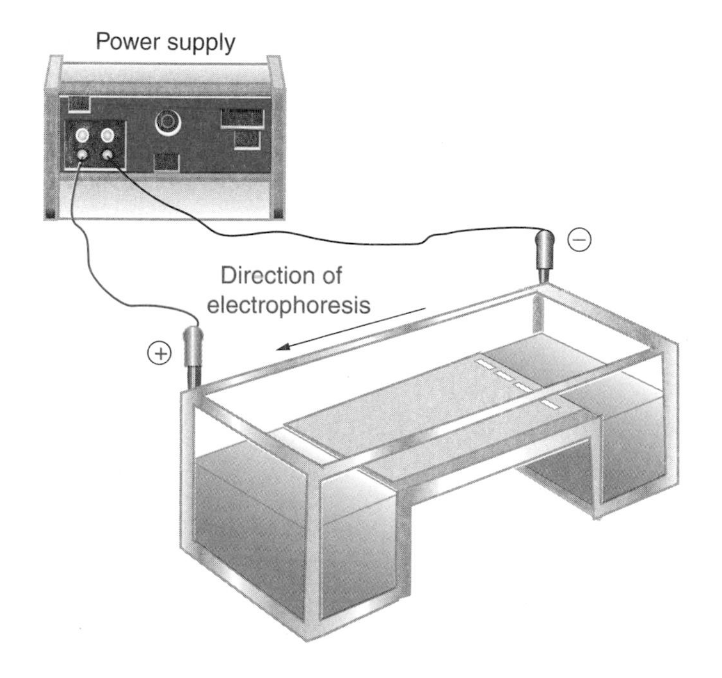

Figure 1.12 Horizontal agarose gel electrophoresis can be used to separate DNA molecules in the range of 0.2–10 kb; it is performed using a simple Plexiglas apparatus and an alternating current power supply. Nucleic acids can be visualized with ultraviolet light in agarose gels that have been stained with ethidium bromide.

the late 1980s, scientific journals still printed the actual sequence of cloned genes, and this information was then scanned into the database by National Institutes of Health (NIH) staff members. However, since about 1990, all newly characterized DNA sequences have been electronically deposited directly into GenBank through the Internet. These GenBank sequences are referred to by a specific database file number included in the published article.

There are two basic reasons why a researcher would need to determine the DNA sequence of a cloned DNA segment. First, if the experimental approach is aimed at characterizing genes or contiguous regions of genomic DNA that have not been studied before, the sequence of the unknown DNA would have to be determined so that it could be deposited into GenBank, which also allows other researchers to access the sequence information using a variety of computer algorithms (Chapter 9). Second, recombinant DNA fragments sometimes must be sequenced to (1) confirm the arrangement of cloned segments in a plasmid vector, (2) screen for sequence alterations introduced by mutagenesis, or (3) identify a gene product by comparing the sequence information with the GenBank database. Regardless of the objective, most DNA sequencing is now performed by automated DNA sequencing instruments that are run as an out-service, much like oligonucleotide synthesis services. The sequence information is returned to the user by E-mail or on a computer disk. Some DNA sequencing is still performed manually in the lab, however, it is much more cost-effective and labor-saving to have the sequence determined by a centralized facility using high-throughput instrumentation (Chapter 9).

Biochemical methods for DNA sequencing were developed in the 1970s by two groups. The chemical cleavage reaction was developed by Allan Maxam and Walter Gilbert, and the chain termination method was described by Fred Sanger, the same biochemist who worked out *N*-terminal peptide sequencing in the 1950s. Because the Sanger sequencing strategy is more amenable to automation and can provide more sequencing information per reaction than the Maxam and Gilbert method, this enzyme-based method has become the standard procedure. Both methods are based on producing a pool of single-strand DNA molecules that all have the same 5′ end, but differ in length by one nucleotide owing to random 3′ ends that have been generated in vitro. Because base-specific reactions are used, it is possible to determine the DNA sequence of the starting material by separating the fragments on an acrylamide gel using electrophoresis. Radioactive- or fluorescent-labeling methods are utilized to identify DNA molecules that share the same 5′ end. Figure 1.13 illustrates how a set of chain termination reactions utilizing dideoxynucleotides (ddNTPs) lacking the necessary 3′ OH group required for elongation are used to determine the complementary sequence of a DNA template. Chapter 9 describes the principle of automated DNA sequencing, which is based on the use of fluorescently labeled dideoxynucleotides to produce a pool of truncated DNA molecules that can be detected by their emission spectra following laser excitation.

Membrane Blotting and Hybridization of Nucleic Acids

Perhaps one of the most important and universal molecular genetic techniques to be developed over the past 25 years has been the use of solid support membranes to analyze DNA sequence similarity by nucleic acid hybridization. In 1975, Ed Southern published a paper in the *Journal of Molecular Biology* describing a technique to analyze DNA sequences bound to nitrocellulose membranes. This method became known as the

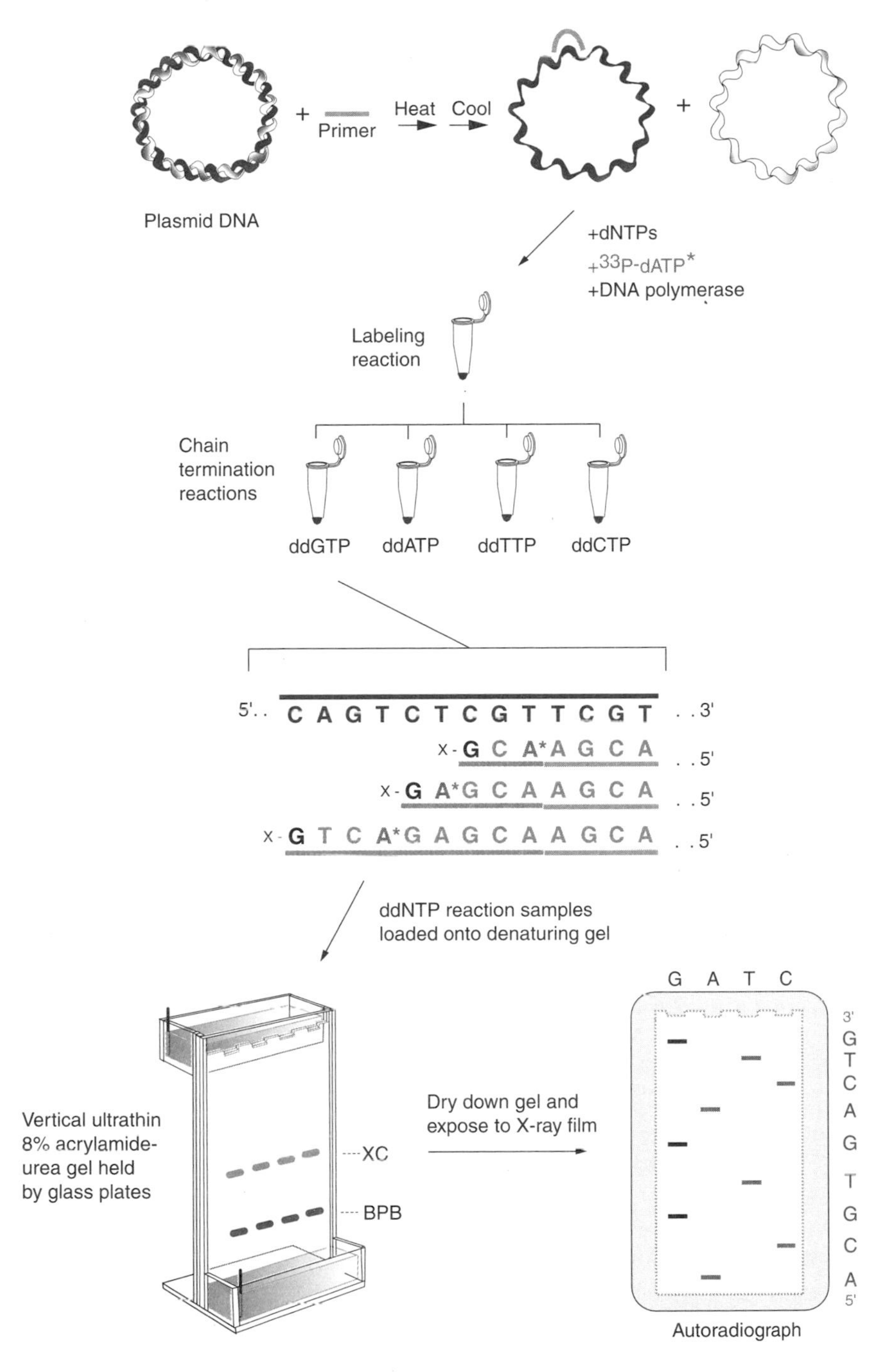

Figure 1.13 Sanger dideoxynucleotide DNA sequencing produces a pool of extended oligonucleotides that have random 3′ ends but the same 5′ terminus. DNA synthesis is initiated at room temperature by adding a modified T7 bacteriophage DNA polymerase, called Sequenase™, in the presence of all four dNTPs, one of which is radioactively labeled, most commonly [α-^{33}P]dATP or [α-^{35}S]dATP. The labeling reaction is then split into four tubes, each of which contains a different ddNTP, and incubated for a few minutes at 37°C. Chain termination in the ddGTP reaction produces a pool of products that all end with ddGMP. Because the ddNTP chain terminators are present at a ~100-fold lower concentration than the dNTPs, chain elongation is able to proceed until a ddNTP is incorporated. The chain termination products are separated by denaturing polyacrylamide gel electrophoresis using tracking dyes (xylene cyanol and bromophenol blue), and the composite DNA sequence is determined by identifying which reaction contains the predominant truncated product at each position in the gel using autoradiography.

"Southern blot" and has been instrumental not only in understanding genome organization, but also in inspiring the development of powerful gene isolation techniques based on the principles of nucleic acid membrane blotting and hybridization.

The basic principle of the Southern blot, as it is done currently, is illustrated in Figure 1.14. Agarose gel electrophoresis is used to separate DNA fragments generated by restriction enzyme digestion, and then the gel is photographed to record the migration of DNA molecular weight markers. The gel is soaked in an alkaline solution (1*N* NaOH) for 30 minutes to denature the DNA, and then the gel is neutralized in Tris buffer. The gel is mounted onto a simple DNA transfer system that is assembled by stacking various materials in the following order: a glass plate placed over a reservoir containing a high salt solution called 20X SSC (3*M* NaCl/0.3*M* sodium citrate), a paper wick, the agarose gel containing denatured DNA, and a piece of nylon membrane. When absorbent paper such as a stack of paper towels is placed on top of the nylon membrane, it allows the 20X SSC solution to be drawn from the reservoir through the gel and into the absorbent paper. This process causes directional diffusion of the DNA out of the gel and transfers it directly onto the nylon membrane maintaining the DNA separation pattern seen on the gel. Following DNA transfer (~12 hours), the membrane is exposed to ultraviolet light to attach the single-strand DNA covalently to the nylon membrane by cross-linking. Detection of specific DNA fragments is accomplished by hybridizing a single-strand radioactive probe to the nylon filter under conditions that promote DNA reassociation. After the excess unhybridized probe is removed by washing, the nylon membrane is analyzed by autoradiography.

Other applications of DNA–DNA hybridizations are the DNA-based screening of genomic libraries (Chapter 3) and restriction-fragment-length polymorphism analysis (Chapter 4). Several useful variations of the original Southern blotting technique have subsequently been developed. The most closely related technique is called the "Northern blot." Northern blots use RNA gels and are processed in essentially the same way as Southern blots (Fig. 1.14) with the exception that organic denaturants (formaldehyde or glyoxal) are used to denature the RNA fully, rather than NaOH, which hydrolyzes RNA. Not to be outdone, immunologists and biochemists developed a protein-based blotting system called the "Western blot," which involves the electrophoretic transfer of proteins from a polyacrylamide gel onto a nylon membrane. These membranes are incubated (not hybridized!) with antibodies under conditions that allow the detection of specific proteins. Two somewhat esoteric variations on the antibody-based Western blot are the "Southwestern blot," which uses double-strand radioactive DNA probes to identify putative DNA binding proteins, and the "Farwestern blot" which relies on high affinity protein–protein interactions, similar to those found in yeast two-hybrid screens (Chapter 4), to identify candidate binding proteins.

Laboratory Practicum 1. *Identifying the transcriptional start site of a gene transcript*

Research Objective

A molecular endocrinologist is interested in estrogen-regulated gene expression in mammary epithelial cells. She has recently isolated a nearly full-length cDNA clone and the corresponding 5′ genomic DNA sequences for a gene that is induced 10-fold by estrogen treatment of human mammary epithelial cells in culture. Her research objective is to identify the 5′ end of the gene transcript in order to facilitate future studies aimed at investigating estrogen-regulated expression of this gene in normal and tumorigenic mammary epithelial cells.

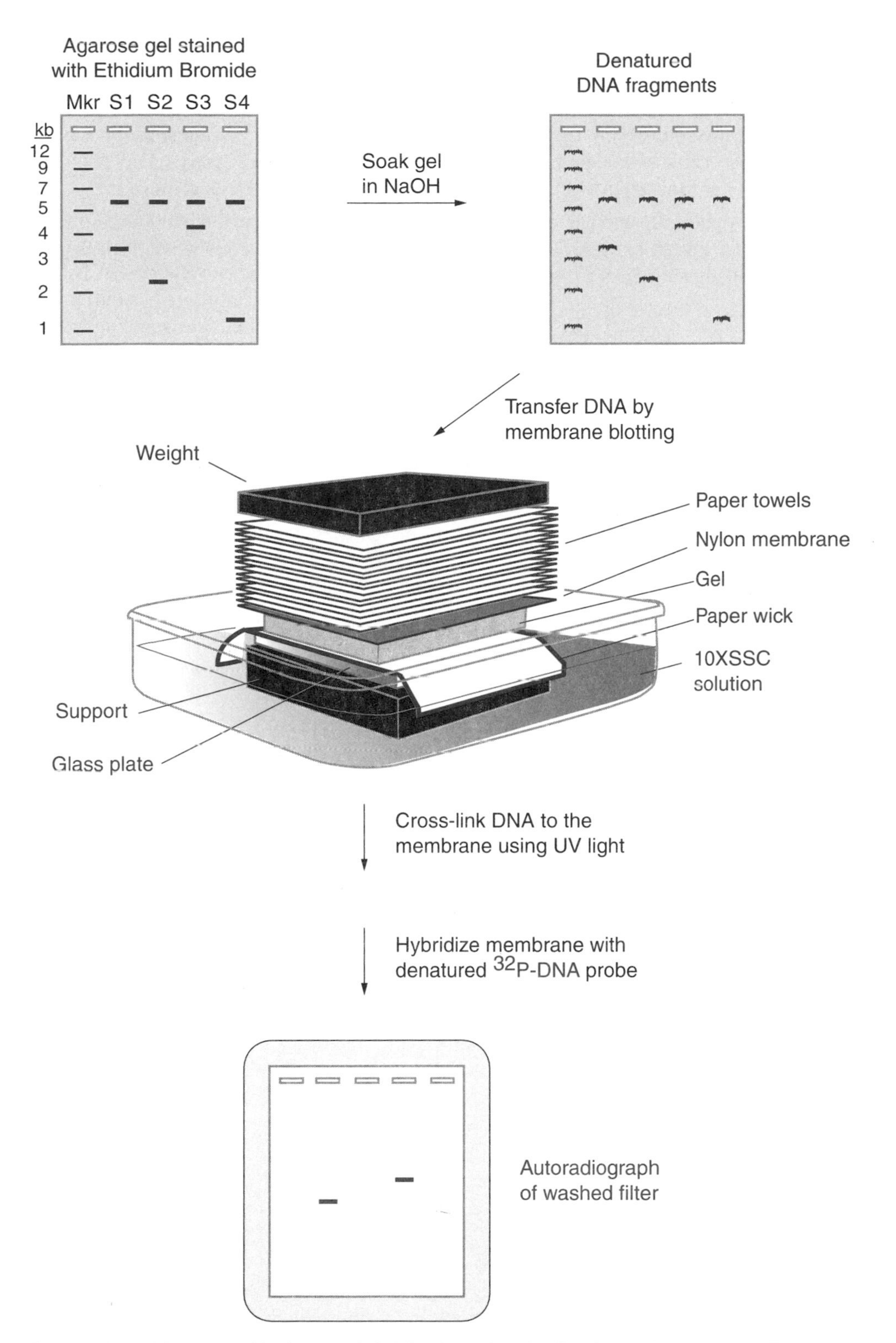

Figure 1.14 Membrane blotting and hybridization using the Southern blot technique. The use of nylon membranes, rather than nitrocellulose as originally done by Southern, has increased the utility of Southern blotting by allowing the more durable nylon membranes to be rehybridized multiple times with different DNA probes. The four DNA samples (S), and molecular weight markers (Mkr), are visualized by staining the gel with ethidium bromide.

Available Information and Reagents

1. Based on the cDNA sequence, and the estimated size of the gene transcript from Northern blots, she predicts that the 5′ end of the transcript is 50–150 nucleotides farther upstream of the sequence in her longest cDNA clone.
2. A plasmid subclone of genomic DNA has been constructed that corresponds to a 2 kb region that overlaps with the most 5′ cDNA sequence and therefore is likely to contain the gene promoter.
3. An antisense oligonucleotide has been designed that is 24 nucleotides long and has a 3′ end that is located 10 nucleotides downstream of the 5′ terminus of the cloned cDNA fragment.
4. RNA from untreated and estrogen-treated mammary epithelial cells has been isolated and shown by Northern blots to contain a 10-fold difference in steady-state levels of the new gene transcript.

Basic Strategy

There are two methods commonly used to map the 5′ end of mRNA transcripts when both the cDNA and genomic DNA corresponding to the 5′ region of the gene have been cloned. The first method is shown in Figure 1.15 and is called RNase mapping. This method uses in vitro RNA synthesis to produce radioactively labeled complementary RNA that includes sequences upstream and downstream of the predicted 5′ terminus of the mRNA. Following solution hybridization between the complementary RNA probe and total cellular RNA, the reassociated heteroduplexes are treated with RNases that degrade single-strand (unhybridized) RNA. The products of RNase digestion are separated on a polyacrylamide gel and the sizes of the undigested and digested RNA probe are determined. The second method is called primer extension, which utilizes an end-labeled oligonucleotide that serves as a primer for cDNA synthesis using the enzyme reverse transcriptase (Fig. 1.16). In the presence of this gene-specific primer, dNTPs, and cellular RNA, reverse transcriptase synthesizes cDNA from any primer that is annealed to template RNA. The length of the longest end-labeled cDNA products should correspond to the total number of nucleotides between the 5′ end of the primer on the antisense strand and the 5′ terminus of the mRNA template. For both the RNase mapping and primer extension methods, the pattern of product formation from parallel reactions using RNA from either untreated or estrogen-treated cells would be used to confirm specific mapping of the estrogen-induced gene.

Comments

The availability of in vitro transcription systems using bacteriophage-specific promoters led to the development of the RNase mapping method by facilitating the synthesis of single-strand high specific activity probes. Both T7 and SP6 bacteriophage RNA polymerase-dependent in vitro transcription systems have been developed. Titration of the RNase digestion conditions is done as an initial experiment to avoid under- or overdigestion of the heteroduplex substrate (Fig. 1.15). In addition to mapping 5′ ends of gene transcripts, RNase mapping can be used to identify RNA splice sites and to measure quantitatively steady-state levels of RNA under different physiological conditions. The primer extension technique (Fig. 1.16) is a reliable method to confirm RNase mapping studies because it relies on product synthesis rather than substrate degradation. A third approach, not shown here, is called S1 nuclease mapping, which is similar to RNase mapping except that a single-strand end-labeled DNA probe is used.

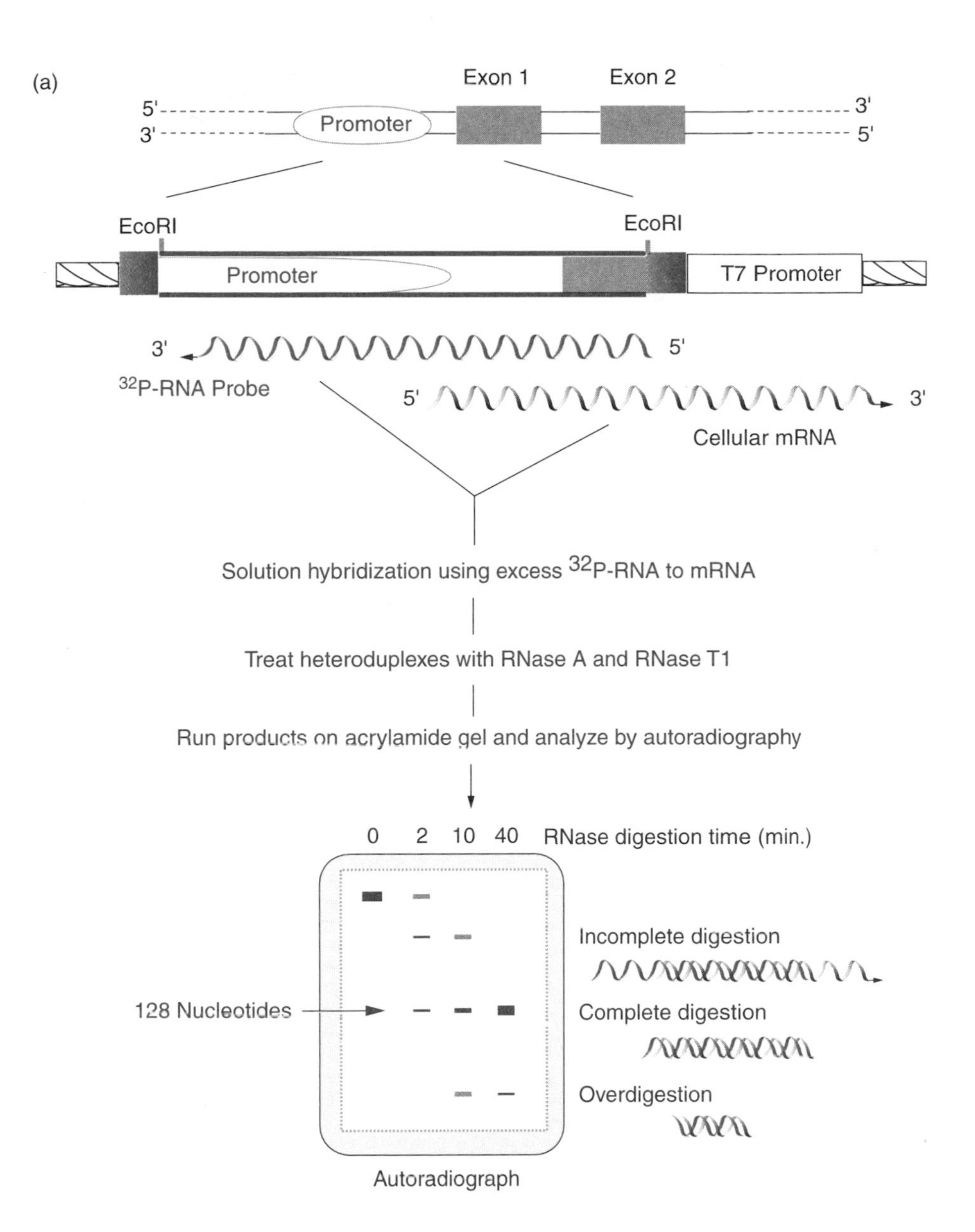

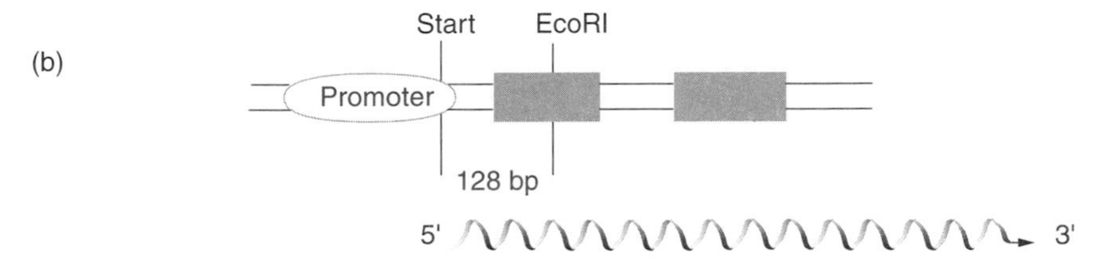

Figure 1.15 Identification of the 5′ end of a gene transcript using RNase mapping. (*a*) The radioactive antisense RNA probe is synthesized in vitro using T7 bacteriophage RNA polymerase and T7 promoter sequences that flank the multiple cloning site. By titrating the RNase digestion time, it is possible to identify the major protected fragment. Discrete anomalous bands arise from preferential digestion of the RNA probe near regions of secondary structure. (*b*) Results from the autoradiograph would be used to predict that the transcriptional start site is located 128 nucleotides upstream (5') of the EcoRI site in exon 1.

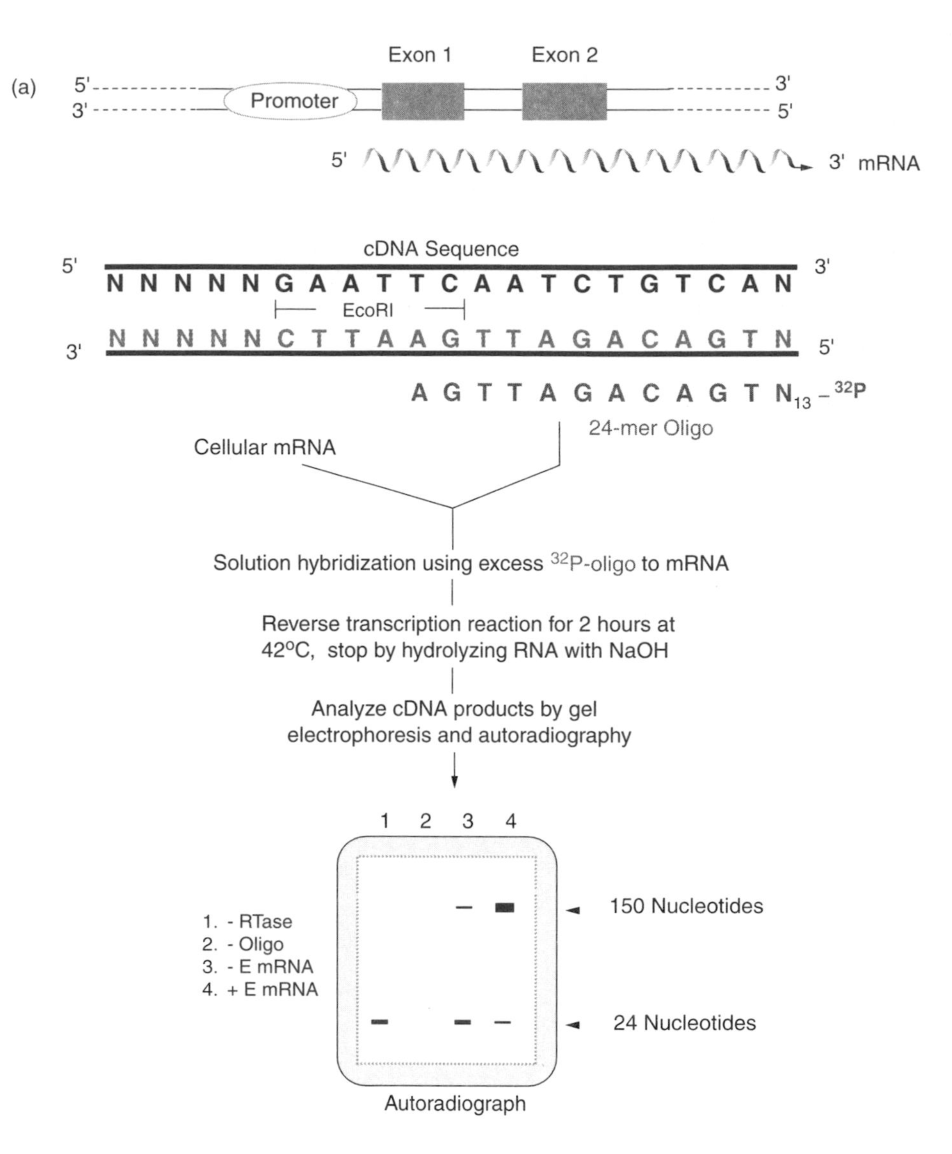

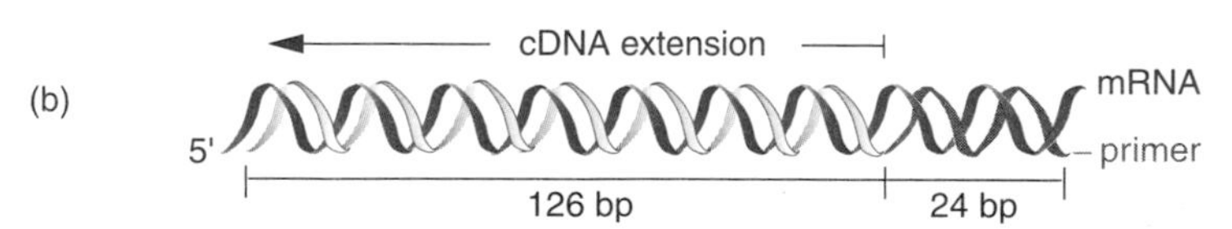

Figure 1.16 Primer extension can be used to identify the 5' end of a gene transcript. (*a*) A 24 nucleotide (24-mer) antisense oligonucleotide was designed based on the cDNA sequence. This 24-mer includes two nucleotides from the EcoRI site at its 3′ end and is radiolabeled with T4 polynucleotide kinase and [γ-^{32}P]ATP at the 5′ end. Following primer hybridization to RNA from either estrogen-free (–E) or estrogen-treated (+E) cells, reverse transcriptase and dNTPs are added to initiate cDNA synthesis. A product of 150 nucleotides would corroborate the RNase mapping studies. Note that there is a ~10-fold difference in the amount of 150 nucleotide product, depending on whether the RNA was isolated from –E or +E cells (compare lanes 3 and 4). Moreover, cDNA synthesis should depend on the presence of reverse transcriptase (lane 1) and inclusion of the antisense 24-mer oligonucleotide in the reaction (lane 2). (*b*) The observed 150 nucleotide long extended cDNA product predicts that the 5′ mRNA terminus is 126 nucleotides upstream of the primer 3′ end.

Prospective

Once the 5′ end of a gene transcript is localized within the context of a genomic sequence, it becomes possible to test functionally for gene promoter activity using sequences within the first ~200 nucleotides upstream of the transcriptional start site. In this example, the researcher could construct a reporter plasmid (Chapter 4) that contains the putative estrogen-regulated promoter and test its activity in normal and tumorigenic mammary cells that have been treated with estrogens. Subsequent promoter mapping experiments could then be done to determine if this gene is a primary target of estrogen action, which may be important to understanding its regulation in normal and neoplastic mammary cells. In vitro transcription studies could also be performed to map the transcriptional start site. This would be done using truncated versions of the cloned genomic DNA as a template in reactions containing nuclear cell extracts and [α-^{32}P]UTP.

REFERENCES

Flow of genetic information

Alberts, B., Bray, D., Johnson, A., Lewis, J., Raff, M., Roberts, K., Walter, P. Essential cell biology: An introduction to molecular biology of the cell. Garland Publishing, New York, 630 pp., 1998.

*Avery, O., MacLeod, C., MacCarty, M. Studies on the chemical nature of the substance inducing transformation of pneumococcal types. J. Exp Med. 79:137–158, 1944.

*Crick, F.H.C. Codon-anticodon pairing: The wobble hypothesis. J. Mol. Biol. 19:548–555, 1966.

*Meselson, M., Stahl, F. The replication of DNA in *Escherichia coli*. Proc. Natl. Acad. Sci. USA 44:671–682, 1958.

*Watson, J.D. The involvement of RNA in the synthesis of proteins. Science 140:17–26, 1963.

Nucleic acid biochemistry

*Britten, R., Kohne, D. Repeated sequences in DNA. Science 161:529–540, 1968.

Mattencci, M., Caruthers, M. Synthesis of deoxyoligonucleotides on a polymer support. J. Am Chem. Soc. 103:3185–3191, 1980.

*Schildkraut, C., Marmur, J., Doty, P. The formation of hybrid DNA molecules and their use in studies of DNA homologies. J. Mol. Biol. 3:595–617, 1961.

*Watson, J.D., Crick, F.H.C. Molecular structure of nucleic acids. A structure for deoxyribose nucleic acid. Nature 171:737–738, 1953.

Wetmur, J.G., Davidson, N. Kinetics of renaturation of DNA. J. Mol. Biol. 31:349–370, 1986.

Wing, R., Drew, H., Takano, T., Brodka, C., Tanaka, S., Itakura, K., Dickerson, R.E. Crystal structure analysis of a complete turn of B-DNA. Nature 287:755–758, 1980.

DNA metabolizing enzymes

Chaconas, G., van de Sande, J.H. 5′-^{32}P labeling of RNA and DNA restriction fragments. Methods Enzymol. 65:75–88, 1980.

Challberg, M.D., Englund, P.T. Specific labeling of 3′ termini with T4 DNA polymerase. Methods Enzymol. 65:39–43, 1980.

Doublie, S., Tabor, S., Long, A., Richardson, C., Ellenberger, T. Crystal structure of a bacteriophage T7 DNA replication complex at 2.2 A resolution. Nature 391:251–258, 1998.

Engler, M.J., Richardson, D.C. DNA ligases. In The Enzymes, Vol. 15B (P.D. Boyer, ed.). Academic Press, New York, pp. 3–30, 1982.

* Landmark papers in applied molecular genetics.

Fuchs, R., Blakesley, R. Guide to the use of type II restriction endonucleases. Methods Enzymol. 100:3–38, 1983.

*Jacobsen, H., Klenow, H., Overgaard-Hansen, K. The N-terminal amino-acid sequences of DNA polymerase I from *Escherichia coli* and of the large and the small fragments obtained by a limited proteolysis. Eur. J. Biochem. 45:623–627, 1974.

Kovall, R., Matthews, B. Structural, functional, and evolutionary relationships between λ-exonuclease and the type II restriction endonucleases. Proc. Natl. Acad. Sci. USA 95:7893–7897, 1998.

*Lawyer, F.C., Stoffel, S., Saiki, R.K., Myambo, K., Drummond, R., Gelfand, D.H. Isolation, characterization, and expression of Escherichia coli of the DNA polymerase gene from *Thermus aquaticus*. J. Biol. Chem. 264:6427–6437, 1989.

Lehman, I.R. DNA polymerase I of Escherichia coli. In The Enzymes, Vol. 14A (P.D. Boyer, ed.). Academic Press, New York, pp. 16–38, 1981.

*Linn, S, Arber, W. Host specificity of DNA produced by Escherichia coli, X. In vitro restriction of phage fd replicative form. Proc. Natl. Acad. Sci. USA 59:1300–1306, 1968.

*Modrich, P., Lehman, I.R. Enzymatic joining of polynucleotides. J. Biol. Chem. 245: 3626–3631, 1975.

Newman, M., Strzelecka, T., Dorner, L.F., Schildkraut, I., Aggarwal, A. Structure of restriction endonuclease BamHI and its relationship to EcoRI. Nature 368:660–664, 1994.

Reid, T.W., Wilson, I.B. Escherichia coli alkaline phosphatase. In The Enzymes, Vol. 4 (P.D. Boyer, ed.). Academic Press, New York, pp. 373–415, 1971.

Richardson, C.C. Bacteriophage T4 polynucleotide kinase. In The Enzymes, Vol. 14A (P.D. Boyer, ed.). Academic Press, New York, pp. 299–314, 1981.

Roberts, R. Restriction and modification enzymes and their recognition sequences. Nucl. Acids Res. 11:r135–r167, 1983.

Rong, M., He, B., McAllister, W., Durbin, R. Promoter specificity determinants of T7 RNA polymerase. Proc. Natl. Acad. Sci. USA 95:515–519, 1998.

*Tabor, S., Richardson, C.C. DNA sequencing analysis with a modified bacteriophage T7 DNA polymerase. Proc. Natl. Acad. Sci. USA 84:4767–4771, 1987.

Verma, I.M. Reverse transcriptase. In The Enzymes, Vol. 14A (P.D. Boyer, ed.). Academic Press, New York, pp. 87–104, 1977.

Biochemical methods to study DNA and RNA

Arnheim, N., Southern, E.M. Heterogeneity of the ribosomal genes in mice and men. Cell 11:363–370, 1977.

Ausubel, F., Brent, R., Kingston, R., Moore, D., Seidman, J., Smith, J., Struhl, K. Current protocols in molecular biology. John Wiley & Sons, New York, 1996.

Birnboim, H.C. A rapid alkaline extraction method for the isolation of plasmid DNA. Methods Enzymol. 100:243–255, 1983.

Chrambach, A., Rodbard, D. Polyacrylamide gel electrophoresis. Science 172:440–451, 1971.

*Denhardt, D.T. A membrane-filter technique for the detection of complementary DNA. Biochem. Biophys. Res. Commun. 23:641–646, 1966.

Doel, M.T., Houghton, M., Cook, E.A., Carey, N.H. The presence of ovalbumin mRNA coding sequences in multiple restriction fragments of chicken DNA. Nucl. Acids Res. 4:3701–3713, 1977.

Helling, R.B., Goodman, H.M., Boyer, H.W. Analysis of R. EcoRI fragments of DNA from lambdoid bacteriophages and other viruses by agarose gel electrophoresis. J. Virol. 14: 1235–1244, 1974.

Holmes, D.S., Quigley, M. A rapid boiling method for the preparation of bacterial plasmids. Anal. Biochem. 114:193–197, 1981.

Khandjian, E.W. Optimized hybridization of DNA blotted and fixed to nitrocellulose and nylon membranes. Bio/Technology 5:165–167, 1987.

Lis, J.T., Schleif, R. Size fractionation of double-stranded DNA by precipitation with polyethylene glycol. Nucl. Acids Res. 2:383–389, 1975.

*Marmur, J. A procedure for the isolation of deoxyribonucleic acid from microorganisms. J. Mol. Biol. 3:208–218, 1961.

*Maxam, A., Gilbert., W. A new method of sequencing DNA. Proc. Natl. Acad. Sci. USA 74: 560–564, 1977.

*McDonell, M.W., Simon, M.N., Studier, F.W. Analysis of restriction fragments of T7 DNA and determination of molecular weights by electrophoresis in neutral and alkaline gels. J. Mol. Biol. 110:119–146, 1977.

Nygaard, A.P., Hall, B.D. A method for the detection of RNA-DNA complexes. Biochem. Biophys. Res. Commun. 12:98–104, 1963.

*Radloff, R., Bauer, W., Vinograd, J. A dye-buoyant-density method for the detection and isolation of closed circular duplex DNA: The closed circular DNA in HeLa cells. Proc. Natl. Acad. Sci. U.S.A. 57:1514–1521, 1967.

Sambrook, J., Fritsch, E.F., Maniatis, T. Molecular Cloning: A Laboratory Manual, 2nd ed., Cold Spring Harbor Laboratory, Cold Spring Harbor, New York, 1989.

*Sanger, F., Nicklen, S., Coulson, A. DNA sequencing with chain termination inhibitors. Proc. Natl. Acad. Sci. USA 74:5463–5467, 1977.

*Sharp, P., Sugden, B., Sambrook, J. Detection of endonuclease activities in *Hemophilus parainfluenza* using analytical agarose-ethidium bromide electrophoresis. Biochemistry 12: 3055–3062, 1973.

*Smith, L.M., Sanders, J., Kaiser, R., Hughes, P., Dodd, C., Connel, C., Heriner, C., Kent, S., Hood, L.E. Fluorescence detection in automated DNA sequence analysis. Nature 321: 674–679, 1986.

*Southern, E.M. Detection of specific sequences among DNA fragments separated by gel electrophoresis. J. Mol. Biol. 98:503–517, 1975.

*Wahl, G.M., Stern, M., and Stark, G.R. Efficient transfer of large DNA fragments from agarose gels to diazobenzyloxymethyl-paper and rapid hybridization by using dextran sulfate. Proc. Natl. Acad. Sci. USA 76:3683–3687, 1979.

Identifying the transcriptional start site of a gene transcript

*Berk, A.J., Sharp, P.A. Sizing and mapping of early adenovirus mRNAs by gel electrophoresis of S1 endonuclease-digested hybrids. Cell 12:721–732, 1977.

Graves, B.J., Eisenberg, S.P., Coen, D.M., McKnight, S.L. Alternate utilization of two regulatory domains within the Moloney murine sarcoma virus long terminal repeat. Mol. Cell. Biol. 5:1959–1968, 1985.

Jones, K.A., Yamamoto, K.R., Tjian, R. Two distinct transcription factors bind to the HSV thymidine kinase promoter in vitro. Cell 42:559–572, 1985.

*McKnight, S.L., Kingsbury, R. Transcription control signals of a eukaryotic protein-encoding gene. Science 217:316–324, 1982.

*Melton, D.A., Krieg, P.A., Rebagliati, M.R., Maniatis, T., Zinn, K., Green, M.R. Efficient in vitro synthesis of biologically active RNA and RNA hybridization probes from plasmids containing a bacteriophage SP6 promoter. Nucl. Acids Res. 12:7035–7056, 1984.

Miesfeld, R., Arnheim, N. Identification of the in vivo and in vitro origin of transcription of human ribosomal DNA. Nucl. Acids Res. 10:3933–3949, 1982.

Tremea, F., Batistuzzo de Medeiros, S.R., ten Heggeler-Bordier, B., Germond, J.-E., Seiler-Tuyns, A., Wahli, W. Identification of two steroid-responsive promoters of different strength controlled by the same estrogen-responsive element in the 5′-end region of the *Xenopus laevis* vitellogenin gene A1. Mol. Endocrinol. 3:1596–1609, 1989.

*Zinn, K., DiMaio, D., Maniatis, T. Identification of two distinct regulatory regions adjacent to the human β-interferon gene. Cell 34:865–879, 1983.

CHAPTER

2

LABORATORY TOOLS FOR MOLECULAR GENETIC APPLICATIONS

The discovery of DNA as the genetic material of cells, and the subsequent elucidation of its chemical structure, gave early molecular geneticists an opportunity to manipulate genetics using the principles of nucleic acid biochemistry. However, as is often the case, the ability of these molecular genetic pioneers to explore and test new approaches was limited by the lack of suitable laboratory techniques. Indeed, much of the research effort in the early years of molecular genetics was devoted to devising novel methods to study and manipulate DNA in vitro and to introduce biologically active DNA back into cells. A turning point came when bacterial geneticists began to use their knowledge of bacterial plasmids and bacteriophage biology to investigate mechanisms of DNA transfer. Restriction enzymes were discovered, and soon after, biochemists provided purified enzymes for the first in vitro recombinant DNA experiments.

Beginning in the late 1960s and through the early 1980s, innovative laboratory researchers added to the growing repertoire of molecular genetic methodologies, until in 1982, a landmark laboratory protocol book called *Molecular Cloning* was written by Tom Maniatis, Ed Fritsch and Joe Sambrook and published by Cold Spring Harbor Press. The availability of the Cold Spring Harbor protocols set off a rapid expansion in the number of life science labs that utilized molecular genetic applications to study biological problems at the molecular level. As a result, a larger and more collaborative laboratory protocol book entitled *Current Protocols in Molecular Biology* was published in 1987 (with subsequent quarterly updates), and it soon became one of the most often cited laboratory references in applied molecular genetics.

Before going on to describe some of the more complex molecular genetic strategies presented in other chapters, we first need to examine several key components one would expect to find in a virtual molecular genetic "toolbox." These basic tools include (1) special *E. coli* strains to propagate DNA, (2) plasmid DNA cloning vectors, (3) bacterial DNA transformation methods, and (4) bacteriophage systems for gene cloning. These topics are followed by a brief summary of the essential laboratory resources required for most molecular genetic applications. Laboratory practicum 2 describes how these basic laboratory tools can be utilized to express a mammalian protein in *E. coli* using an inducible gene expression system.

ESCHERICHIA COLI K-12: THE BACTERIAL HOST FOR ALL OCCASIONS

The choices researchers make regarding model cell types for biological studies often reflect a historical perspective. The brewers' yeast *Saccharomyces cerevisiae*, for example, was intensely studied for centuries because of its central role in alcohol fermentation. To take advantage of this prior knowledge, geneticists chose this same strain of yeast for cell cycle and gene regulation studies. Similarly the reason Stanley Cohen and Herb Boyer chose *Escherichia coli* strain K-12 for their groundbreaking molecular genetic experiments in the early 1970s, was also related to the history of this laboratory strain. *E. coli* K-12 was the bacterial strain of choice in biochemistry labs because it was easy to grow and amenable to metabolic studies. These same properties also made it the primary microorganism for bacterial geneticists to study. Molecular geneticists now use this same strain of *E. coli* for routine procedures because it turned out to be an extremely good host for a variety of molecular genetic applications. Moreover, during the past 25 years, *E. coli* K-12 has proved to be an innocuous biological host for the propagation of recombinant DNA molecules. The attenuated *E. coli* K-12 strain does not thrive outside of the laboratory environment and it is unable to compete against the more genetically robust *E. coli* serotypes normally found in the human intestine.

Biology of *E. coli* K-12

E. coli is a Gram-negative, rod-shaped bacterium about 2.5 μm long that contains flagella and a genome of 4,639,221 bp encoding at least 4000 genes (see Chapter 4). The K-12 strain was first isolated in 1921 from the stool of a malaria patient and it has been maintained in laboratory stocks as a pure strain for the past 75 years. By virtue of their natural habitat in the intestine, *E. coli* are able to grow in simple media containing glucose, ammonia (or more commonly peptide digests), phosphate, and salts (Na^+, K^+, Ca^{2+}, and Mg^{2+}). *E. coli* contain both an outer and inner membrane that are separated by a protein-rich periplasmic space containing peptidoglycan which gives the bacterium osmotic strength. The outer membrane contains proteins that function in the import and export of metabolites and that also provide binding sites for infectious bacteriophage. Like all Gram-negative bacteria, *E. coli* have no nuclear membrane and the chromosome is a large circular duplex molecule with a membrane attachment site and a single origin of replication. As described in the next section, *E. coli* cells can support the replication of DNA plasmids, many of which encode antibiotic-resistance genes. *E. coli* cells undergo cell division every 20 minutes when grown under optimal aerobic conditions, and they are facultative anaerobes, which allows them to be grown to high density using fermentation devices. Single *E. coli* colonies are isolated on standard bacterial agar plates and individual strains can be stored anaerobically in agar slabs or frozen as glycerol stocks.

Although *E. coli* K-12 is a safe, nonpathogenic bacterium, it should be emphasized that sterile laboratory practices are always warranted when working with this strain, and that specific biosafety guidelines must be followed when using *E. coli* K-12 for recombinant DNA research.

The *lac* Operon

There are a number of well-characterized bacterial operons (gene clusters) in *E. coli* that serve to illustrate various principles of prokaryotic gene regulation. One of these classic examples, the *lac* operon, has been used to develop several important molecular genetic tools. We first review the biology of the *E. coli lac* operon and then describe two applications that utilize various components of this genetic unit.

The *lac* operon encodes three enzymes required for lactose metabolism in *E. coli*. The first gene in the operon is the *lacZ* gene encoding β-galactosidase, an enzyme that cleaves lactose to produce galactose and glucose. When glucose levels are low and lactose levels are high, transcription of the *lac* operon is maximally induced through a combined mechanism of transcriptional derepression and activation which requires the *lac* repressor and catabolite activator protein (CAP), respectively. Figure 2.1 shows the basic components of the *lac* operon, including the *lac* promoter–operator region, the coding sequences of *lacZ*, *lacY*, and *lacA*, and the *lac* repressor protein that is encoded by the nearby *lacI* gene. For the purposes of this discussion, we focus our attention on the mechanism by which lactose metabolites bind to the *lacI* repressor protein and cause transcriptional derepression of the *lac* operon.

The *lac* repressor is a ~38 kDa protein that binds to the *lac* operator site as a tetramer (two dimers). *lac* repressor binding to the operator sequence is extremely tight ($K_d = 10^{-13}M$) and specific (the K_d for nonspecific DNA binding is only $10^{-6}M$), causing transcriptional repression through a mechanism involving repressor interference with RNA polymerase binding to the *lac* promoter. When an effector molecule such as the nonmetabolizable allolactose analogue, isopropylthiogalactoside (IPTG), binds to

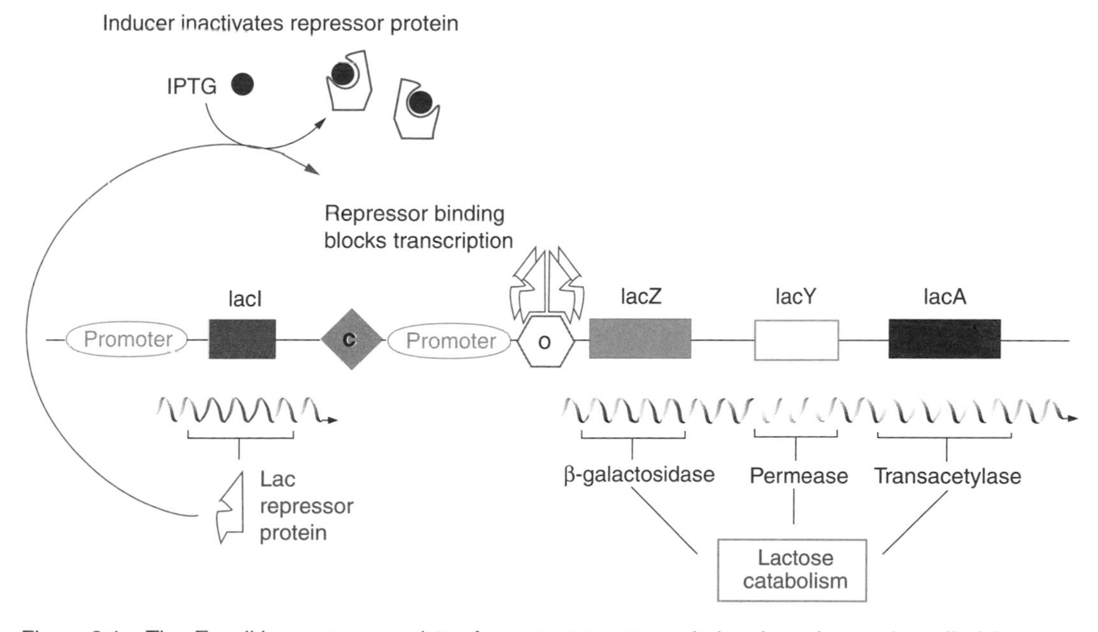

Figure 2.1 The *E. coli lac* operon consists of a promoter, a transcriptional regulatory site called the operator (o), a CAP binding site (c), and three structural genes (*lacZ*, *lacY*, and *lacA*) that are transcribed as a single polycistronic mRNA. Transcription of the *lac* operon is regulated by the *lac* repressor protein (*lacI*), which is encoded on a gene physically linked to the *lac* operon. *lac* operon inducers, such as IPTG, inactivate the *lac* repressor protein resulting in transcriptional derepression of the *lac* operon.

the *lac* repressor protein, its DNA binding affinity for the *lac* operator decreases by three orders of magnitude and transcription of the *lac* operon is derepressed (Fig. 2.1).

Numerous *E. coli* mutants have been isolated that display altered regulation of the *lac* operon. The *E. coli lacI*^{q} mutation, for example, overexpresses the *lac* repressor gene and exhibits a higher degree of *lac* promoter repression because of the elevated levels of *lac* repressor protein in the cell. A second class of mutations is in the *lac* promoter, and although these mutations do not alter repression of the *lac* operon, they do result in increased promoter activity. One example of this type is the *lacUV5* promoter mutation. By combining the *lacI*^{q} mutation with the *lacUV5* mutation in the same strain, it is possible to obtain >100-fold induction of the *lac* operon with IPTG (Fig. 2.2).

Using β-galactosidase activity as a measurement of gene activity is a second molecular genetic application that arose from studies of the *E. coli lac* operon. β-Galactosidase

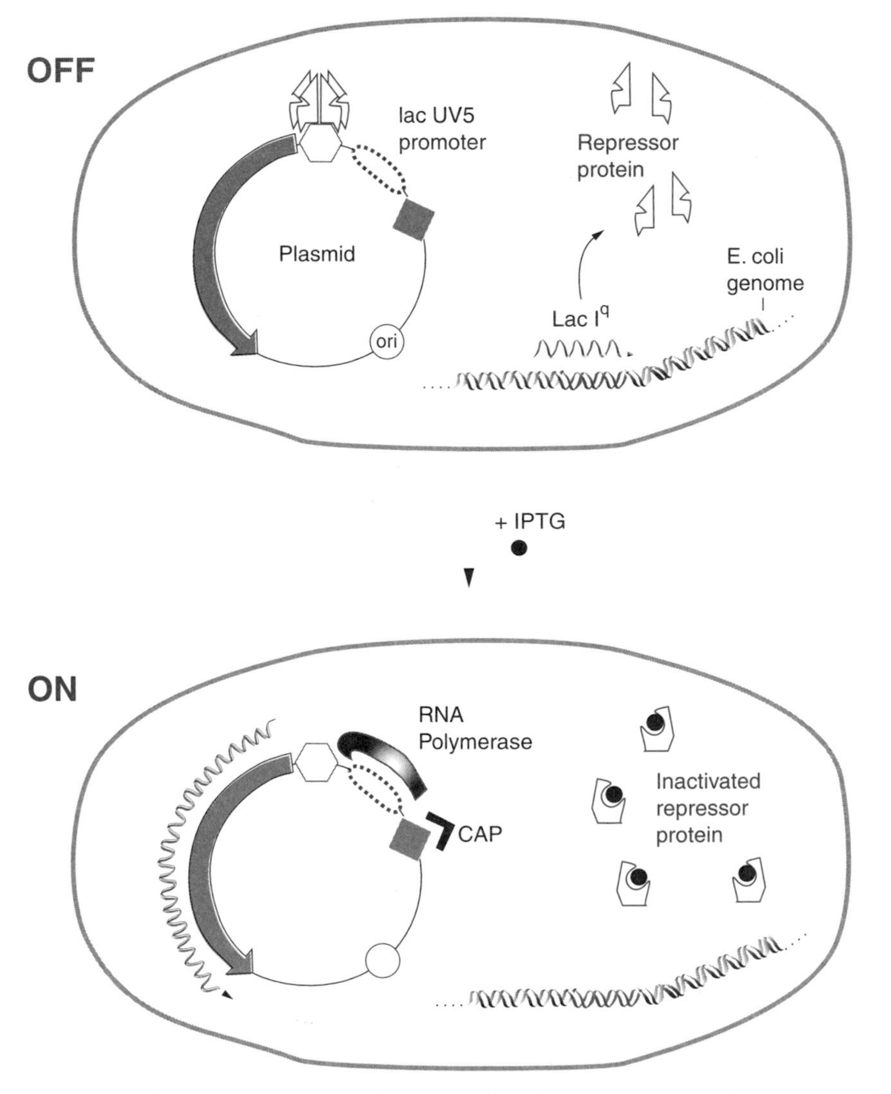

Figure 2.2 An IPTG-inducible expression system has been developed using components of the *lac* operon. Expression of cloned genes can be regulated using plasmids containing the *lacUV5* promoter that are grown in *E. coli* strains containing the *lacI*^{q} receptor overexpression mutation.

activity can be measured in live cells using the chromogenic substrate 5-bromo-4-chloro-3-indolyl-β-D-galactoside (X-gal). In wild-type *E. coli* cells, cleavage of X-gal produces a blue color that can be visualized as a blue colony on agar plates. The X-gal assay can be used to identify *lac* operon mutants, which appear as white colonies against a blue background. Another way the X-gal assay has been exploited is by taking advantage of β-galactosidase mutants that function through a mechanism called α complementation. Characterization of these mutants showed that β-galactosidase activity could be reconstituted in *E. coli* cells that expressed both nonfunctional *N*-terminal (α fragment) and *C*-terminal (ω fragment) β-galactosidase polypeptides. α complementation has been used to monitor insertional cloning into plasmids containing restriction sites within the ~150 amino acid coding sequence of the α fragment (Fig. 2.3). When these plasmids are grown in *E. coli* strains expressing the ω fragment on an F′ plasmid, it is

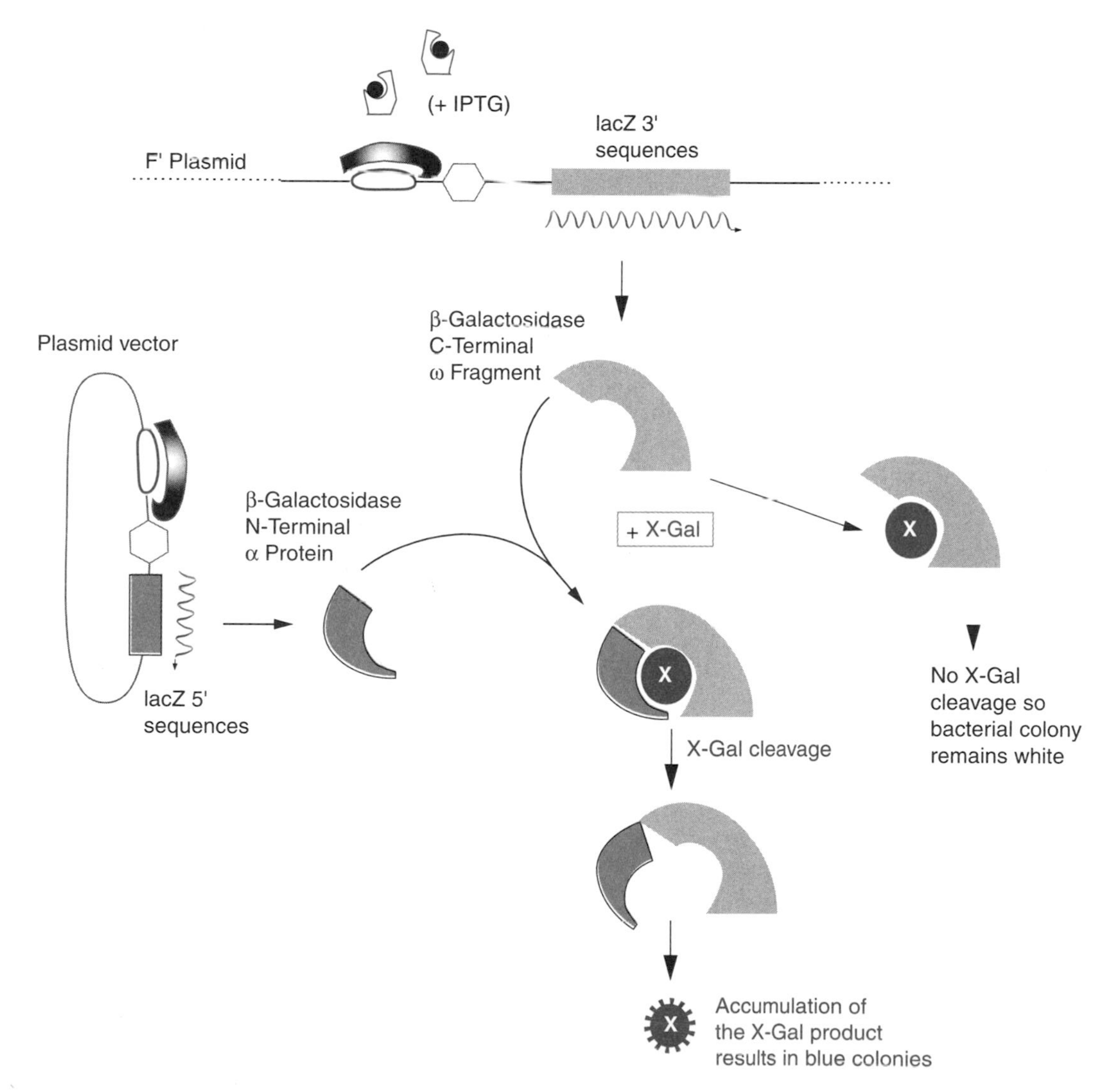

Figure 2.3 β-Galactosidase activity can be used as an indicator of gene function. Plasmids that express the *N*-terminal α fragment of β-galactosidase can be used in *E. coli* strains expressing the *C*-terminal ω fragment of β-galactosidase on an F′ plasmid.

possible to score antibiotic-resistant cells that contain the plasmid cloning vector and are either blue (no gene insertion) or white (coding sequences are disrupted). The advantage of using α complementation for DNA cloning is that plasmid vectors are limited by the size of insert DNA that can be accommodated and the α fragment (~450 bp) is small compared to the entire *lacZ* gene (~3500 bp).

Genetic Variants of *E. coli* K-12 Used for Gene Cloning

Wild-type *E. coli* K-12 have several natural defenses against foreign DNA that can make them very difficult to use for gene cloning strategies. Some of these defenses involve chemical modification of the foreign DNA, either by endonuclease cleavage or by DNA recombination, whereas other properties are related to efficiency of plasmid DNA propagation or enhanced ability to perform specific cloning selection schemes. Fortunately, through a combination of luck and directed genetic screening efforts, a variety of *E. coli* K-12 derivatives are available that have greatly improved gene cloning efficiencies. Some of these properties are described here and others listed in Appendix E.

Bacterial restriction modification systems pose the biggest problem for gene cloning. There are four restriction modification systems known to be present in *E. coli* K-12, and these can be divided into two general categories. The first is similar to that described in Chapter 1, in which unmethylated DNA is degraded by site-specific restriction enzymes very shortly after entering the cell. The *E. coli* K-12 restriction system (EcoK) requires three proteins encoded by the *hsdR*, *hsdM,* and *hsdS* genes that function together as a protein complex both to cleave and to methylate DNA. The EcoK restriction system recognizes the sequence 5′-AAC$(N)_5$GTGC-3′, and if it is unmethylated, cleaves the DNA at a variable distance from this recognition site. The *hsdR* gene encodes the endonuclease of the system and therefore *hsdR*$^-$ strains of *E. coli* K-12 are often used for gene cloning. In addition, some strains contain mutations in *hsdS,* which encodes the specificity determinant of the hsdRMS protein complex. *hsdR* mutations eliminate only the nuclease activity, whereas mutations in *hsdS* abolish both the nuclease and methylation functions of the hsdRMS complex.

The second type of restriction modification system that causes problems is one that degrades inappropriately methylated DNA, for example, mammalian DNA that is methylated at cytosine and guanosine residues. Mutations in the *E. coli* genes *mcrA*, *mcrB,* and *mrr* are required in *E. coli* K-12 strains used for cloning genomic DNA fragments from human or murine sources. Interestingly, *D. melanogaster* and *S. cerevisiae* do not contain methylated bases, although higher plants and some lower eukaryotes do. Figure 2.4 summarizes the function of these restriction systems and emphasizes the importance of eliminating them from *E. coli* K-12 strains used for gene cloning. Note that the *dam* and *dcm* genes encode two other *E. coli* DNA methylases, which together with the *hsdM* gene account for the known DNA methylation patterns found in *E. coli*.

Rearranged or deleted insert DNA is equally deleterious to successful gene cloning experiments. Most *E. coli* K-12 strains used for cloning contain mutations in the *recA* gene, which encodes a well-characterized recombination protein. *recA* mutations also enhance the biosafety of *E. coli* K-12 because *recA*$^-$ strains are sensitive to ultraviolet light. Other *E. coli* genes involved in recombination can cause problems, especially when repetitive sequences are present in genomic DNA fragments. To overcome this cloning obstacle, various *E. coli* K-12 strains have been developed with mutations in one or more of the *recB*, *recC*, *recJ,* or *sbcC* recombination genes. Inverted DNA repeats

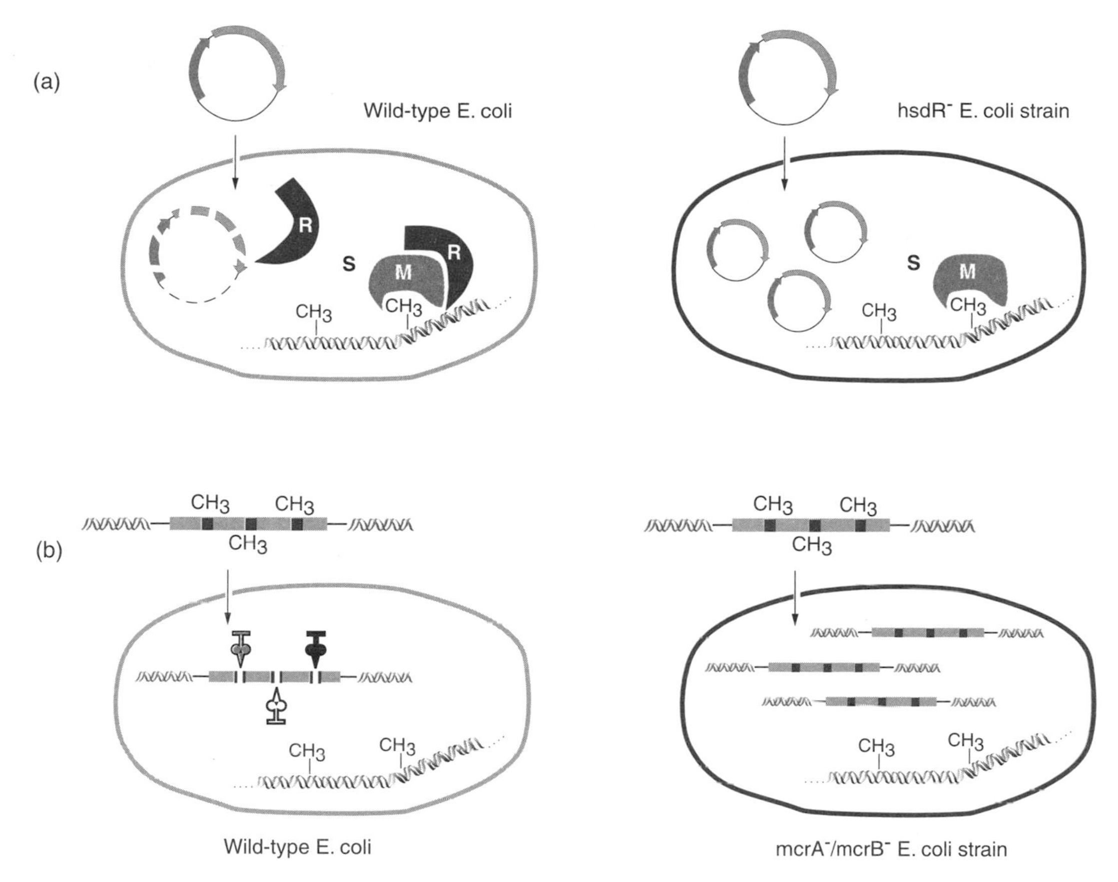

Figure 2.4 Several types of restriction modification systems must be eliminated from *E. coli* K-12 strains used for gene cloning. (*a*) Unmethylated foreign DNA is degraded by the hsdR nuclease and therefore *hsdR⁻* strains must be used. (*b*) Foreign DNA that is methylated in positions different from *E. coli* genomic DNA is degraded by the mdrA/mcrB proteins. Genomic DNA libraries made from mammalian cells and most plants must be carried in *E. coli* K-12 strains that have *mcrA/mcrB* mutations.

are also unstable sequence motifs in *E. coli*, and therefore strains with mutations in the DNA repair pathway genes *uvrC* and *umuC* are often useful.

Most *E. coli* K-12 strains contain a mutation in the *endA* gene encoding DNA specific endonuclease I. Loss of this endonuclease greatly increases plasmid DNA yields and improves the quality of DNA that is isolated using standard biochemical preparations. Finally, as described in the preceding section, various components of the *lac* operon have been utilized for molecular genetic applications involving gene expression in *E. coli*. In these systems, *E. coli* strains must be used that are defective in the endogenous *lacZ* gene so that the host strain cannot metabolize X-gal and forms white colonies in the presence of IPTG. In addition, the $lacI^q$ mutation is often present in these strains, as well as a selectable F′ plasmid containing an *N*-terminal deletion of the *lacZ* gene (*lacZΔM15*) encoding the carboxy terminal ω polypeptide that is required for α complementation.

In summary, the standard bacterial host for most recombinant DNA cloning experiments is an *E. coli* K-12 strain with mutations in the endogenous restriction modification system (*hsdR*⁻), homologous DNA recombination function (*recA*⁻) and endonuclease I activity (*endA*⁻). Depending on the specific objectives of a molecular genetic strategy, a number of additional *E. coli* K-12 mutations may also be desirable.

BACTERIAL PLASMIDS: THE BIOLOGICAL CURRENCY OF APPLIED MOLECULAR GENETICS

Plasmids are freely replicating multicopy extrachromosomal DNA elements that were initially found in strains of bacteria displaying enhanced pathological phenotypes. The three general classes of plasmids are (1) virulence plasmids encoding toxin genes, (2) drug-resistance plasmids that confer resistance to antibiotics, and (3) plasmids that encode genes required for bacterial conjugation. Basically, all molecular genetic methods involving the manipulation of specific DNA fragments utilize some aspect of plasmid DNA biology as an integral prerequisite, or essential, step in the experimental protocol.

Plasmid Biology

Plasmids range in size from 1 to 200 kb and depend on host proteins for maintenance and replication functions. Most of the plasmid-encoded genes that have been characterized impart some growth advantage to the bacterial host, and the extrachromosomal, multicopy nature of plasmids is important to this function. The movement of gene fragments between the bacterial chromosome and other plasmids is what led bacterial geneticists to the discovery of these autonomously replicating genetic elements. One of the first bacterial plasmids to be discovered and characterized was the *E. coli* F factor. This extrachromosomal genetic element is a ~95 kb bacterial plasmid that was named for its role in bacterial "fertility." The F plasmid is present at one copy per cell and encodes ~100 genes, some of which are required for F plasmid transfer during conjugation (*tra* genes). As a result of integration into the *E. coli* chromosome, and then an imprecise excision resulting in the reformation of a circular plasmid, the F plasmid is able to transfer portions of the *E. coli* genome into another bacterium via conjugation. The F plasmid contained in some *E. coli* K-12 strains carries the tetracycline-resistance gene and a mutated copy of the *lac* operon that is utilized for α complementation. F plasmid-containing *E. coli* strains are also required for molecular genetic applications involving M13 filamentous bacteriophage. The F plasmid carries the sex pili gene which encodes the cell surface binding site for M13 phage.

Studies of plasmid replication have shown that initiation of DNA synthesis begins at a single origin site on the molecule that can be defined genetically as *ori*. Most plasmids used in molecular genetic applications have a "relaxed" mode of DNA replication, meaning that they can initiate replication independent of the host cell cycle, although host-encoded replication proteins are required. The best studied plasmid replication system is that of ColE1, which is a small *E. coli* plasmid encoding antibacterial proteins called colicins. ColE1 replication has been shown to involve transcription of a DNA sequence near the origin, which creates the RNA primer required for initiation of DNA synthesis. ColE1 plasmid replication control is mediated through this RNA synthesis event. The inability to maintain stably two plasmids utilizing the same plasmid replication mechanism is called plasmid incompatibility. For example, two ColE1 plasmid molecules with different genotypes are not stably inherited in the absence of independent genetic selections. In contrast, a ColE1-based plasmid vector can co-exist in cells carrying the F plasmid because the replication control mechanism is different for these two plasmids. Processes that determine plasmid partitioning during cell division also contribute to plasmid incompatibility.

Antibiotic-Resistance Genes are Often Encoded on Plasmids

R plasmids are a type of bacterial plasmid that encode antibiotic resistance genes. R plasmids are similar to F plasmids in that they can be transferred between bacteria by conjugation. The widespread use of antibiotics in the clinic has led to a rapid rise in bacterial strains that are cross-resistant to a variety of antibiotics. For example, the R1 plasmid of *Salmonella paratyphi* encodes antibiotic-resistance genes for chloramphenicol, streptomycin, sulfonamide, ampicillin, kanamycin, and neomycin. *E. coli* strains containing R plasmids with multi-antibiotic resistance genes have also been isolated from clinical samples. Many of the antibiotic-resistance genes are contained within bacterial transposons. Transposable genetic elements such as *Tn*10, which encodes the tetracycline-resistance gene, tet^r, have special DNA sequences that facilitate DNA recombination by an insertional mechanism. Antibiotic-resistance genes present on R plasmids likely arose from transposition events occurring between sites in the bacterial chromosome and extrachromosomal plasmids.

Antibiotics are antimicrobial agents that can be classified based on their mode of action. These agents were initially discovered by testing for antibacterial activity using extracts prepared from microorganisms. Penicillin, for example, was first isolated from the soil fungus *Penicillium*. Most antibiotics in use today are synthetic derivatives of some previously identified natural compound. Two examples are ampicillin and amoxicillin, both of which are potent antibiotics related to the parent compound 6-aminopenicillanic acid. Bacterial antibiotic resistance genes encode proteins that inactivate these agents, either by preventing their accumulation in the cell or by inactivating them once they have been imported. By utilizing these antibiotic-resistance genes as dominant genetic markers in plasmid cloning vectors, it is possible to select for *E. coli* cells that have maintained high copy replication of plasmid DNA molecules. Table 2.1 lists examples of commonly used antibiotics and the corresponding bacterial antibiotic-resistance genes that have been exploited for specific molecular genetic applications.

Plasmid DNA Cloning Vectors

The first gene cloning experiments were performed using plasmid DNA molecules derived from one of several naturally occurring plasmids. These plasmids were classified on the basis of their DNA replication origin (replicon) and the encoded antibiotic-resistance gene(s). The most common plasmids used today contain the ColE1 *ori* which has a relaxed mode of replication. The term plasmid "vector" refers to the use of these biological reagents as transporting vehicles that carry foreign DNA from the test tube to the host cell. In addition to the plasmid vectors described in this section, various other DNA cloning vectors are presented below that have been derived from bacteriophage (λ cloning vectors), yeast (2μ and YAC vectors), insect viruses (baculovirus vectors), and animal viruses (retroviral vectors). The important features of a useful cloning vector are as follows:

1. Ability to replicate in host cells.
2. Genetic marker to select for host cells containing the vector.

TABLE 2.1 Antibiotics and antibiotic-resistance genes are important tools in applied Molecular Genetics

Antibiotic	Mode of action	Resistance gene	Application
Ampicillin	Inhibits cell wall synthesis by disrupting peptidoglycan cross-linking	β-Lactamase (*amp*^{r}) gene product is secreted and hydrolyzes ampicillin	*amp*^{r} gene is included on plasmid vectors as a positive selection marker
Tetracycline	Inhibits binding of aminoacyl tRNA to the 30S ribosomal subunit	*tet*^{r} gene product is membrane bound and prevents tetracycline accumulation by an efflux mechanism	*tet*^{r} gene is a positive selection marker on some plasmids (e.g., pBR322, F′ derivatives)
Kanamycin	Inactivates translation by interfering with ribosome function	Neomycin or aminoglycoside phosphotransferase (*neo*^{r}) gene product inactivates kanamycin by phosphorylation	*neo*^{r} gene is a positive selection marker on plasmids commonly used in eukaryotic molecular genetics
Bleomycin	Inhibits DNA and RNA synthesis by binding to DNA	The *bla*^{r} gene product binds to bleomycin and prevents it from binding to DNA	*bla*^{r} gene is a positive selection marker on plasmids and also used as a marker in eukaryotic cells (*zeo*)
Hygromycin B	Inhibits translation in prokaryotes and eukaryotes by interfering with ribosome translocation	Hygromycin-B-phosphotransferase (*hph* or *hyg*^{r}) gene product inactivates hygromycin B by phosphorylation	*hyg*^{r} gene is used as a positive selection marker in eukaryotic cells that are sensitive to hygromycin B
Chloramphenicol	Binds to the 50S ribosomal subunit and inhibits translation	Chloramphenicol acetyl transferase (*CAT* or *CM*^{r}) gene product metabolizes chloramphenicol in the presence of acetyl CoA	*CAT/CM*^{r} gene is used as a selectable marker, and as transcriptional reporter gene of promoter activity in eukaryotic cells

3. Unique restriction enzyme sites for insertional cloning.
4. Minimum amount of nonessential DNA.

In the mid 1970s, Herb Boyer and colleagues developed the first versatile plasmid cloning vector suitable for gene cloning and gave it the name pBR322. A feature of pBR322 that was important at the time to the public acceptance of recombinant DNA technology was that it lacked sequences required for plasmid mobilization and therefore could not be transferred from one bacterium to another. The pBR322 plasmid vector has a ColE1 replicon and is maintained in *E. coli* at ~20 copies per cell. pBR322 also contains both the *amp*^{r} and *tet*^{r} antibiotic-resistance genes. The total length of this double-strand circular plasmid is 4362 bp. As shown in Figure 2.5, foreign DNA can be cloned into unique restriction sites present in pBR322, the most useful of which are contained within the coding regions of either the *amp*^{r} or *tet*^{r} genes.

During the 1980s, numerous improvements were made in plasmid cloning vectors to enhance their suitability for gene cloning. Most notably, plasmid vectors were developed by molecular biology companies that specialized in providing purified DNA

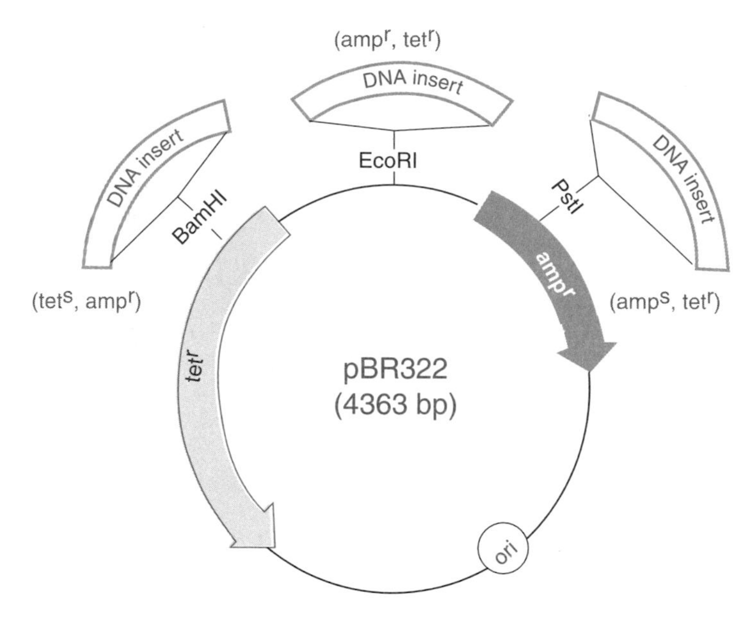

Figure 2.5 pBR322 was the first plasmid cloning vector that had all tho features required for gene cloning. Several restriction enzyme sites are shown that could be used as insertion sites for foreign DNA. Depending on the cloning strategy, the appropriate antibiotic-resistance phenotype was used to identify *E. coli* K-12 cells that contained the replicating plasmid. By replica plating the bacteria onto agar plates containing either tetracycline or ampicillin, it was possible to identify bacteria harboring recombinant plasmids with inserts into the BamHI or PstI sites, as indicated. Cloning strategies utilizing the EcoRI site required a plasmid purification screening step to identify *amp^r*, *tet^r* isolates that contained plasmids larger than the vector itself (>4.4 kb). Screening antibiotic-resistant colonies on the basis of plasmid size is also commonly used as a means to identify recombinant expression vector plasmids.

metabolizing enzymes. Some of the important features of these new and improved plasmid vectors follow:

1. Mutations in the ColE1 *ori* control sequences that result in a >10-fold increase in plasmid copy number (~500 copies/cell).
2. Addition of multiple unique insertional cloning sites to expand the number of strategies available for gene cloning.
3. Inclusion of the *N*-terminal coding sequence of the β-galactosidase α fragment to enable screening for gene insertions using X-gal indicator plates.
4. Incorporation of bacteriophage promoter sequences flanking the insertional cloning sites to facilitate in vitro RNA synthesis of foreign DNA using purified bacteriophage RNA polymerases.
5. Addition of the M13 bacteriophage replication origin to allow production of single-strand circular DNA by infection of the *E. coli* host cell with special helper phage.

Figure 2.6 illustrates the important features of a modern plasmid cloning vector.

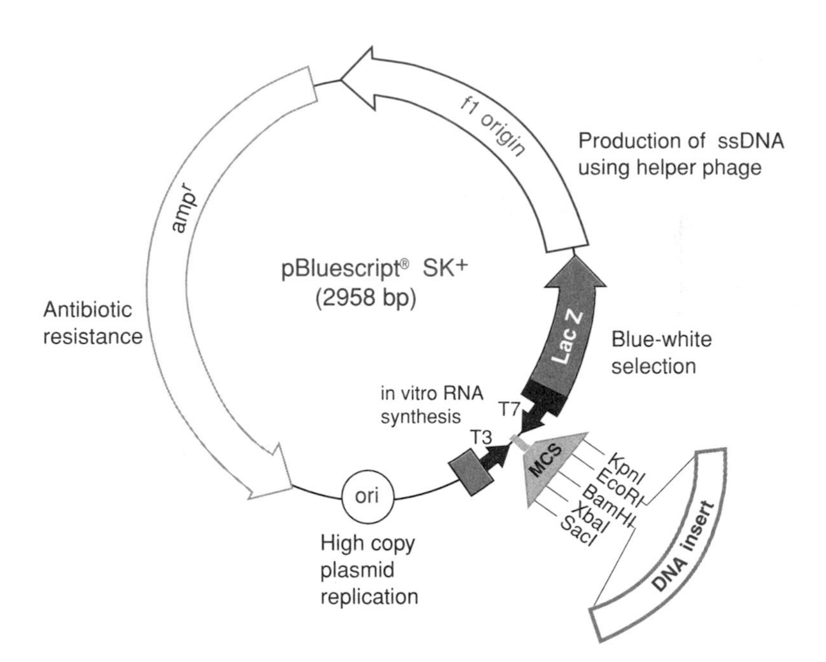

Figure 2.6 The Bluescript® SK+ cloning vector from Stratagene Cloning Systems (La Jolla, California) is an example of a commercially available plasmid. The essential features of this convenient plasmid vector include a modified ColE1 replication origin (ori) that increases plasmid copy number, the inclusion of a multiple cloning site (MCS) to facilitate the insertion of gene fragments with a variety of restriction enzyme site termini, the α fragment coding sequence of β-galactosidase flanking the MCS to permit blue-white colony screening on X-gal plates, T3 and T7 bacteriophage promoter sequences on either side of the multiple cloning site to direct the in vitro synthesis of RNA, and the f1 origin sequence to allow the production of single strand circular DNA following infection of the host bacterial cell with M13 helper phage.

Transferring DNA into *E. coli* by Transformation

The process of transferring exogenous DNA into cells, called "transformation," refers to any application in molecular genetics in which purified DNA is actively or passively imported into host cells using nonviral methods. A technical breakthrough in molecular genetics came in 1970 when a simple DNA transformation protocol was developed that permitted the transfer, and stable inheritance, of exogenous λ bacteriophage and bacterial genomic DNA into *E. coli* cells. Soon afterward, plasmids encoding antibiotic resistance genes were used in DNA transformation studies, and within 3 years, experiments based on recombinant DNA technology were reported.

There are basically two general methods for transforming bacteria, as illustrated in Figure 2.7. The first is a chemical method utilizing $CaCl_2$ and heat shock to promote DNA entry into cells. A second method is called electroporation; it uses a short pulse of electric charge to facilitate DNA uptake. The major difference between these two techniques is transformation efficiency. Electroporation is several orders of magnitude more efficient than transformation by $CaCl_2$.

The chemical method uses bacteria that are incubated with DNA on ice in a solution containing $CaCl_2$ followed by a brief heat shock at 42°C. These cells are allowed to recover for a short time in growth media, and then plated on agar plates containing

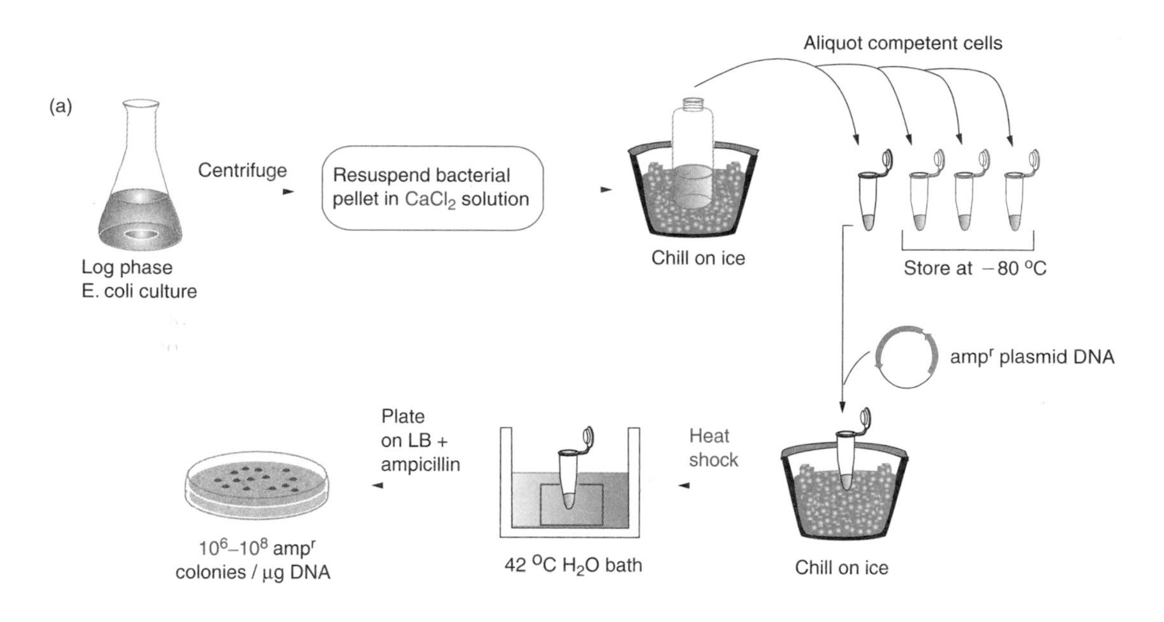

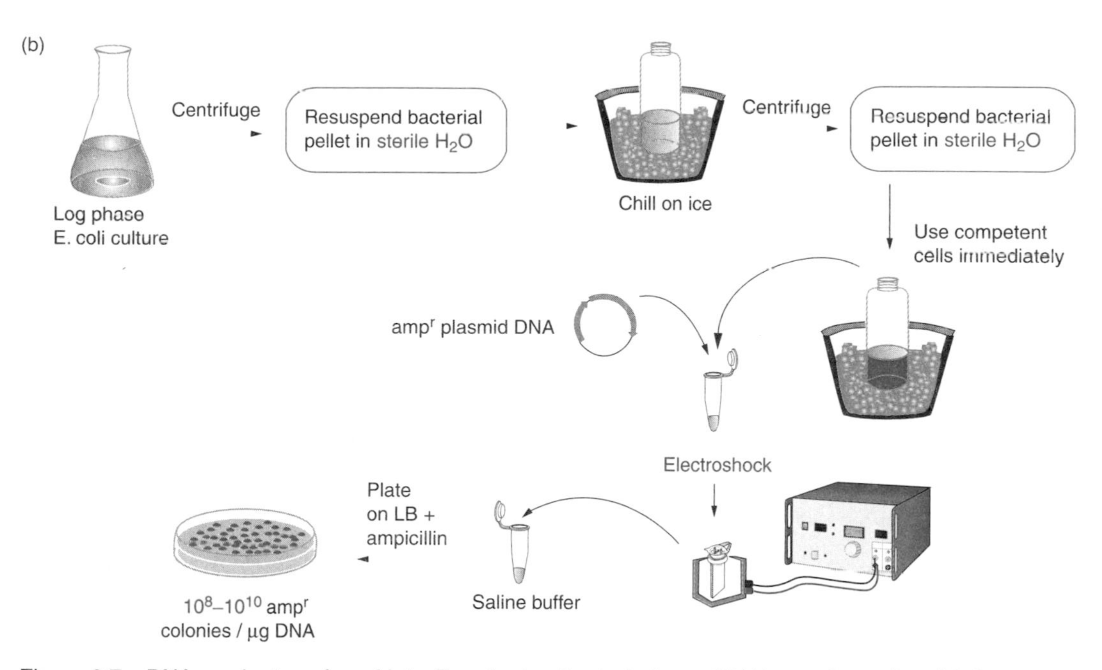

Figure 2.7 DNA can be transferred into *E. coli* using the technique of DNA transformation. (*a*) Chemical transformation utilizes $CaCl_2$ and heat to stimulate DNA uptake. This method is simple and convenient but transformation efficiencies are relatively low. (*b*) DNA transformation by electroporation is done by delivering a short electric pulse to cells that are suspended in low ionic strength solution in a special cuvette. Transformation frequencies using electroporation are 10^2 – 10^3 higher than with $CaCl_2$.

the appropriate antibiotic. In most cases, it is possible to obtain ~10^6 – 10^8 antibiotic-resistant colonies per microgram of circular plasmid DNA using this method. In typical gene cloning experiments, however, the number of transformed bacterial cells is much lower (~10^2 colonies per transformation). This is because the in vitro recombination frequencies are suboptimal and linear plasmid DNA is not replicated nor efficiently recircularized inside the cell. Nevertheless, this quick procedure is quite satisfactory, especially when the research objective is limited to forming uncomplicated recombinant molecules using purified insert and vector DNA. The mechanism of bacterial DNA transformation is not known, but some studies have suggested that the DNA enters the cell through protein pores that are activated by both calcium and heat. Numerous modifications of the original $CaCl_2$ procedure have been developed in an attempt to increase the transformation efficiency of various *E. coli* strains. For example, addition of hexamminecobalt chloride or dimethyl sulfoxide to the $CaCl_2$ solution was found to increase transformation efficiencies by 10- to 100-fold.

Electroporation was first applied to mammalian cells by molecular geneticists to overcome limitations of chemically based transformation methods that do not work well for unattached cells such as lymphocytes. By increasing the field strength using high voltage power sources, electroporation parameters were found that worked exceptionally well for bacteria. The mechanism of DNA transformation by electroporation is also poorly understood, but it is thought that the short pulse (~3 milliseconds) of electric charge (2.5 kilovolts, ~200 ohms, ~30 microfaradays) may cause a transitory opening in the cell wall, or perhaps activate membrane channels. The major advantages of electroporation are that transformation frequencies of 10^8 – 10^{10} transformants/μg plasmid DNA are routinely obtained, and once optimized, electroporation works equally well with most bacterial strains.

BACTERIOPHAGE CLONING VECTORS

Bacteriophages (phages) are bacterial viruses that infect their hosts by attaching to specific receptor proteins on the outside of the cell. Phages generally gain entry into the host cell by injecting their DNA directly into the cell or by being internalized by host cell processes. In most cases, once the phage genome is inside the host cell, phage proteins and phage DNA are synthesized and infectious particles are released through lysis or membrane budding. The ability to transfer DNA from the phage genome to specific bacterial hosts during the process of bacterial infection gave molecular geneticists the idea that specially designed phage vectors would be useful tools for certain types of gene cloning applications. Two of the most common phage-based methods are the use of gene cloning vectors constructed from components of the temperate bacteriophage λ and the production of single-strand DNA using cells infected with vectors based on the filamentous bacteriophage M13.

Biology of Bacteriophage λ

The life cycle of phage λ has been extensively studied as a model system of genetic control. λ infection of *E. coli* can result in either phage replication and cell lysis leading to the release of ~100 infectious particles, or the λ DNA can integrate into the *E. coli* genome as a prophage and remain dormant, creating a lysogenic host cell. Factors that control this genetic decision have been well characterized and shown to reflect transcriptional control of λ-encoded genes that are expressed shortly after infection. The primary event in this genetic "switch" between lysis and lysogeny is governed by the level of λ encoded proteins required for the lytic pathway. If one set of the λ immediate early genes is preferentially transcribed, then the lytic pathway is favored, however, if a transcriptional repressor protein encoded by the λ *cI* gene inhibits this transcription, then lysogeny is favored. Figure 2.8 illustrates how early transcriptional events during λ infection determine whether the outcome will be phage production and cell lysis, or lysogeny resulting in λ DNA integration into the bacterial chromosome.

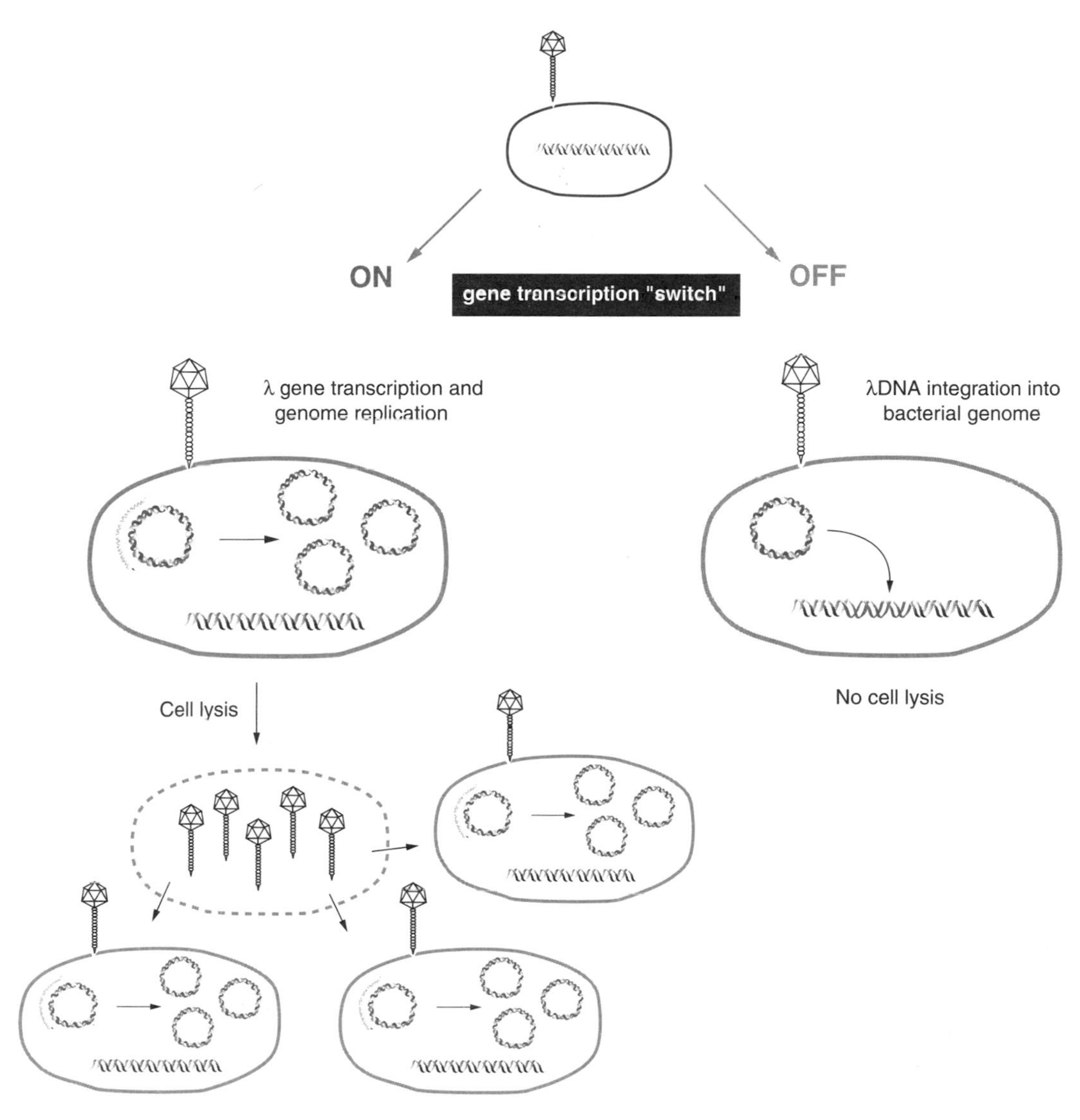

Figure 2.8 Infection of *E. coli* by λ phage results in the induction of the lytic cycle, or lysogeny, depending on transcriptional regulatory events controlling λ gene expression. If λ gene transcription is activated, then lytic infection occurs leading to phage release and infection of nearby *E. coli* cells. Alternatively, if λ gene transcription is repressed, then the infection is lysogenic, resulting in λ DNA integration into the *E. coli* genome.

The λ phage particle contains two major structural components called a head and a tail. The λ genome is a double-strand DNA molecule of 48,502 bp that is packaged within the phage head. The tail proteins are required for phage attachment to the *E. coli* outer membrane and injection of phage DNA into the host cell. The receptor protein for phage attachment is the product of the *E. coli lamB* gene which encodes a maltose transport protein. To maximize the efficiency of λ infection, *E. coli* cells are grown on maltose as the primary carbon source to induce expression of the *lamB* gene. Following phage attachment, the λ genome is injected into the host cell as a linear molecule that quickly forms a circle through the homologous base pairing of a 12-nucleotide single-strand sequence located at DNA termini. These λ DNA "cohesive" ends are joined together to form the *cos* site, which plays an important role in λ late gene transcription and DNA packaging in the lytic cycle.

Initially, λ DNA replication begins at the *ori* and proceeds bidirectionally to form a theta structure; however, as bidirectional replication continues, a rolling circle mechanism begins to produce linear forms of double-strand λ genomic DNA. Staggered cleavage of this large DNA concatamer at the regularly spaced *cos* sites is mediated by the λ terminase protein. Packaging of the newly replicated λ genomic DNA into phage head particles is coincident with *cos* site cleavage. Physical constraints on the amount of DNA that will fit into the phage head also control DNA packaging. Although the wild-type λ genome is 48 kb, DNA that is at least 38 kb but no more than 52 kb can be packaged provided that *cos* site cleavage occurs.

It is possible to isolate identical strains of λ phage genetically using the plaque formation assay as shown in Figure 2.9. By mixing a low ratio of phage particles to bacterium during the adsorption step, it ensures that each bacterium will be infected by only a single phage particle. This infection ratio is called the multiplicity of infection (MOI).

λ Phage Cloning Vectors

Studies of λ phage mutants showed that as much as 18 kb of the genome was dispensable for lytic growth, primarily by eliminating genes required specifically for lysogeny. This meant that portions of the λ genome could be replaced with foreign DNA using in vitro recombinant DNA methods. Moreover, because the amount of DNA that can be packaged in a phage head is somewhat flexible (38–52 kb), the exact length of insert DNA in the recombinant λ vector is not critical. Initially, λ DNA was isolated from a mutant phage that had lost the EcoRI sites normally located in genes required for lysis, but that had retained an EcoRI site within the dispensable region. By inserting foreign DNA into this single EcoRI site, and then transforming an appropriate *E. coli* strain, it was possible to recover infectious phage particles containing the recombinant λ vector.

The utility of λ vectors for gene cloning was significantly enhanced following the development of two major improvements in the general strategy. First, to overcome the low efficiency of Ca Cl_2-mediated DNA transformation (0.1% of the cells), as compared to phage infection (10% of the cells), protein extracts were developed that permitted ligated DNA to be packaged in vitro into infectious phage particles. Following the ligation of foreign DNA to the purified λ vector arms, the ligation reaction is then carefully mixed with an enriched protein fraction containing phage packaging proteins. Through an energy-independent association process involving highly ordered protein–protein interactions, the λ DNA is cleaved at appropriately spaced *cos* sites by λ terminase and packaged into infectious particles. Commercially available packaging extracts are commonly used to achieve phage titers of ~10^7 plaque forming units/microgram vector DNA.

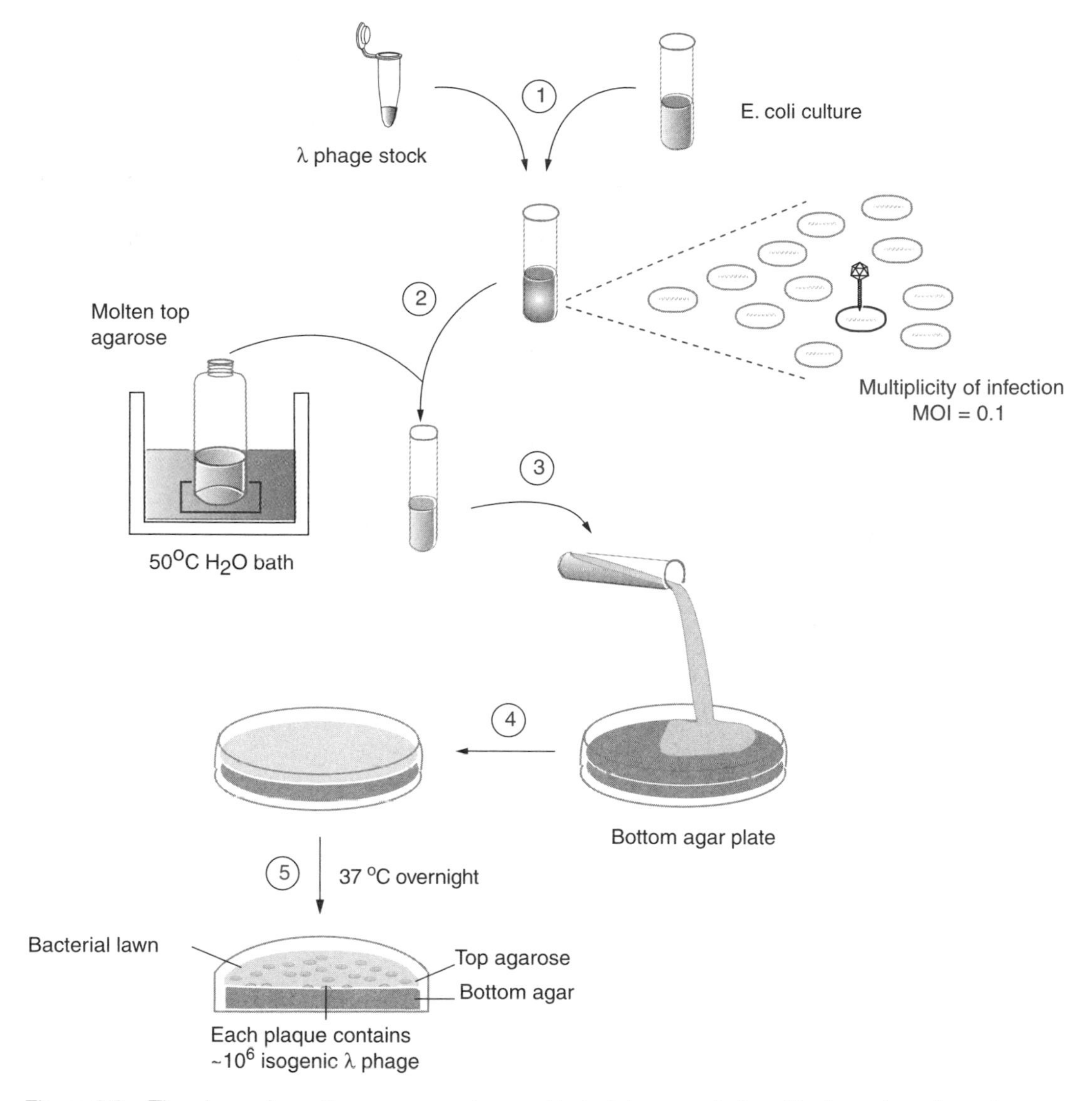

Figure 2.9 The plaque formation assay can be used to isolate pure strains of λ phage by using a low multiplicity of infection (MOI). Phage particles are combined with *E. coli* at a MOI of less than 1 and allowed to adsorb for a short period. Molten agarose is added to this mixture and quickly poured onto an LB agar plate. Once the agarose has solidified, the plate is incubated at 37°C. Plaques can be visualized as "holes" in the bacterial lawn corresponding to areas where *E. coli* lysis has occurred. Approximately 10^6 isogenic phage can be isolated from each plaque.

A second improvement in λ cloning systems led to an increase in the ratio of recombinant to nonrecombinant phage in the pool of infectious particles. This was a problem with the first generation of EcoRI-based λ vectors because uncleaved λ vector, or re-ligation of vector without an insert, resulted in high numbers of nonrecombinant phage. Therefore, several genetic strategies were devised to select genetically against nonrecombinant λ genomes. Two of these selection schemes are illustrated in Figure 2.10. One involves small DNA insertions (usually cDNA) into the λ *cI* gene, which controls the lysogenic pathway, and the other relies on replacement of vector "stuffer" DNA

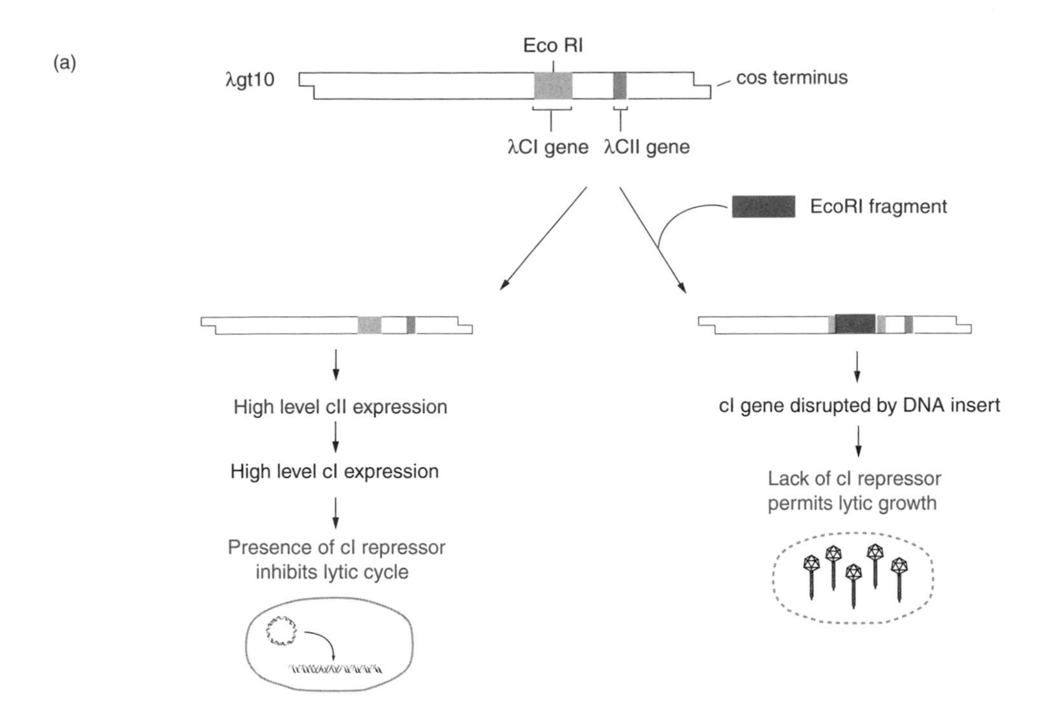

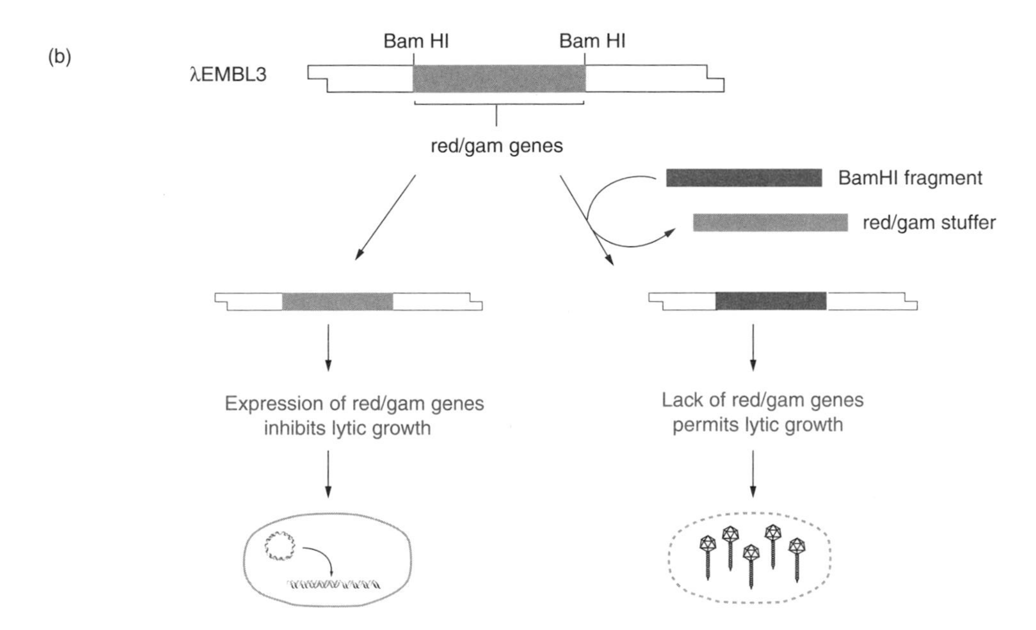

Figure 2.10 Genetic strategies are used to increase the number of recombinant λ vectors in a population of ligated molecules. (*a*) Insertional cloning into the *cI* gene of the λgt10 cDNA cloning vector (DNA inserts of ~1–5 kb) can be selected in *hfl* (high frequency of lysogeny) mutant strains of *E. coli*. In *hflA* strains of *E. coli*, expression of the λ *cII* gene is elevated, resulting in transcriptional induction of the λ *cI* repressor gene, which promotes lysogeny. Disruption of the λ *cI* coding sequence by DNA insertion into the unique EcoRI site of the λgt10 cDNA cloning vector blocks the lysogenic pathway leading to cell lysis and plaque formation. (*b*) Replacement of λ DNA containing the *red* and *gam* genes in the λEMBL3 genomic DNA cloning vector with BamHI compatible DNA inserts of ~10–20 kb permits lytic growth of recombinant phage in *E. coli* strains containing the P2 bacteriophage lysogen. Non-recombinant λEMBL3 vector DNA is not large enough to be efficiently packaged.

encoding two λ genes that promote lysogeny (*red*, *gam*), with large fragments of foreign DNA (genomic DNA 15–20 kb in size).

M13 Bacteriophage Cloning Vectors

Joachim Messing and his colleagues developed cloning vectors from the M13 isolate of an *E. coli* filamentous phage in the early 1980s, primarily as a means to obtain single-strand DNA for use as templates in dideoxy DNA sequencing reactions. Filamentous bacteriophage infect *E. coli* through the bacterial sex pili normally used for bacterial conjugation. Proteins required for sex pili formation are encoded on the *E. coli* F plasmid and therefore this genetic element must be present in the host cell (also referred to as a "male" cell). The M13 phage particle contains a single strand circular DNA genome of 6407 nucleotides that is released into the host cell as a result of phage absorption.

Circular (+) strand M13 DNA is replicated by host proteins shortly after infection to form the double-strand replicative form (RF). Circular RF DNA is replicated bidirectionally from a single origin to produce ~20–40 RF molecules/cell. Viral proteins required for M13 phage packaging are synthesized from mRNA transcribed off the (–) strand of the RF molecule. Late in the infection cycle, some of the RF molecules are nicked at a specific site by the M13 gene II encoded nuclease and rolling circle replication is initiated to produce large numbers of circular (+) strand progeny DNA. Single strand M13 DNA is packaged at the host cell membrane through interactions with phage proteins leading to the release of infectious particles. Although infection of *E. coli* cells with M13 inhibits bacterial cell division by about twofold, M13 viral particles are still produced at the rate of a ~100 particles/cell/hour.

The advantage of using M13 based cloning vectors to produce single strand templates for DNA sequencing reactions may be offset by the difficulty of obtaining sufficient amounts of the double-strand DNA (RF type) required for recombinant DNA reactions. In addition, the rolling circle mechanism of M13 replication can be affected by insert sequence and size, causing underrepresentation of certain recombinant molecules following infection. To overcome these problems, hybrid M13 phage/plasmid cloning vectors called "phagemids" were constructed in the early 1990s. The original phagemid vectors were derived from the high copy pUC18/pUC19 plasmids first developed by Messing. These phagemids include the minimal M13 sequences required for DNA replication and packaging, a multiple cloning site, the ColE1 replicon, and the *amp*^{r} gene. The Bluescript® SK+ cloning vector shown in Figure 2.6 is an example of an M13-based phagemid.

Figure 2.11 illustrates how single (+) strand phagemid DNA can be isolated from *E. coli* infected with a replication deficient M13 helper phage. To obtain both the (+) and (–) strands of the insert DNA as single strand templates, it is necessary to construct phagemids that differ in the orientation of the insert relative to the M13 replication sequences.

ESSENTIAL LABORATORY RESOURCES FOR MOLECULAR GENETIC RESEARCH

It should be clear from the topics discussed so far that both biochemistry, especially nucleic acid and protein biochemistry, and bacteriology are essential laboratory resources needed at some point for most all applied molecular genetic methods. How-

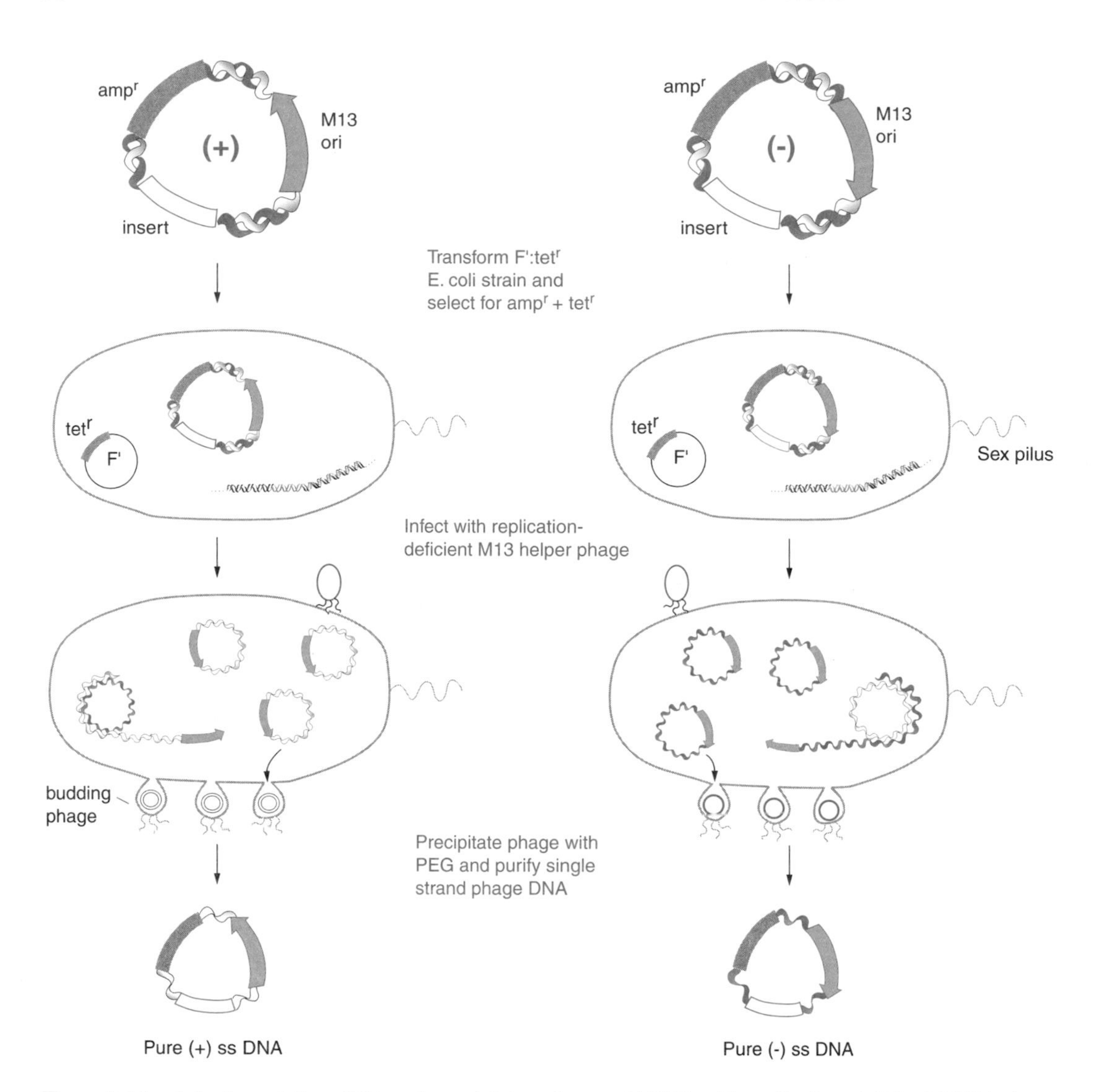

Figure 2.11 Infection of *E. coli* F+ cells containing phagemid DNA with replication-deficient M13 helper phage results in the packaging of single strand phagemid DNA. The orientation of the M13 replication origin (+ or –), relative to the insert DNA, in the phagemid vector determines which strand of the insert DNA (coding or noncoding) is contained in the packaged phage. Because the bacterial sex pilus proteins are encoded on the F′ plasmid, and the pilus is required for M13 phage attachment, *E. coli* strains have been developed that include a transposon-encoded antibiotic-resistance gene on the F′ plasmid (*tet^r* or *kan^r*). Rolling circle replication and phage packaging by helper virus proteins result in the production of recombinant M13 phage that can be used as a source of (+) or (–) strand circular DNA for subsequent strand-specific sequencing reactions or in vitro mutagenesis.

ever, these laboratory skills alone would not suffice for many of the advanced molecular genetic strategies discussed in later chapters. Two other resources are also required. The first is access to commercial products marketed under the heading "molecular biology reagents." The second resource, which is becoming increasingly more important, is access to computer facilities, or at least an ability to retrieve and store information downloaded from the Internet. Each of these four laboratory resources is summarized below:

1. *Laboratory resources for biochemical techniques.* These includes supplies required for wet lab inorganic chemistry and preparative and analytical biochemistry. For example, most protocols utilize some type of low ionic strength buffered solution to stabilize DNA and proteins. Essential equipment includes a top loading digital scale, pH meter, 37°C water bath, rotors and centrifuges, and cold storage (4°C and –20°C) and gel electrophoresis equipment. Another important resource is a radioisotope storage and work area; however, this may change in the future owing to biohazard concerns that have resulted in the development of nonisotopic detection methods. Resources for collecting biochemical data often include a need for photographic and spectrophotometric or fluorometric equipment. Data analysis may require specialized instrumentation, such as liquid scintillation counters to quantify radioactive samples, or imaging equipment, for example, a phosphorimager to document and quantify gel electrophoresis results. These analytical instruments, as well as equipment used for large-scale biochemical sample preparation, can be very expensive and are therefore usually purchased as shared equipment.
2. *Skills and reagents needed for bacteriological studies.* Knowledge of sterile culture techniques and methods for the isolation and maintenance of clonal isolates of *E. coli* are necessary; however, practical experience in this area is easy to obtain with minimal hands-on training. Supplies are needed to prepare standard bacterial growth media and agar plates, which require sterilization with an autoclave. Special equipment includes a 37°C oven for incubating bacterial plates, a 37°C shaking incubator to grow liquid cultures, and a tabletop and preparative centrifuge to harvest bacterial cells. With the exception of antibiotics and specific reagents required for the growth of certain *E. coli* strains, most other reagents would be stocked as standard biochemical supplies.
3. *Access to "molecular biology" reagents.* It would be impossible to perform even the most rudimentary molecular genetic application without access to purified restriction enzymes. Therefore, every lab using molecular genetic methods must have a -20°C enzyme freezer to store ~30 different restriction enzymes that are needed for gene cloning and analysis procedures. In addition, various DNA polymerases, DNA ligases and alkaline phosphatase enzymes must also be purchased and stored. Other reagents in this category include partial or complete "kits" for constructing DNA libraries and for preparing nucleic acid probes for membrane hybridization. Special DNA products are also routinely needed, for example, custom oligonucleotides, cloning and expression vectors, and molecular weight markers. Although many of the special biochemical and bacteriological reagents needed for applied molecular genetic research are commercially available, most can be prepared more economically in the lab.
4. *Computer-based resources.* Computers are an essential resource in any research lab because of their utility in data management and analysis. Many applied molecular genetic methods have an additional need for computer resources that support DNA sequence analysis software, specifically with regard to accessing DNA sequence databases. As described in Chapter 9, DNA sequence analysis is a valuable tool which can aid in the interpretation and planning of molecular genetic experiments. The type of computer resource needed for these analyses can be as simple as a personal computer with modem access to the Internet, or it may include an on-site license for DNA sequence analysis software. Although we do not discuss molecular modeling software in this book, powerful computer

graphics work stations are also becoming integral resources for a variety of molecular biological and biochemical investigations.

Laboratory Practicum 2. *Regulated expression of a cloned gene product in* E. coli

Research Objective

A plant biologist is interested in studying the calcium binding protein calreticulin in the plant *Arabidopsis thaliana* to determine its possible function in floral tissues. As a first step, he wants to generate antibodies that specifically recognize *Arabidopsis* calreticulins in order to determine which floral cell types express these proteins using immunocytochemistry. His plan is to express an *Arabidopsis* calreticulin coding sequence in *E. coli* to allow isolation of calreticulin antigen for polyclonal antibody production in rabbits. He searched the *Arabidopsis* expressed sequence tag (EST) database over the Internet using a known human calreticulin sequence and identified several candidate cDNA clones that were available from the *Arabidopsis* Biological Resource Center at Ohio State University.

Available Information and Reagents

1. A 1.4 kb *Arabidopsis* cDNA was obtained from the Resource Center and its complete DNA sequence was determined. It was found to encode a 425 amino acid protein with high amino acid homology to human calreticulin.
2. A pET$(His)_6$ bacterial expression vector, the pLysS plasmid encoding T7 lysozyme, and the BL21(DE3) *E. coli* λDE3 lysogen strain containing the T7 RNA polymerase gene under *lac* control have been obtained from a commercial source.
3. Biochemical supplies are available for purifying proteins containing a polyhistidine sequence from bacterial cell extracts using metal chelation chromatography.
4. Resources are available for polyclonal antibody production in rabbits using purified antigen. The plant biologist has expertise in plant immunocytochemistry and biochemistry.

Basic Strategy

The researcher has chosen the pET bacterial expression system because it has been shown to generate large amounts of protein suitable for antibody production. This IPTG-inducible expression system is based on the finding that bacteriophage T7 RNA polymerase is highly specific and efficient in transcribing genes containing a T7 promoter. An ampicillin-resistant plasmid vector, called pET, has been developed that contains a multiple cloning site downstream of a strong T7 promoter. By transforming a recombinant calreticulin pET vector into an *E. coli* strain containing the T7 RNA polymerase gene under *lac* control, it will be possible to induce calreticulin expression by treating the cells with IPTG (Fig. 2.12). Because low level expression of the T7 RNA polymerase gene occurs in this system in the absence of IPTG, and appreciable levels of the plant calreticulin protein may inhibit bacterial growth, a compatible plasmid containing the T7 lysozyme gene will also be introduced into the host cells. T7 lysozyme is a bifunctional protein that cleaves peptidoglycan linkages in the bacterial cell wall and also stoichiometrically inhibits the function of T7 RNA polymerase. Therefore, low level expression of T7 lysozyme inhibits the activity of T7 RNA polymerase prior to IPTG treatment.

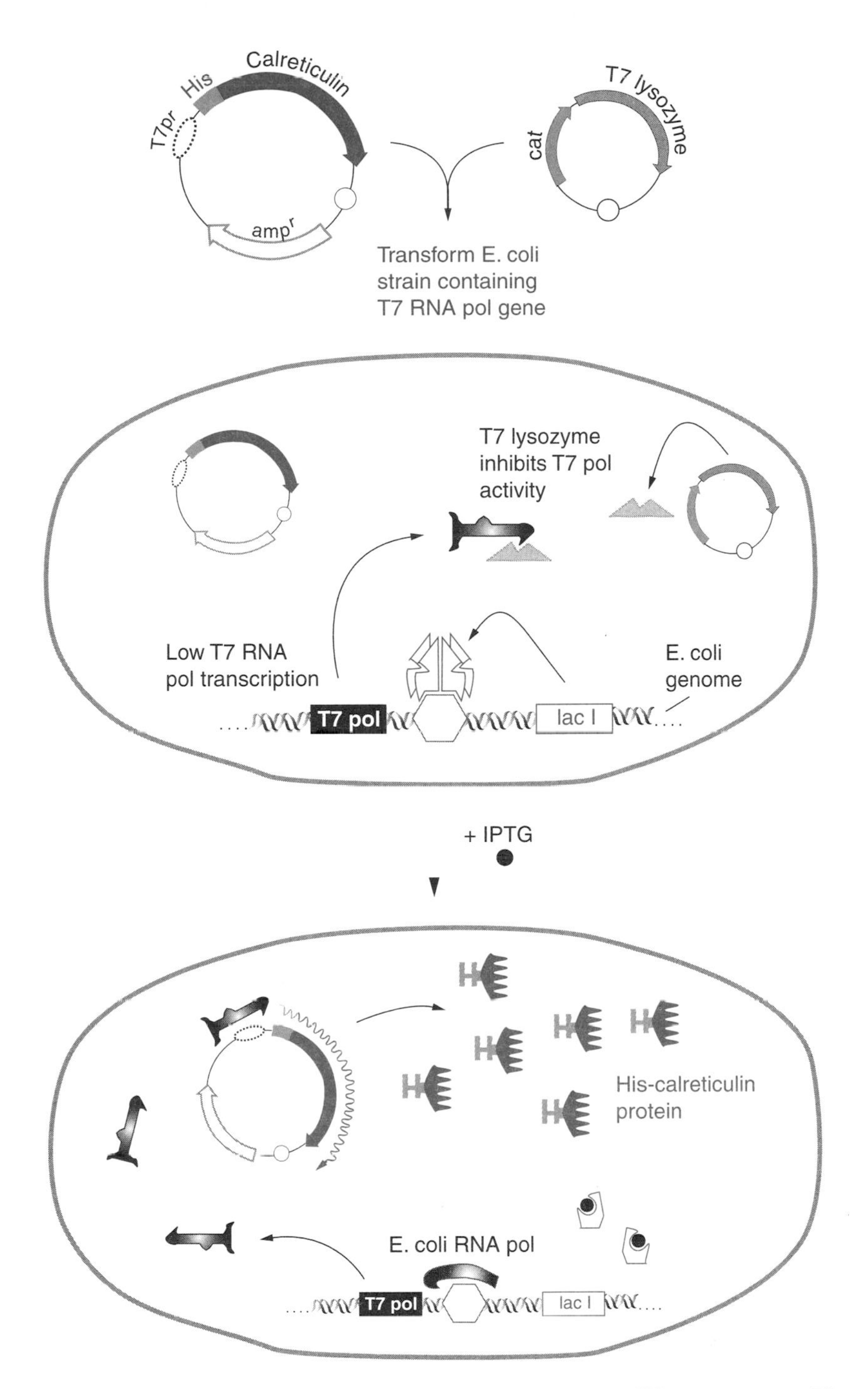

Figure 2.12 Regulated expression of genes in *E. coli* using the pET vector system. The two essential components of this T7 bacteriophage RNA polymerase-based expression system are the *amp*^{r} pET plasmid containing the coding sequence of a DNA insert downstream of the T7 promoter (Tpr), and an *E. coli* strain containing a genomic copy of the T7 RNA polymerase (T7 pol) gene under the control of the lac UV5 promoter. Also shown in this example is a T7 lysozyme encoding plasmid containing the chloramphenicol-resistance gene (*cat*). T7 lysozyme inhibits T7 RNA polymerase activity that is present even in the absence of IPTG. High level expression of the histidine-tagged calreticulin protein is obtained by blocking lacI repressor function with the addition of IPTG to the media.

The plan is to subclone the calreticulin cDNA obtained from the Biological Resource Center into the pET(His)$_6$ vector, and then establish an *E. coli* strain that has IPTG-inducible levels of calreticulin protein. Proteins with the (His)$_6$ polyhistidine sequence can be readily purified by affinity chromatography using divalent cations (Ni^{2+}) immobilized on a solid support. The purified calreticulin will be used as an antigen for polyclonal antibody production. Once these anti-calreticulin antibodies are

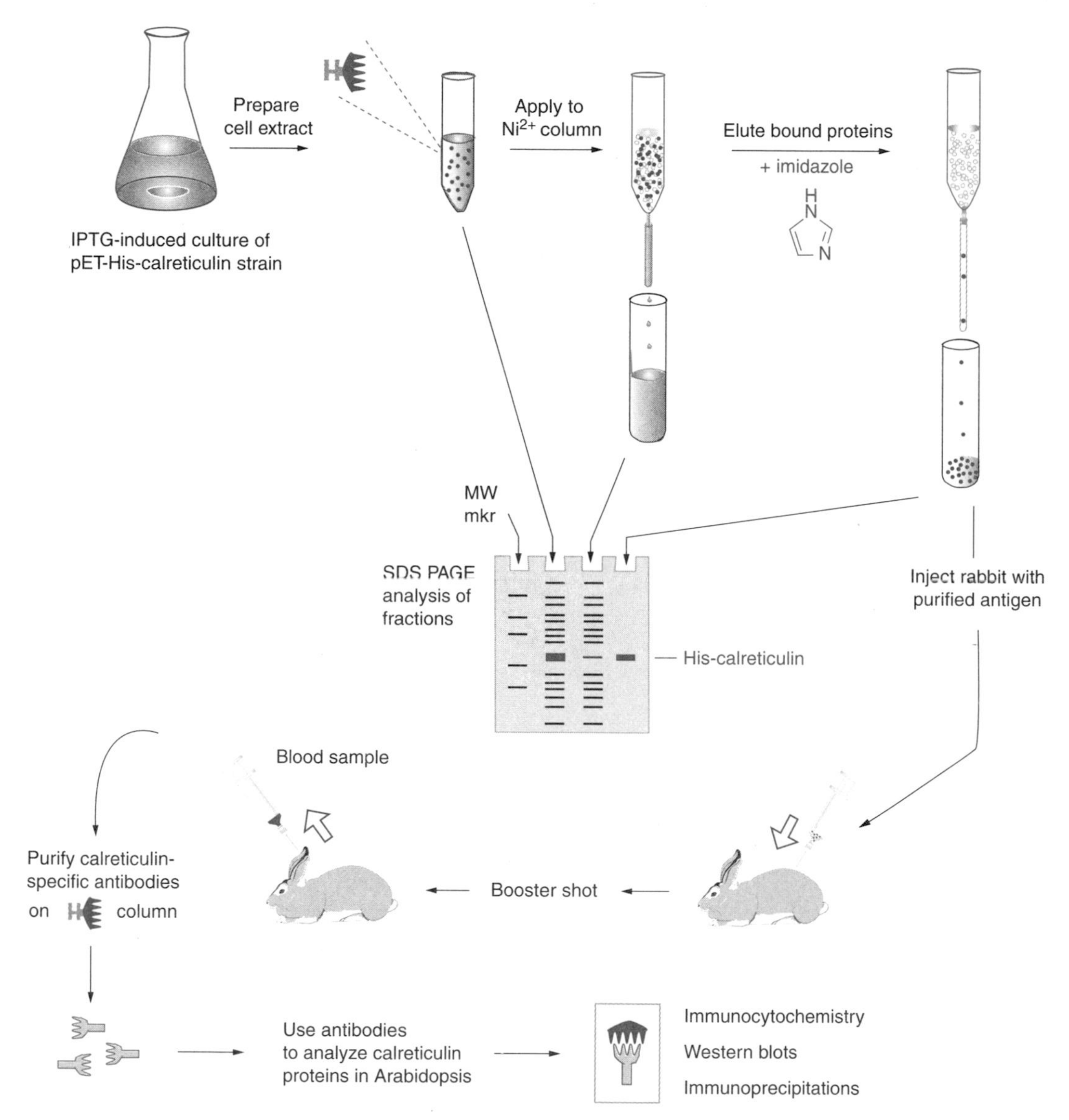

Figure 2.13 Flow scheme for calreticulin antibody production using purified protein from bacterial extracts. The polyhistidine tag contained within the recombinant calreticulin protein can be used to purify soluble protein from cell extracts using a Ni^{2+} affinity column. The bound calreticulin protein is eluted from the column with imidazole, which outcompetes the histidines for Ni^{2+} binding. Protein purification is monitored by gel electrophoresis (SDS-PAGE). In this example, calreticulin-specific polyclonal antibodies are produced in a rabbit and immunopurified with a calreticulin affinity column that is made using the calreticulin antigen protein from the *E. coli* extracts. The immunopurified calreticulin-specific antibodies could be useful as reagents in Western blots, immunocytochemical studies, and immunoprecipitation experiments.

available, it will be possible to characterize calreticulin protein in floral tissues by immunocytochemistry and protein blots (Western blots). Figure 2.13 illustrates the flow scheme for these experiments.

Comments

E. coli is a convenient host for expression of recombinant proteins and a number of vector systems have been developed. The two main features of these bacterial expression systems are regulated gene expression and protein purification by affinity chromatography. The $(His)_6$ fusion proteins are in general easy to purify but this added polypeptide sequence can sometimes interfere with protein structure or function. To avoid these problems, a protease cleavage site that is recognized by thrombin or enterokinase can be engineered into the cloning vector to permit removal of the $(His)_6$ moiety from the purified protein. The production of antigen in *E. coli* is the least demanding application of bacterial expression systems because denaturing conditions can be used to aid in protein solubilization. In fact, recombinant proteins expressed in *E. coli* are often sequestered in the periplasmic space as insoluble "inclusion" bodies. On the one hand this subcellular localization of expressed protein can be a useful aid in the purification of denatured protein from cell extract fractions enriched for inclusion bodies, but on the other hand, it can be very difficult to recover soluble recombinant proteins after a renaturation step that retain full biochemical function. To circumvent this protein insolubility problem encountered in *E. coli* expression systems, a number of alternative expression systems in prokaryotic and eukaryotic host cells have been developed (Chapter 7).

The production of polyclonal antibodies using recombinant proteins has become a standard procedure. Many researchers use commercial services that specialize in antibody production. Protein samples are sent to the service by express mail, and 7–10 weeks later, serum is sent back for analysis. Because the antigen is produced in *E. coli* and therefore available in large quantities, antigen-specific antibodies can be purified from the serum using affinity chromatography. Depending on the results obtained using polyclonal antibodies, the production of antigen-specific monoclonal antibodies may also be warranted.

Prospective

The plant biologist found that the anti-calreticulin antibodies detected high levels of the protein in developing seeds, a result consistent with its role as a chaperone protein in tissues with elevated rates of protein synthesis. During seed development, large amounts of storage proteins are synthesized that are later degraded as part of the germination process. It would be interesting to determine if calreticulin is required for seed development. This could be done, for example, by disrupting calreticulin function using cell-specific antisense genetics (Chapter 7) and by characterizing calreticulin in *Arabidopsis* mutants containing defects in seed development. Calreticulin function in the seeds could also be studied by isolating seed proteins that interact with the bacterial expressed calreticulin$(His)_6$ protein following absorption to a Ni^{2+} affinity column.

REFERENCES

Laboratory protocols in molecular genetics

Ausubel, F., Brent, R., Kingston, R., Moore, D., Seidman, J., Smith, J., Struhl, K. Current protocols in molecular biology. John Wiley & Sons, New York, 1996.

* Landmark papers in applied molecular genetics.

Glover, D.M., Hames, B.D. DNA cloning: A practical approach, Vols. 1–4. Oxford University Press, Oxford, 1995.

Sambrook, J., Fritsch, E.F., Maniatis, T. Molecular Cloning: A Laboratory Manual, 2nd ed., Cold Spring Harbor Laboratory, Cold Spring Harbor, N.Y., 1989.

Escherichia coli K-12: The bacterial host for all occasions

Bachmann, B.J. Linkage map of *Escherichia coli* K-12, edition 7. Microbiol. Rev. 47:180–230, 1983.

*Beckwith, J. A deletion analysis of the *lac* operator region in *E. coli*. J. Mol. Biol. 8:427–440, 1964.

*Berg, P., Baltimore, D., Brenner, S., Roblin, R., Singer, M. Asilomar conference on recombinant DNA molecules. Science 188:991–994, 1975.

Boyer, H.W. and Roulland-Dussoix, D. A complementation analysis of the restriction and modification of DNA in *Escherichia coli*. J. Mol. Biol. 41:459–472, 1969.

*Cohen, S., Chang, H., Boyer, H., Helling, R. Construction of biologically functional bacterial plasmids in vitro. Proc. Natl. Acad. Sci. USA 70:3240–3244, 1973.

*Jackson, D., Symons, R., Berg, P. Biochemical method for inserting new genetic information into DNA of simian virus 40: Circular SV40 DNA molecules containing lambda phage genes and the galactose operon of *E. coli*. Proc. Natl. Acad. Sci. USA 69:2904–2909, 1972.

Marinus, M.G. DNA methylation in *Escherichia coli*. Ann. Rev. Genet. 21:113–132, 1987.

Miller, J.H. The *lacI* gene: Its role in lac operon control and its uses as a genetic system. In The operon (J. Miller, ed.). Cold Spring Harbor Laboratory, Cold Spring Harbor, NY, pp. 31–88, 1978.

Oishi, M., Cosloy, S. The genetic and biochemical basis of the transformability of *E. coli* K12. Biochem. Biophys. Res. Commun. 49:1568–1575, 1972.

Raleigh, E.A., Murray, N.E., Revel, H., Blumenthal, R.M., Westaway, D., Reith, A.D., Rigby, P.W.J., Elhai, J., Hanahan, D. *McrA* and *McrB* restriction phenotypes of some *E. coli* strains and implications for gene cloning. Nucl. Acids Res. 16:1563–1575, 1988.

Ullman, A., Jacob, F., Monod, J. Characterization by in vitro complementation of a peptide corresponding to an operator-proximal segment of the β-galactosidase structural gene of *Escherichia coli*. J. Mol. Biol. 24:339–343, 1967.

Woodcock, D.M., Crowther, P.J., Doherty, J., Jefferson, S., De Cruz, E., Noyer-Weidner, M., Smith, S.S., Michael, M.Z., Graham, M.W. Quantitative evaluation of *Escherichia coli* host strains for tolerance to cytosine methylation in plasmid and phage recombinants. Nucl. Acids Res. 17:3469–3478, 1989.

Bacterial plasmids: the biological currency of applied molecular genetics

Balbas, P., Soberon, X., Merino, E. et. al. Plasmid vector pBR322 and its special purpose derivatives: A review. Gene 50:3–18, 1986.

*Bolivar, F., Rodriguez, R.L., Greene, P.J., Betlach, M.C., Heynecker, H.L, Boyer, H.W. Construction of useful cloning vectors. Gene 2:95–113, 1977.

*Calvin, N.M., Hanawalt, P.C. High-efficiency transformation of bacterial cells by electroporation. J. Bacteriol. 170:2796–2801, 1988.

Cesareni, G., Banner, D.W. Regulation of plasmid copy number by complementary RNAs. Trends Biochem. Sci. 10:303–308, 1985.

Chung, C.T., Niemela, S.L., Miller, R.H. One-step preparation of competent *Escherichia coli*: Transformation and storage of bacterial cells in the same solution. Proc. Natl. Acad. Sci. U.S.A. 86:2172–2175, 1989.

*Cohen, S., Chang, A., Hsu, L. Nonchromosomal antibiotic resistance in bacteria: Genetic transformation of *E. coli* R-factor DNA. Proc. Natl. Acad. Sci. USA 69:2110–2114, 1972.

Dagert, M., Ehrlich, S.D. Prolonged incubation in calcium chloride improves competence of *Escherichia coli* cells. Gene 6:23–28, 1974.

Foster, T.J. Plasmid-determined resistance to antimicrobial drugs and toxic metal ions in bacteria. Microbiol. Rev. 47:361–409, 1983.

*Hanahan, D. Studies on transformation of *Escherichia coli* with plasmids. J. Mol. Biol. 166: 557–580, 1983.

Kahn, M., Kolter, R., Thomas, C., Figurski, D., Meyer, R., Remaut, E., Helinski, D.R. Plasmid cloning vehicles derived from plasmids ColE1, F, R6K, and RK2. Methods Enzymol. 68: 268–280, 1979.

*Mandel, M., Higa, A. Calcium-dependent bacteriophage DNA infection. J. Mol. Biol. 53: 159–162, 1970.

*Melton, D.A., Krieg, P.A., Rebagliati, M.R. Maniatis, T., Zinn, K., Green, M.R. Efficient in vitro synthesis of biologically active RNA and RNA hybridization probes from plasmids containing a bacteriophage SP6 promoter. Nucl. Acids Res. 12:7035–7056, 1984.

Moazed, D., Noller, H.F. Interaction of antibiotics with functional sites in 16S ribosomal RNA. Nature 327:389–394, 1987.

*Sutcliffe, J.G. Complete nucleotide sequence of the *Escherichia coli* plasmid pBR322. Cold Spring Harbor Symp. Quant. Biol. 43:77–90, 1978.

Waxman, D.J., Strominger, J.L. Penicillin-binding proteins and the mechanism of action of β-lactam antibiotics. Ann. Rev. Biochem. 52:825–851, 1983.

*Weinstock, G.M., Rhys, C., Berman, M.L., Hampar, B., Jackson, D., Silhavy, T.L., Weisemann, J., Zweig, M. Open reading frame expression vectors: A general method for antigen production in *Escherichia coli* using protein fusions to β-galactosidase. Proc. Natl. Acad. Sci. USA 80:4432–4436, 1983.

Bacteriophage cloning vectors

Albright, R., Matthews, B. How Cro and λ-repressor distinguish between operators: The structural basis underlying a genetic switch. Proc. Natl. Acad. Sci. USA 95:3431–3436, 1998.

*Blattner, F.R., Williams, B.G., Blechl, A.E., et al., O. Charon phages: Safer derivatives of bacteriophage lambda for DNA cloning. Science 196:161–169, 1977.

Enquist, L.W., Skalka, A. Replication of bacteriophage λ DNA is dependent on the function of host and viral genes. I. Interaction of *red*, *gam* and *rec*. J. Mol. Biol. 75:185–198, 1973.

Hendrix, R. Bacteriophage DNA packaging: RNA gears in a DNA transport machine. Cell 94: 147–150, 1998.

*Herskowitz, I. Control of gene expression in bacteriophage lambda. Ann. Rev. Genet. 7: 289–323, 1973.

*Hohn, B., Murray, K. Packaging recombinant DNA molecules into bacteriophage particles in vitro. Proc. Natl. Acad. Sci. USA 74:3259–3264, 1977.

Johnson, A.D., Meyer, B.J., Ptashne, M. Interactions between DNA bound repressors govern regulation by lambda phage repressor. Proc. Natl. Acad. Sci. USA 76:5061–5065, 1979.

*Leder, P., Tiemeier, P., Enquist, L. EK2 derviatives of bacteriophage lambda useful in cloning DNA from higher organisms: The λgtWES system. Science 196:175–179, 1977.

*Muller-Hill, B., Crapo, L., Gilbert, W. Mutants that make more lac repressor. Proc. Natl. Acad. Sci. USA 59:1259–1263, 1968.

*Norrander, J., Kempe, T., Messing, J. Construction of improved M13 vectors using oligonucleotide-directed mutagenesis. Gene 26:101–106, 1983.

*Ptashne, M. A genetic switch: Gene control and phage lambda. Blackwell Scientific Publications, Oxford, 1986.

Rambach, A., Tollais, P. Bacteriophage lambda having EcoRI endonuclease sites only in the nonessential region of the genome. Proc. Natl. Acad. Sci. USA 71:3927–3932, 1974.

*Short, J.M., Fernandez, J., Sorge, J., Huse, W. λZap: A bacteriophage λ expression vector with in vivo excision properties. Nul. Acids Res. 16:7583–7590, 1988.

*Thomas, M., Cameron, J., Davis, R.W. Viable molecular hybrids of bacteriophage lambda and eukaryotic DNA. Proc. Natl. Acad. Sci. USA 71:4579–4584, 1974.

*Yamamoto, K.R., Alberts, B.M., Benzinger, R., Lawhorne, L., Treiber, G. Rapid bacteriophage sedimentation in the presence of polyethyleneglycol and its application to large scale virus production. Virology 40:734–743, 1970.

Yanisch-Perron, C., Vieira, J., Messing, J. Improved M13 phage cloning vectors and host strains: Nucleotide sequences of the M13mp18 and pUC19 vectors. Gene 33:103–119, 1985.

Regulated expression of a cloned gene product in E. coli

Chen, W., Tabor, S., Struhl, K. Distinguishing between mechanisms of eukaryotic transcriptional activation with bacteriophage T7 RNA polymerase. Cell 50:1047–1055, 1987.

*de Boer, H.A., Comstock, L.J., Vasser, M. The tac promoter: A functional hybrid derived from the trp and lac promoters. Proc. Natl. Acad. Sci. USA 80:21–25, 1983.

*Edman, J.C., Hallewell, R.A., Valenzuela, P., Goodman, H.M., Rutter, W.J. Synthesis of Hepatitis B surface and core antigens in *E. coli*. Nature 291:503–506, 1981.

Nakai, H., Richardson, C.C. Interactions of the DNA polymerase and gene 4 protein of bacteriophage T7. J. Biol. Chem. 261:15208–15216, 1986.

Nelson, D.E., Glausinger, B., Bohnert, H.J. Abundant accumulation of the calcium-binding molecular chaperone calreticulin in specific floral tissues of *Arabidopsis thaliana*, Plant Physiol. 114:29–37, 1997.

*Studier, F.W., Moffatt, B.A. Use of bacteriophage T7 RNA polymerase to direct selective high-level expression of cloned genes. J. Mol. Biol. 189:113–130, 1986.

*Studier, F.W., Rosenberg, A.H., Dunn, J.J., Dubendorff, J.W. Use of T7 RNA polymerase to direct the expression of cloned genes. Methods Enzymol. 185:60–89, 1990.

*Tabor, S., Richardson, C.C. A bacteriophage T7 RNA polymerase/promoter system for controlled exclusive expression of specific genes. Proc. Natl. Acad. Sci. U.S.A. 82:1074–1078, 1985.

CHAPTER

3

OVERVIEW OF GENE ANALYSIS METHODS

This chapter describes how novel gene sequences are initially isolated from DNA libraries through a process called "library screening." In the first part of this chapter, an overview is given of the general steps involved in constructing and screening a λ phage DNA library. This is followed by a description of methods used for gene characterization based on restriction enzyme mapping and plasmid subcloning. In the last section of this chapter, strategies are presented that allow the researcher to accelerate evolution in a test tube by performing in vitro mutagenesis. This important experimental approach is often the first step required to characterize functionally a newly isolated gene sequence. Laboratory practicum 3 describes how to use an in vitro mutagenesis strategy, called alanine scanning, to map systematically amino acid residues required for biological function.

DNA LIBRARIES CONTAIN A COLLECTION OF GENE SEQUENCES

There are two basic types of gene sequences represented in DNA libraries. First, there are contiguous DNA segments representing the exact nucleotide sequence of a chromosomal region. These are called genomic DNA fragments and they can contain gene coding sequences found in exons, noncoding gene segments in introns, untranscribed 5′ and 3′ regions of the gene, and transcriptional regulatory regions. A second type of gene sequence used to construct DNA libraries is complementary DNA (cDNA), which is derived from converting single-strand cellular mRNA into double-strand DNA using the enzyme reverse transcriptase (Chapter 5). Because cDNA libraries contain only gene sequences from transcribed regions of the genome, and gene expression is temporally and spatially controlled in a highly regulated manner, a large number of cDNA libraries must be constructed to represent the many different cell types of a multicellular organism. For example, cDNA libraries from mouse liver, heart, brain, and kidney have been constructed as a means to capture mRNA sequences present in these different cell types. In contrast, a single mouse genomic library made from mouse liver genomic DNA is

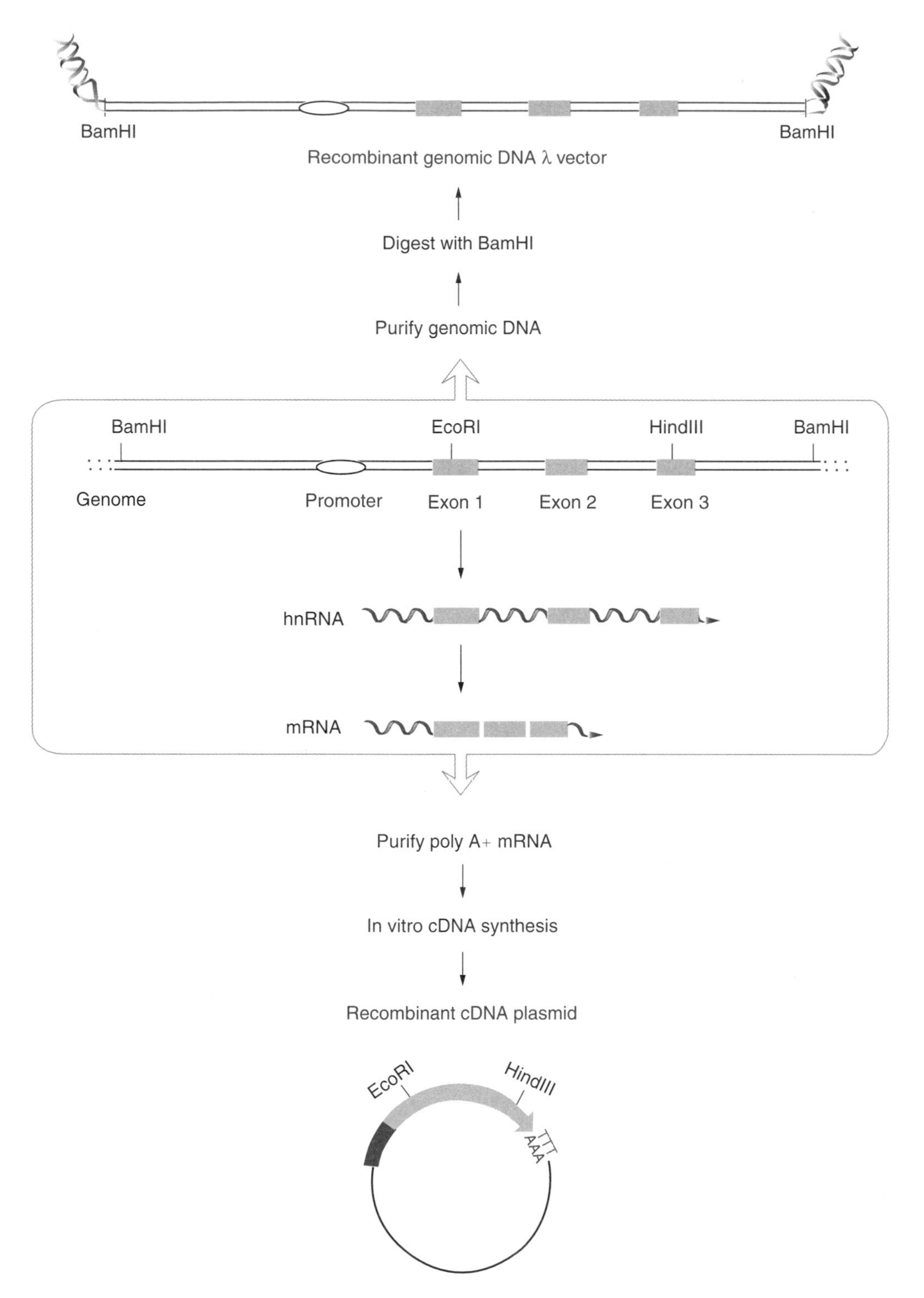

Figure 3.1 Genomic DNA libraries contain linear segments of the genome and cDNA libraries contain double strand DNA copies of transcribed gene sequences. Purifying genomic DNA from a cell provides material that can be cloned directly into a λ phage vector using restriction enzyme sites present in the endogenous gene. The genomic DNA insert is identical to the corresponding segment of DNA present in the chromosome. In contrast, cDNA is an in vitro synthesized product that is generated from purified preparations of polyadenylated (A+) mRNA. The cDNA insert contains only those sequences present in the processed mRNA specifically lacking gene regulatory elements and intronic regions.

sufficient to capture the entire mouse genome. Figure 3.1 illustrates the major difference between genomic and cDNA libraries.

Representative DNA Libraries

A DNA library is considered "representative" if the combined set of DNA segments contained within the library corresponds to all possible DNA sequences present in the original sample. Moreover, the relative abundance of DNA segments should be maintained when converting the sample material into a library. For example, it should be possible to isolate DNA segments from a representative genomic DNA library that together cover an entire chromosome. The term representative also applies to cDNA libraries. In this case, the relative abundance of each cDNA clone reflects the steady-state level of RNA transcripts in the original sample used to synthesize the cDNA. Because the steady state level of albumin mRNA in a liver cell is >100 times higher than that of glucocorticoid receptor mRNA, a representative mouse liver cDNA library should contain many more clones containing albumin cDNA than glucocorticoid receptor cDNA sequences.

A critical parameter to consider when constructing a representative library is how redundant the DNA segments are with regard to containing overlapping regions of a particular gene. Tom Maniatis was the first to develop a strategy for constructing genomic libraries that contain random overlapping DNA segments. The principle of this approach is shown in Figure 3.2. Cloning vectors contain only a limited amount of insert DNA, and more importantly, the termini of insert DNA must be compatible with the cloning vector. One commonly used method to obtain overlapping genomic DNA segments is to digest the DNA with a restriction enzyme, such as Sau3A, that has the four base recognition sequence GATC. By using reaction conditions that result in incomplete cleavage of the genomic DNA into Sau3A fragments (Figure 3.2), it is possible to generate random fragments with termini that can be ligated into appropriate cloning vectors (Chapter 4). Although the enzymatic manipulations are different, representative cDNA libraries can also be constructed using randomly primed mRNA that results in the synthesis of overlapping cDNA clones (Chapter 5).

Theoretical calculations can be made to determine how many individual clones should be present in a representative library such that a given gene sequence is likely to be included. The number of primary recombinants contained in a library is fixed at the time of construction and reflects the relative efficiency of the combined cloning steps. The following statistical equation can be used to calculate how many primary recombinants should exist in a representative human genomic library;

$$N = \frac{\ln(1 - P)}{\ln(1 - f)}$$

In this equation, P is the desired probability, f is the fraction of the entire genome contained within a single clone, and N is the number of recombinants required. A representative genomic library with a 99% chance of containing any given 20 kb segment of human DNA (human genome size is 3×10^6 kb) should therefore have ~700,000 primary recombinants. It is less reliable to calculate the estimated number of primary recombinants required for a cDNA library because the complexity (relative abundance of different RNA species) of the RNA pool from any given cell type is not known. Nevertheless, based on statistical calculations using conservative estimations of RNA com-

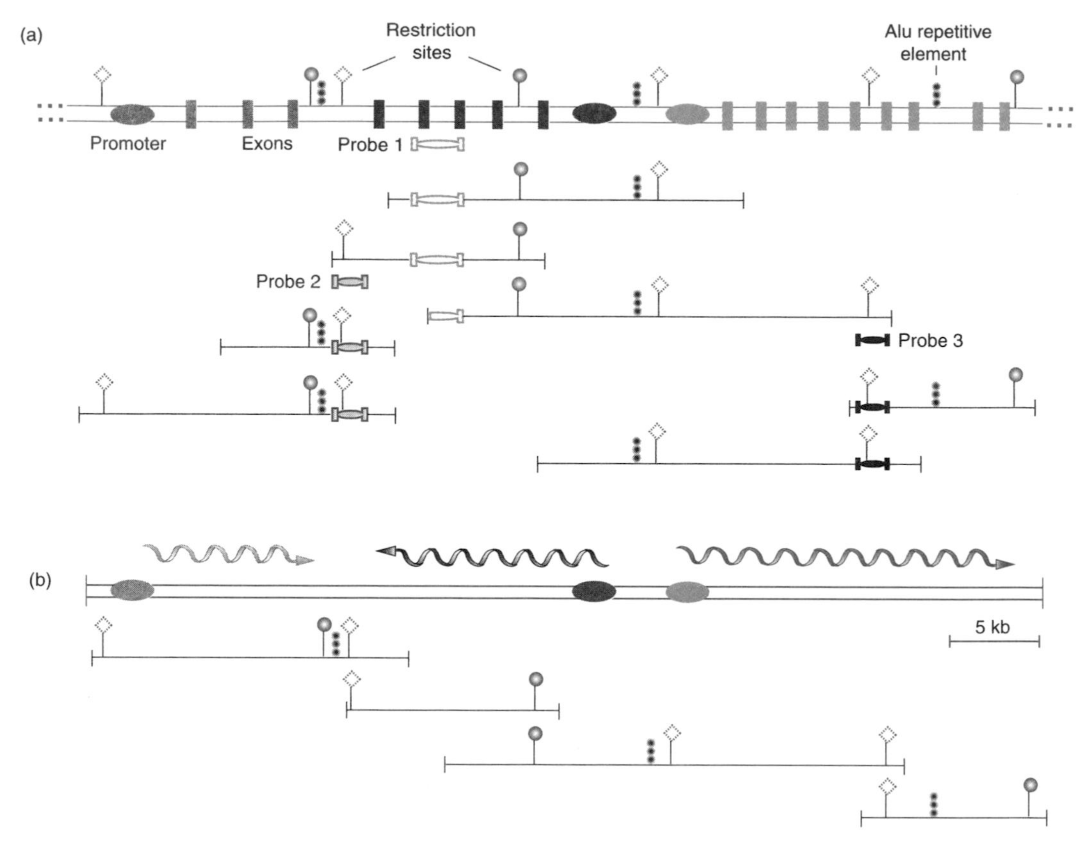

Figure 3.2 A representative genomic library contains multiple clones with overlapping gene segments. (*a*) A collection of gene segments can be isolated from a representative library by screening with radioactive probes derived from previously cloned DNA segments. Recombinant DNA fragments isolated from the first library screening with probe 1 can be used to rescreen the library with probes 2 and 3 to obtain additional overlapping segments. Landmarks such as rare restriction sites and repeated DNA sequences (Alu repetitive element) are used to align the set of isolated genomic DNA clones. (*b*) A composite set of minimal overlapping DNA fragments can be assembled using individual clones that have been isolated from a representative genomic library. In this example, three transcribed genes were found to be contained within the ~50 kb of cloned genomic DNA.

plexity and on empirical determinations, it is generally considered that a representative cDNA library should contain a minimum of ~500,000 primary recombinants.

The last step in constructing a representative DNA library for general use is to amplify uniformly the number of primary recombinants in the library so that the same set of clones can be screened repeatedly using different probes. For λ phage libraries, this is done by plating out the entire library onto a large number of plates and then collecting the phage by diffusion into a phage suspension buffer. Because a phage plaque contains ~10^6 infectious particles, this results in a 10^6 fold amplification of the primary library. Bacterial plasmid and cosmid libraries are amplified in a similar way by recovering bacterial colonies from agar plates. Although library amplification is the best way to preserve a particular set of gene sequences for future use, there is always the danger that not all recombinants will be represented at the same relative abundance after amplification. Because differences in cell division times are minimized by the limitation of nutrients in any given area of the plate, primary DNA libraries are amplified by agar plating, rather than by using liquid cultures.

TABLE 3.1 Common cloning vectors used to construct DNA libraries

Vector type	Insert size (kb)	Representative vectors	Major advantage	Major disadvantage
Genomic Libraries				
λ Phage	9–23	EMBL3/4 λ DASH, λGEM	Easy to propagate and amplify	Relatively small insert size
Cosmid	30–45	pWE15, Lawrist4 SuperCos	Larger insert size than λ vectors	More difficult to screen than λ libraries
P1 phage (PAC)	50–100	pNS358, pNS582	Can accommodate inserts of 100 kb	Packaging of P1 recombinants is variable
F factor (BAC)	50–300	pBAC108L, pBeLoBAC	Large inserts in a bacterial host	Can be difficult to transform bacteria
YAC (yeast)	200–1000	pYAC4, pYACneo	Entire human genome contained in 5000 clones	Large YAC library is stored in ~500 microtiter plates
cDNA Libraries				
λ phage	0.5–7	λgt10, λgt11, λZap	Easy to screen and store the library	Few, if any, disadvantages
Plasmid	0.5–8	pUC, pBS pAD-GAL4, pCDM8	Highly versatile cloning and screening strategies	Can have low transformation efficiency

Cloning Vectors Used for DNA Library Construction

Subdividing genomic DNA into clonable fragments using restriction enzymes is the first step in library construction. However, the choice of enzyme and the reaction conditions used for the digestion determine what type of cloning vector is most suitable for the intended library. An important consideration is the desired size of the insert, for example, 20 kb for λ bacteriophage vectors, all the way up to 1000 kb for yeast artificial chromosome vectors. Similarly, depending on how a cDNA library is going to be screened, for example, by nucleic acid hybridization or protein expression, it is important to choose the most suitable cloning vector prior to library construction. Table 3.1 lists the major cloning vectors used for constructing DNA libraries. The λ bacteriophage vectors are easy to store as a collected pool and the gene isolation procedures (screening) are straightforward. Table 3.1 also lists several specialized genomic DNA cloning vectors that have been developed to take advantage of minimal requirements for autonomous chromosomal replication (BAC and YAC) or λ packaging (cosmids).

SCREENING A λ PHAGE LIBRARY BY DNA HYBRIDIZATION

The basic reason for constructing a DNA library is that it provides a pool of gene sequences that can be accessed using biochemical methods, in a way analogous to finding World Wide Web (WWW) sites on the Internet using a search algorithm. Just as the choice of search parameters in a WWW browsing query influences what sites are found, the choice of reagents and strategy used for library screening determines the type of gene sequences that are isolated. The most common library screening strategy uses a radioactive DNA probe in a membrane hybridization reaction to locate physically plaques or bacterial colonies that contain complementary DNA segments. This method is equivalent to spreading out 500,000 jigsaw puzzle pieces faceup on a very large table, to search for the one piece corresponding to Waldo's red and white striped cap in a Where's Waldo™ puzzle!

The first step in screening a λ library is to infect *E. coli* with the recombinant phage stock at a ratio of phage to bacterium that favors multiple rounds of infection and plaque formation on agar plates. This is usually determined empirically by titering a phage stock with a fixed amount of freshly prepared *E. coli* cells. An infection ratio of ~1:10 (phage:bacterium) is the general range that gives a multiplicity of infection (MOI) of 0.1 (Chapter 2). The infected *E. coli* are then mixed with molten agarose using an amount of cells that will result in approximately 30,000 plaques per large petri plate. By pouring a total of 15–20 plates in this manner, it is possible to separate out ~500,000 recombinant phage as shown in Figure 3.3.

Once the plaques are formed on each agar plate, the library is ready to be screened by DNA hybridization. Figure 3.4 shows the basic protocol used for λ library screening based on the plaque lift method. The most important step in this screening strategy is the ability to realign the exposed film with the master plate in order to identify the precise region of the plate containing the candidate plaque. One way to do this is to use registration marks on the filter that accurately fix the relative position of the filter and the agar plate. Several asymmetric needle punctures through the filter can be used to produce a unique pattern that permits alignment of the master plate, replica filters, and film. Because false-positive signals can arise from imperfections in the top agar or membrane, it is common to make two identical plaque lifts from the same agar master plate using the registration marks as a guide. If the same hybridization signal is observed on both filters, then it is likely that the corresponding plaque on the master plate contains the desired recombinant λ phage (Figure 3.4).

The final step in library screening is to purify the candidate phage by limiting dilution as shown in Figure 3.5. It is at this step of plaque purification where positive signals are confirmed and false-positive phages are discarded. The purified λ phage in the final stage is called "plaque pure," meaning that 100% of the phage in the tertiary plating hybridizes to the probe.

RESTRICTION ENZYME MAPPING AND SUBCLONING

As described in Chapter 1, restriction enzymes are endonucleases that bind to specific DNA sequences and create double strand DNA breaks by enzymatic cleavage. Because restriction enzyme recognition sites are highly specific, it is possible to use commercially available enzyme preparations to map cleavage sites within cloned DNA using agarose gel electrophoresis. The mapping strategy involves several rounds of DNA digestion with infrequent cutting enzymes (recognition site of 6–8 bp), followed by agarose gel electrophoresis to determine fragment lengths. Using this information, a tentative restriction map can be constructed as the basis for designing a second round of experiments using multiple enzyme combinations.

The first task when constructing a restriction map is to identify one or more unique restriction sites in the cloning vector to anchor a relative map position (5′ or 3′ of the insert). Figure 3.6 shows how a restriction map can be constructed for a large genomic DNA insert within the EMBL3A λ cloning vector. Although it is possible to construct a more detailed restriction map using additional enzyme digestions in various combina-

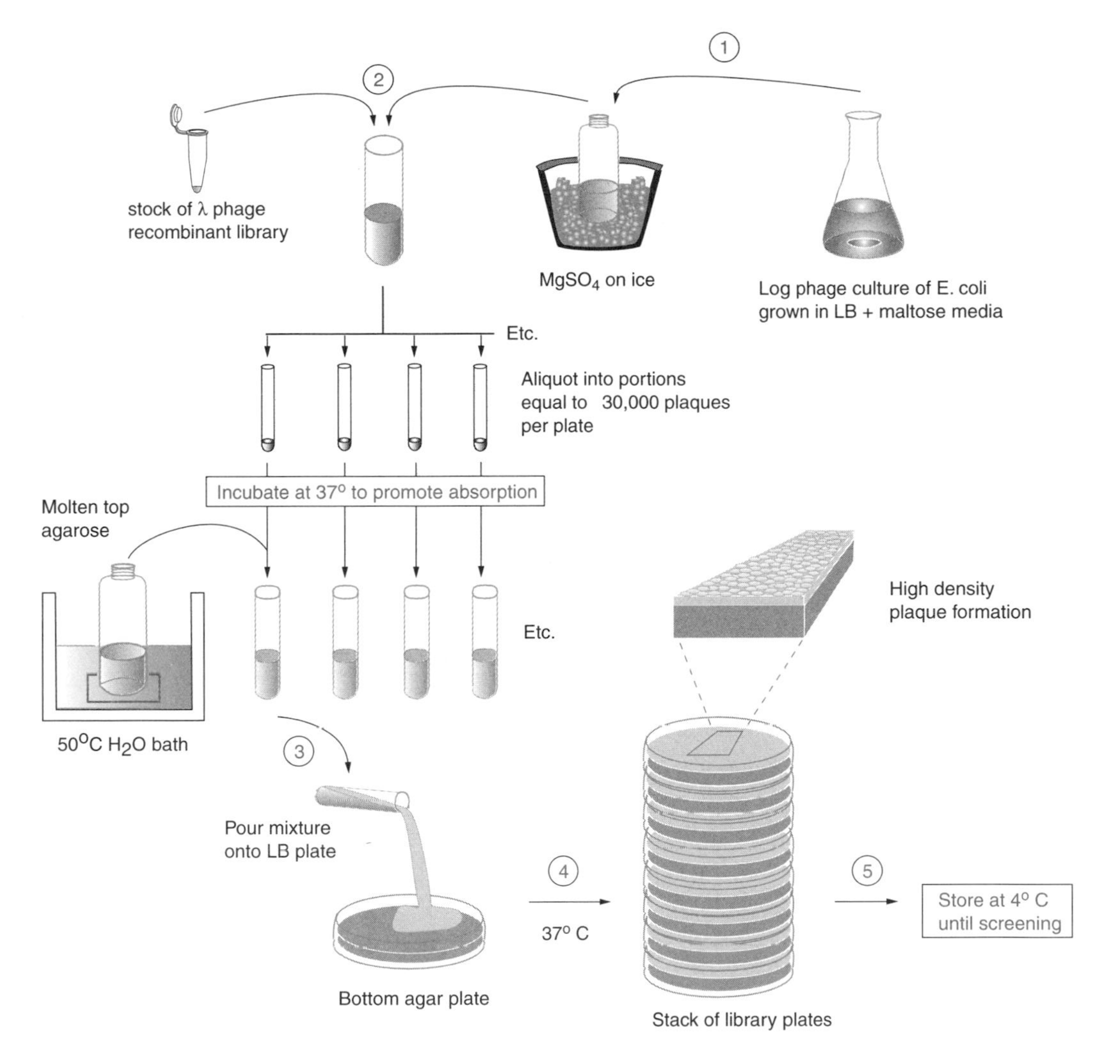

Figure 3.3 General method for plating 5 x 10^5 recombinants from an amplified λ DNA library onto agar plates. *E. coli* cells are grown in LB broth containing maltose to induce expression of the maltose receptor which promotes λ phage attachment as described in Chapter 2. The bacteria are harvested during log phase growth and resuspended in $MgSO^4$ to enhance λ infection further. Predetermined amounts of phage stock and bacterial culture are mixed and aliquoted into portions that correspond to ~30,000 phage/plate. This plaque density is optimized so that each plaque is ~1 mm in diameter and the entire plate is covered with closely packed, but distinct, plaques. After the phage and bacteria have been incubated on ice in $MgSO^4$ to allow phage attachment to the bacteria, a small amount of molten agarose (45–50°C) is added to each tube and then the mixed contents are poured onto a large LB agar plate. The top agarose solidifies at room temperature, fixing the position of the uninfected and infected bacterium in the agarose layer.

tions, and by performing partial enzyme digests with end-labeled DNA fragments, it is now more efficient to use subcloning and automated DNA sequencing to locate restriction enzyme sites precisely.

In order to characterize better a large genomic DNA sequence, it must be subdivided into fragments small enough to be cloned into a plasmid vector suitable for functional analysis and DNA sequencing. Figure 3.7 shows two common strategies for

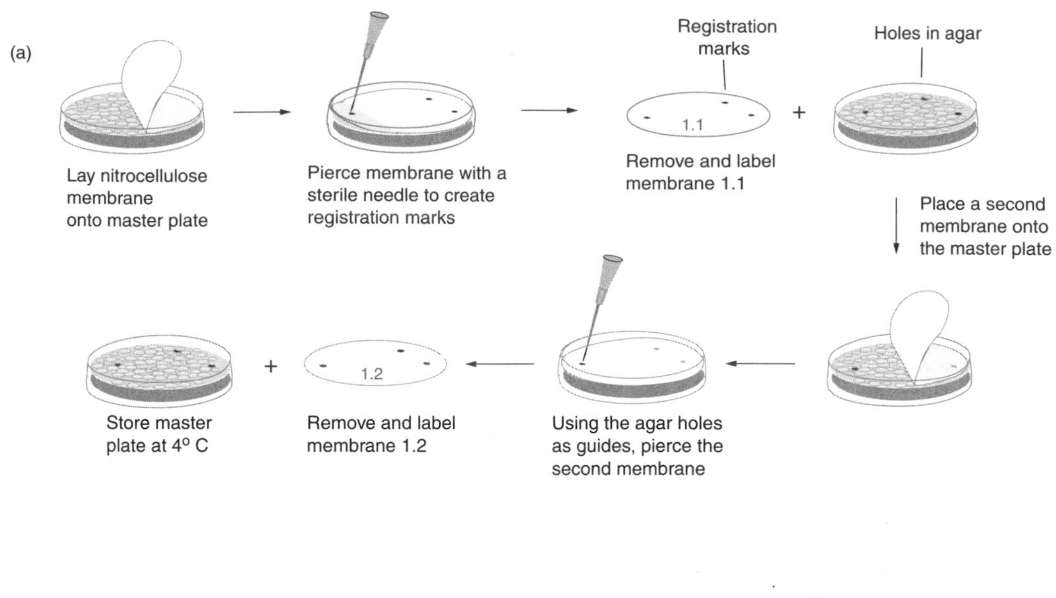

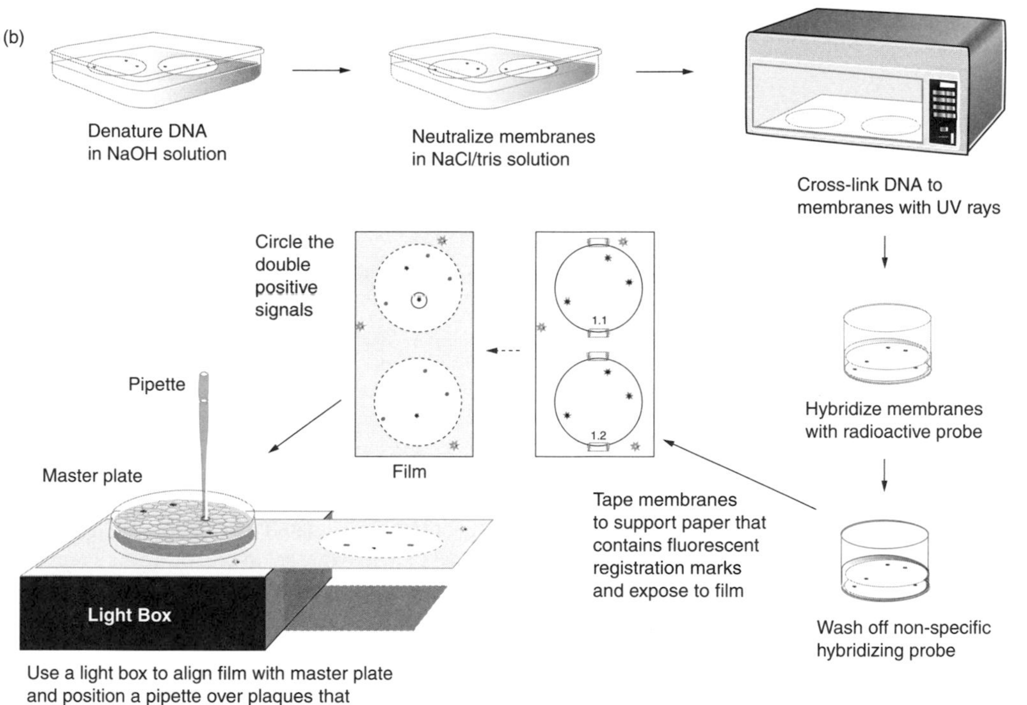

Figure 3.4 Plaque lift hybridzations are used to screen a λ library using radioactive DNA probes. (*a*) The first step in phage library screening is to place carefully a sheet of nitrocellulose membrane directly onto the top agar of each agar master plate. Registration marks are made by piercing the membrane with a syringe needle. The membrane is then peeled off to remove a portion of the phage in each plaque. A second membrane is placed onto the same master plate and registration marks are made to generate a replica filter. (*b*) The phage DNA on the filters is denatured with NaOH, neutralized, and then covalently attached to the filter by untraviolet cross-linking. The methods for probe labeling and filter hybridization are the same as those used for Southern blots (Chapter 1). Once the positive signal on the autoradiographic film is properly aligned with the agar master plate, a glass pipette is used to isolate an agar core from the precise position corresponding to a positive hybridization signal on the aligned film.

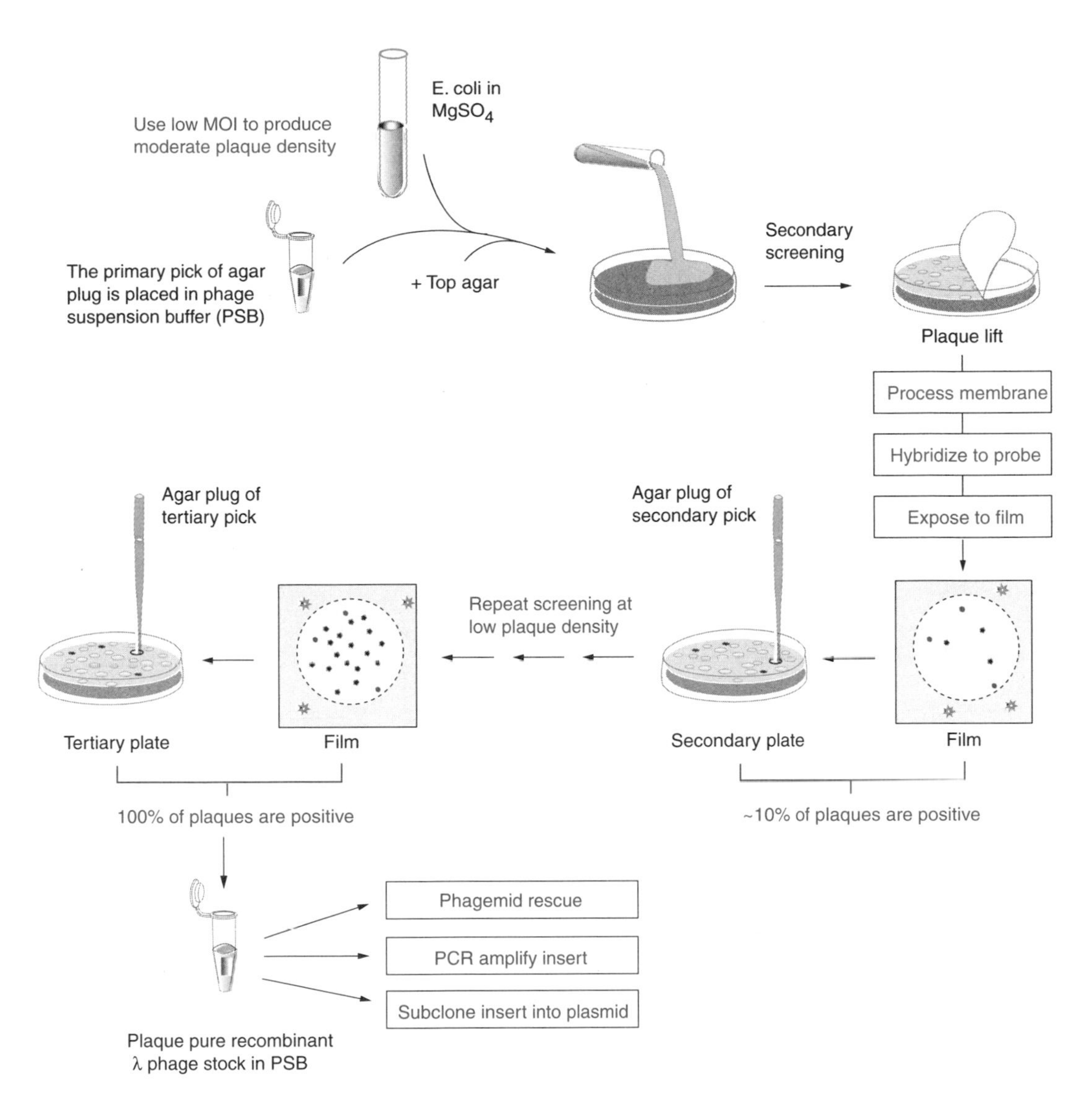

Figure 3.5 A standard method for purifying positive plaques is based on rescreening the primary phage stock using the same hybridization probe and low density plating. Because the phage density on the master plate is saturating, and realignment between the filter membrane, film, and agar plate is not exact, the agar plug from primary pick is actually a mixture of recombinant phage. Therefore, the true positive recombinant phage must be purified by a limiting dilution strategy, usually requiring at least one, and sometimes two more screenings. A λ phage stock is considered plaque pure when 100% of the plaques are positive in the final screening. Insert DNA can be recovered from the recombinant λ phage using phagemid rescue, PCR amplification, or restriction enzyme cleavage and subcloning.

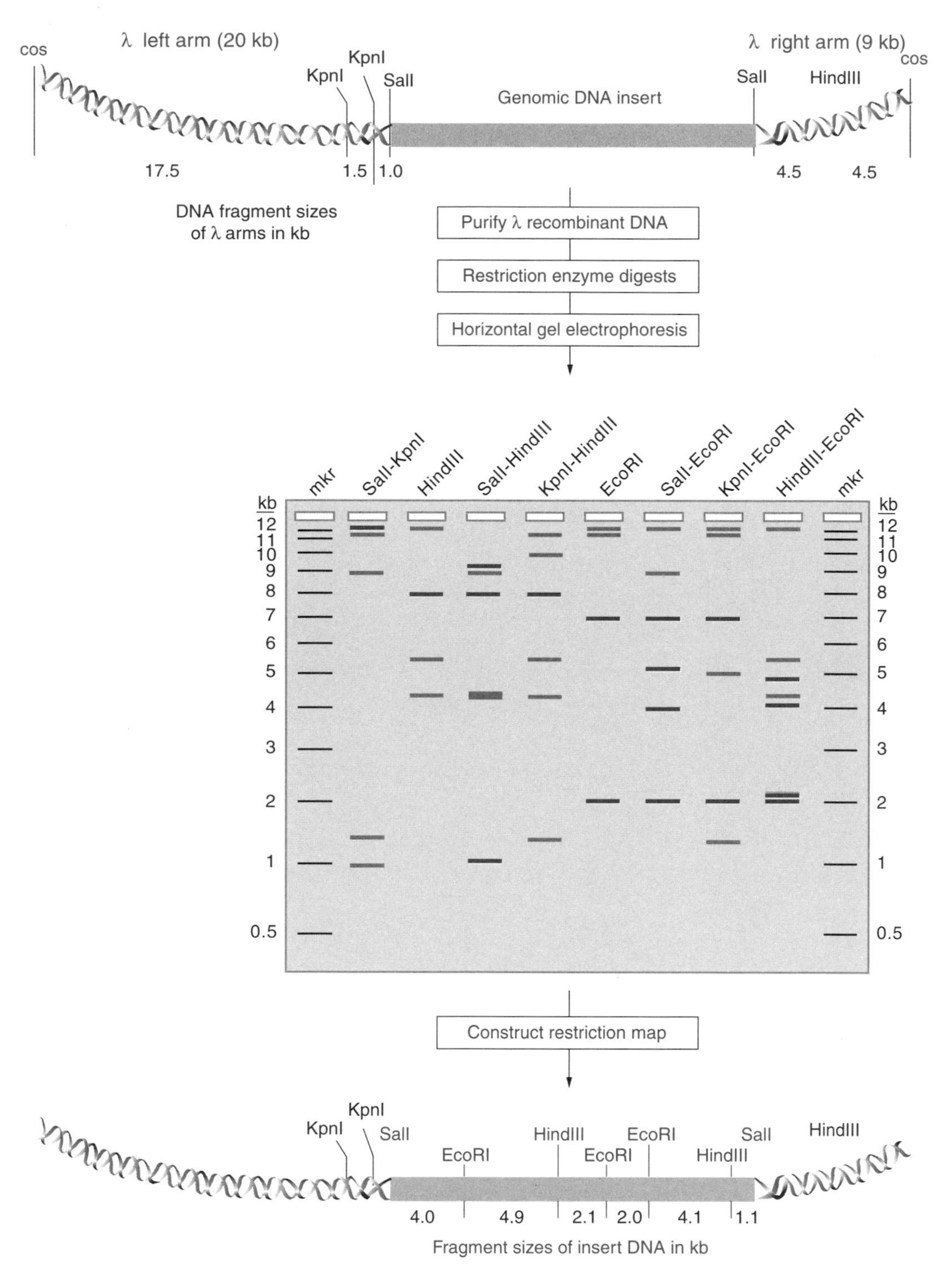

Figure 3.6 Restriction enzyme digests are used to construct physical maps of cloned DNA. A series of restriction enzyme digests analyzed by agarose gel electrophoresis provides the necessary information for a restriction map. Based on the results of single and double enzyme digestions, and by using enzymes that cleave the vector backbone asymmetrically (KpnI and HindIII), it is possible to construct a low resolution map of this SalI insert. The colored restriction fragment bands represent insert DNA lacking attached vector sequences. The DNA marker (mkr) lanes contain DNA fragments of known length that are used to construct a standard curve to calculate the size of experimental DNA fragments.

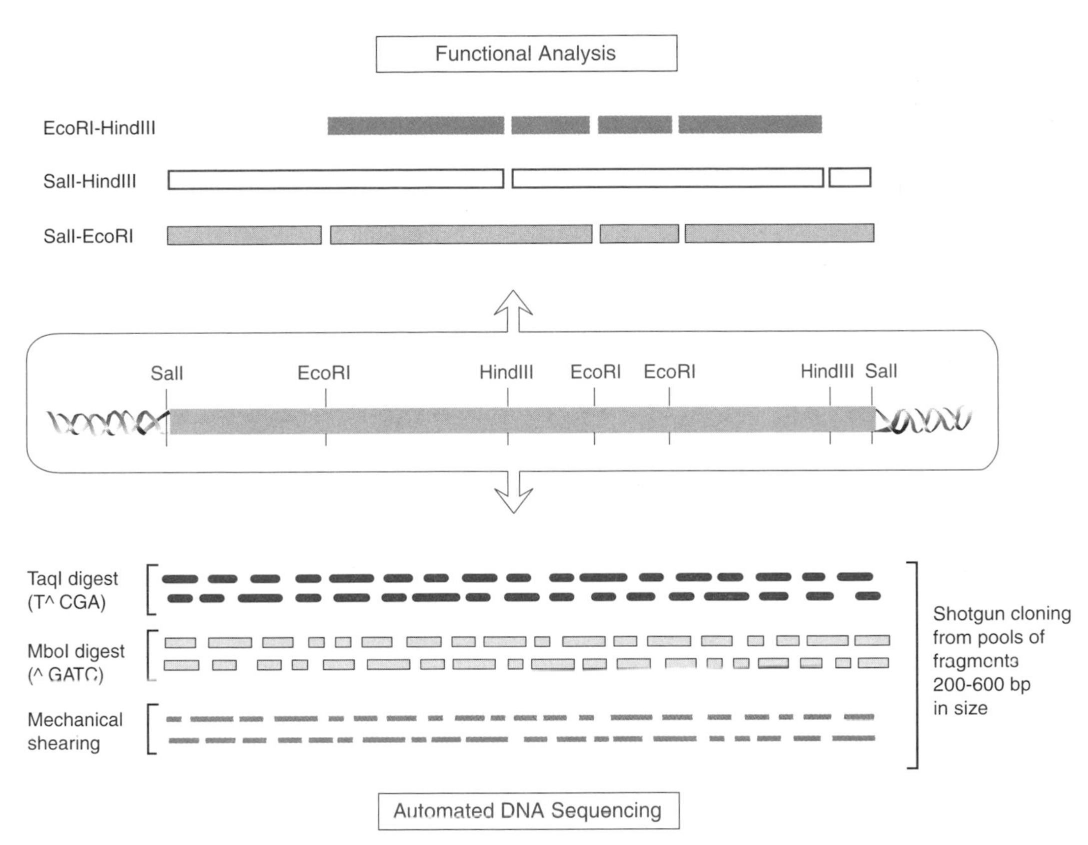

Figure 3.7 Genomic DNA segments contained in λ vectors are subcloned into plasmid vectors to facilitate functional gene analysis and DNA sequencing. Using the restriction enzyme map shown in Figure 3.6, two complementary subcloning strategies are shown. One type is directed subcloning, which is done by digesting the purified recombinant λ DNA with combinations of enzymes that will generate predicted DNA fragments that can be isolated from agarose gels and ligated into an appropriately digested plasmid vector. Alternatively, random, or shotgun, cloning is done using different frequent cutting restriction enzymes, such as TaqI and MboI, which will generate multiple overlapping fragments suitable for DNA sequencing. Small random fragments for DNA sequencing can also be obtained using a mechanical shearing method that produces DNA segments of ~500 bp, which are then repaired by treatment with Klenow DNA polymerase and blunt-ligated to a plasmid vector.

subcloning DNA inserts into a plasmid vector: (1) a directed cloning strategy based on a previously determined restriction enzyme map, and (2) random subcloning using frequent cutting restriction enzymes or sheared DNA. This latter approach is called "shotgun" cloning and is often the first step in high-throughput DNA sequencing using automated instrumentation (Chapter 9). DNA sequence information obtained from shotgun cloning can be analyzed by computer algorithms to compile an aligned sequence that can be used to find all known restriction enzyme sites within that DNA segment.

IN VITRO MUTAGENESIS OF CLONED DNA SEQUENCES

Sometimes the objective of isolating gene sequences from DNA libraries is simply to obtain inserts that can be analyzed by DNA sequencing, for example, when performing comparative studies of the same gene between different species. However, in many cases, the isolation of novel gene sequences requires the use of functional assays to characterize the cloned DNA. As with conventional genetics, a biological comparison between wild-type and mutant phenotypes can often reveal the normal gene function. This same approach can be taken in molecular genetics, except that the researcher is able to choose precisely what types of mutations will likely be the most informative. The ability to perform in vitro mutagenesis as a means to produce desired gene alterations, or to accelerate evolutionary processes, is one of the most powerful tools in applied molecular genetics. In this section, three basic in vitro mutagenesis strategies are described, each of which is used for creating specific types of mutations in cloned DNA sequences.

Deletion Mutagenesis

The most straightforward in vitro mutagenesis strategy is deletion mutagenesis, which can be done by removing nucleotides from the termini or by internal deletions. Information from restriction maps can be used to design specific deletions that allow convenient recloning of the modified DNA segment. Alternatively, exonucleases can be used to degrade DNA nonspecifically from the termini of a linearized plasmid. The extent of exonuclease treatment, usually as a factor of reaction time, determines the amount of deleted DNA. Figure 3.8 illustrates how a plasmid insert can be mutated by restriction enzyme-mediated deletion mutagenesis using compatible restriction enzyme-mediated termini, or by digesting linearized plasmid DNA with exonucleases followed by treatment with Klenow DNA polymerase.

Deletion mutagenesis is sometimes used to create novel proteins lacking one or more functional domains, for example, deleting the hormone-binding domain of a steroid receptor to produce a constitutively active transcription factor. Internal deletions of protein coding sequences are more complicated because two thirds of all deletions result in frame-shift mutations. Perhaps one of the most elegant examples of deletion mutagenesis is linker scanning mutations, first pioneered by Steve McKnight to study the regulatory elements in the thymidine kinase promoter. In this strategy, a collection of 5′ and 3′ deletants is created and the termini are ligated to an oligonucleotide linker. Based on the DNA sequence of individual deletants, paired combinations are chosen and used to create a new DNA fragment in which the linker sequence precisely replaces 8 bp of the original sequence without altering the spacing of surrounding nucleotides (Fig. 3.9).

Oligonucleotide-Directed Mutagenesis

Terminal deletion mutagenesis is often used to define the 5′ and 3′ boundaries of a functional region. However, fine structure mapping is required to delineate the minimal nucleotide region necessary for the function under study. Another in vitro mutagenesis technique is called oligonucleotide- or site-directed mutagenesis. This method is based on the use of short oligonucleotides, containing defined mutations, as strand specific primers in an in vitro DNA synthesis reaction. In order to use this strategy, the nucleotide sequence of the region to be mutated must be known to permit synthesis of a complementary mutagenic oligonucleotide. The mutagenic oligonucleotide is used to prime single strand circular DNA templates for in vitro DNA synthesis as shown in Fig.

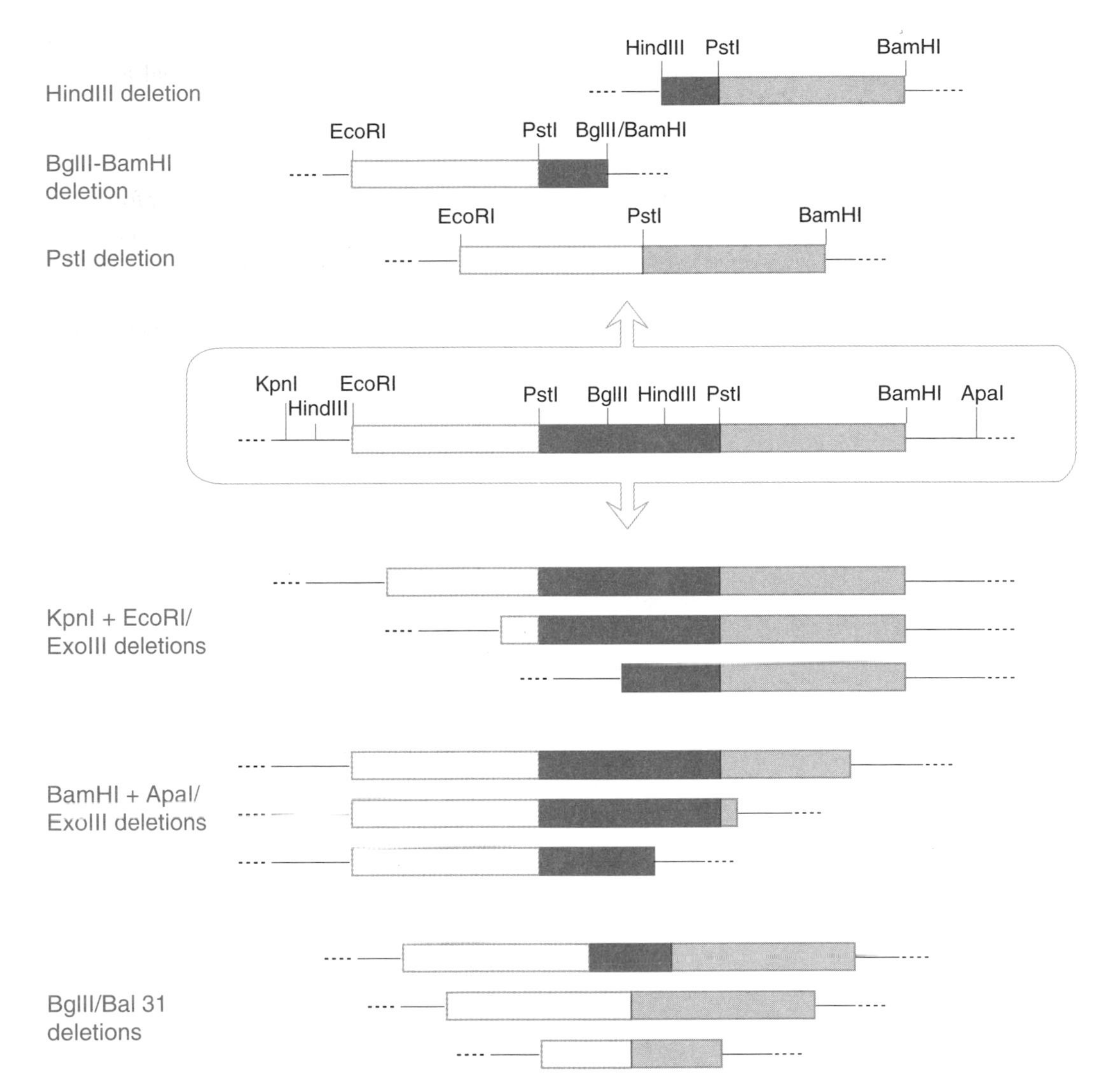

Figure 3.8 Deletion mutagenesis using restriction enzymes and exonucleases. Based on the restriction map of a plasmid DNA insert, it is possible to design a strategy to remove nucleotides from the termini or within the cloned segment, using intramolecular ligation reactions. BglII and BamHI digestion result in compatible termini (GATC). Exonuclease III (Exo III) degrades one strand of double-strand DNA from the 3′ end, but only if the DNA termini contain either a 5′ overhang or a blunt end, but not an extended single-strand 3′ overhang. This property can be exploited to generate unidirectional deletions by digesting target DNA with two restriction enzymes, one that leaves a 5′ overhang (EcoRI or BamHI) and the other a 3′ overhang (KpnI or ApaI). Exo III-deleted DNA must be treated with the single-strand nuclease S1 to remove the undigested 5′ strand. Bal31 exonuclease degrades double-strand DNA to yield bidirectional deletions that can be blunt-ligated without prior treatment with S1 nuclease.

3.10. It is important during the bacterial in vivo replication phase that the directed base pair mismatch not be repaired. This is done by using special *E. coli* strains that have specific defects in DNA repair pathways. Following transformation, single colonies are isolated and the recovered plasmid DNA is sequenced to determine if it contains the desired mutation. The efficiency of site-directed mutagenesis can be as high as 80% depending on the method used to enrich for mutated plasmids (see Table 3.2).

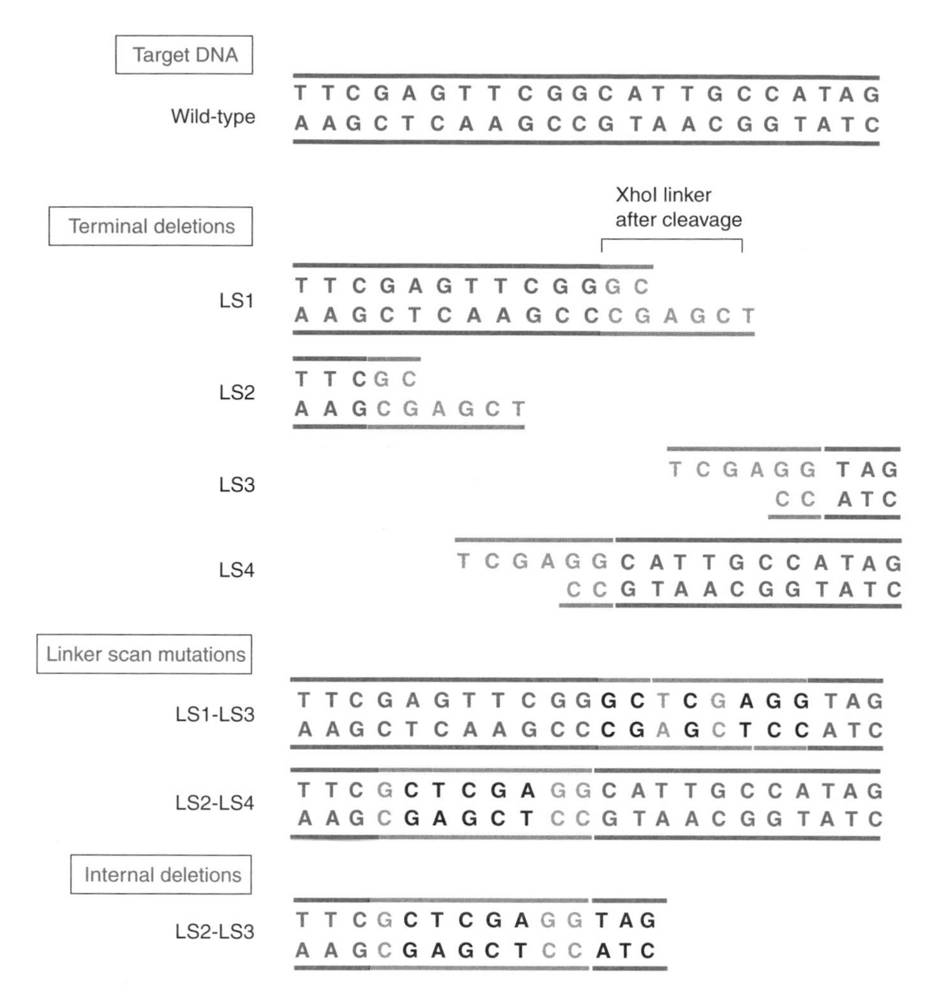

Figure 3.9 Linker scanning mutagenesis can be used to create a series of small internal deletions and clustered point mutations. A collection of 5′ and 3′ terminal deletants is generated, each containing an 8-mer restriction site linker ligated to the end, for example, an XhoI linker (GCTCGAGG) containing the palindromic XhoI recognition sequence (CTCGAG) in the middle. By digesting selected deleted segments with XhoI, followed by ligation, it is possible to created nested point mutations in the target DNA (LS1-LS3 and LS-2-LS-4). Not all linker nucleotide substitutions in the ligated region will be mutations, for some of the base pair positions will fortuitously match the wild-type target sequence. Internal deletion mutations can also be created by ligating two nonadjacent segments (LS2–LS3).

A key strategy in oligonucleotide-directed mutagenesis methods is to increase the relative number of mutated plasmids in the pooled population. The dUTP incorporation strategy shown in Figure 3.10 relies on the degradation and repair of template DNA that has been "marked" with deoxyuridine prior to the in vitro DNA synthesis step. Thomas Kunkel developed the dUTP incorporation strategy by taking advantage of an *E. coli* strain that contains defects in genes responsible for preventing dUTP incorporation into DNA. One of the mutations is in the *ung* gene which encodes uracil-*N*-glycosylase, an enzyme that normally removes uracil bases from DNA that result from cytosine deamination. The second defect is in the *dut* gene (dUTPase) which functions to prevent the buildup of dUTP pools in the cell. Production of single strand M13 phage in an *ung*⁻/

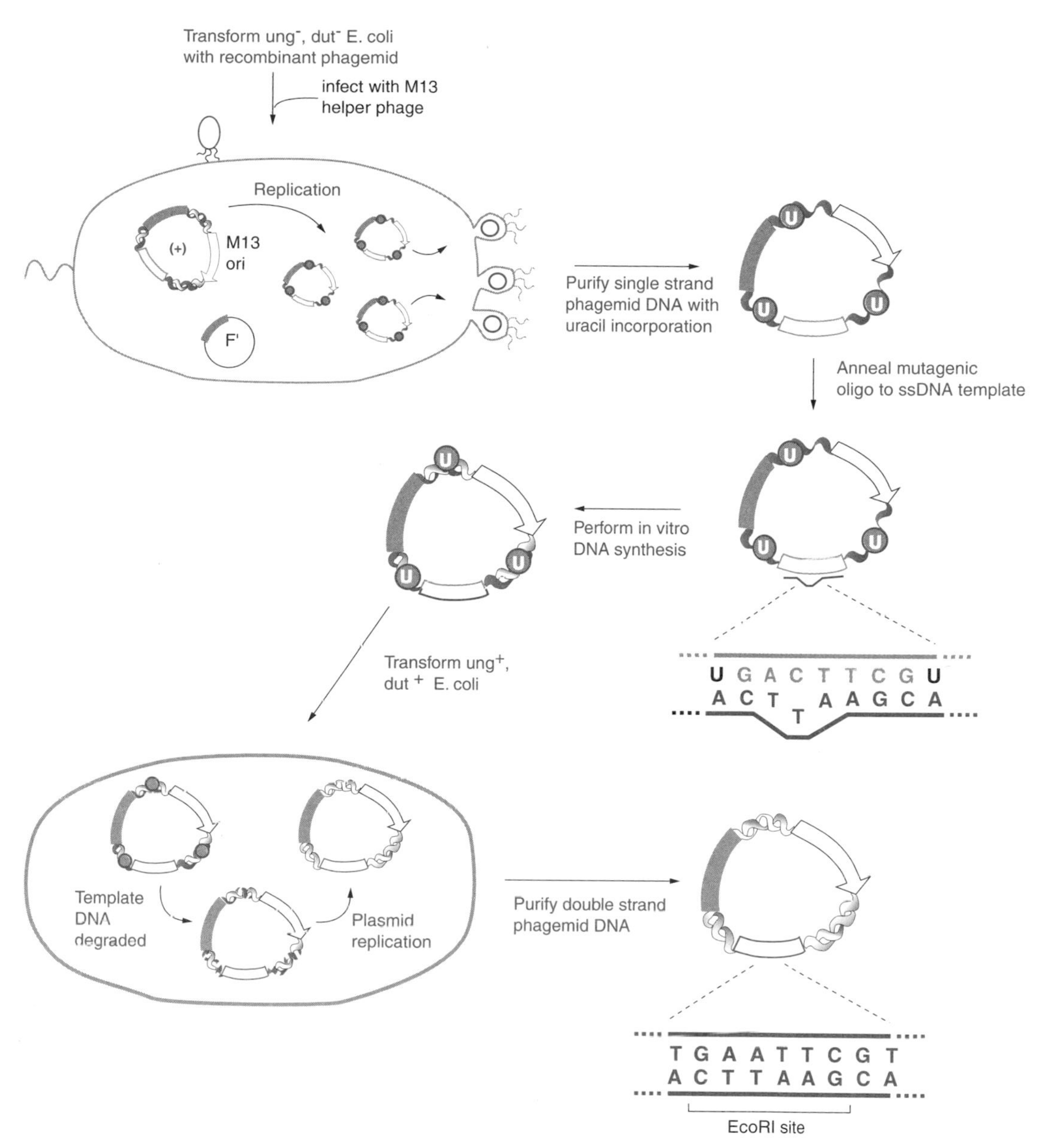

Figure 3.10 Oligonucleotide-directed mutagenesis is an efficient method to generate specific nucleotide alterations in cloned DNA. dUTP incorporation into M13 DNA is done by using a phagemid vector to produce single strand DNA in an *E. coli* strain defective in the two enzymes uracil-*N*-glycosylase (*ung*) and dUTPase (*dut*). The mutagenic oligo is annealed to uracil-containing template DNA (U), and in vitro DNA synthesis is performed to generate double-strand DNA. Following transformation of ung$^+$, dut$^+$ *E. coli*, the template strand is degraded and the surviving double-strand phagemid plasmid is isolated and sequenced. In this example, the mutant oligo was used to engineer a novel EcoRI site into the DNA insert.

dut$^-$ *E. coli* strain results in the synthesis of uracil-containing template DNA (Fig. 3.10). Following primer annealing and in vitro DNA synthesis, the double-strand DNA is transformed into a ung$^+$, dut$^+$ *E. coli* strain. Because the template DNA contains a large number of dU residues in place of dA, it is degraded by the activity of uracil-*N*-glycosylase.

TABLE 3.2 Strategies used to increase the efficiency of oligonucleotide-directed in vitro mutagenesis

Descriptive name	Method to remove template strand	Advantages	Disadvantages
dUTP incorporation	M13 template grown in a DNA repair mutant (dut⁻, ung⁻) contains dUTP in place of dATP	No special selections schemes or restriction enzymes required	Requires the use of a M13 phagemid and appropriate *E. coli* strains
Phosphoro-thioate incorporation	in vitro DNA synthesis incorporates phosphoro-thioates into mutant strand which protects it from enzyme cleavage with NciI	Critical reactions are done in vitro and does not require a special *E. coli* strain	Depends on efficient phosphorothioate incorporation and exonuclease III digestion
Gapped duplex DNA	Mutant strand is marked with both gain of chloramphenicol resistance and loss of *amp*^{r} phenotypes	Reversal of two antibiotic resistant phenotypes allows for calculation of mutation frequency	Requires a special vector and depends on efficient in vitro heteroduplex formation
Site elimination	Mutant strand is marked with a second oligo-directed mutation in a nonessential enzyme site	Nonessential enzyme site can be anywhere on vector	Appropriate enzyme site is required and differences in oligo annealing temperature may affect efficiency
amp^s site reversion	Mutant strand is marked with an oligo-directed reversion of an *amp*^{s} to *amp*^{r} phenotype	Both the *amp*^{r} and *tet*^{r} genes can be used sequentially to obtain mutations	Requires a special *amp*^{s} cloning vector that can be used for reversion
PCR amplification	Methylated template strand is eliminated after PCR using the restriction enzyme DpnI	Requires only one transformation and can be done quickly	PCR reaction can introduce mutations at nontarget sites

The two basic approaches used to distinguish template and mutant DNA (which directly relates to the efficiency of mutagenesis) are those that use prior "history" of the template DNA to mark it for degradation after the in vitro DNA synthesis step, and methods that depend on secondary point mutations that alter the antibiotic resistance or restriction map of either the unmutated or mutated plasmids. Many of the protocols rely on the use of single-strand M13 bacteriophage DNA as the template strand for in vitro DNA replication, whereas others have been developed that utilize heat-denatured double-strand DNA.

Random Point Mutagenesis

It is difficult to know a priori what nucleotides to mutate if functional information is not available. Random mutagenesis can be used in these circumstances to introduce point mutations within a target DNA sequence. Treatment of DNA with the mutagenic compounds ethyl methanesulfonate (EMS) and *N*-nitroso-*N*-methylurea (NMU) causes base modifications that result in the production of point mutations (transitions and transversions) following DNA replication. Random mutations can also be introduced into cloned DNA using PCR amplification of a target DNA sequences under conditions that favor nucleotide misincorporation by *Taq* DNA polymerase. This PCR-based technique is described in Chapter 6.

The biggest problem with random mutagenesis is that it can be very tedious to identify the small number of mutant plasmids within a plasmid pool without the use of a functional in vivo screen. To overcome this limitation, a method was developed using gradient gel electrophoresis to distinguish between heteroduplexes that contain mismatched base pairs and homoduplexes that are unmutated. Rick Myers was the first to describe this application of gradient gel electrophoresis to in vitro mutagenesis when he used it to identify point mutations in the β-*globin* gene regulatory region. However, random mutagenesis strategies are more often used in conjunction with a reliable in vivo screen in yeast or bacteria, or coupled to automated high-throughput DNA sequencing (Chapter 9).

Laboratory Practicum 3. *Alanine-scanning mutagenesis of a gene coding sequence*

Research Objective

Apoptosis is a highly regulated cellular process that mediates biochemical cell death. A cell biologist at a pharmaceutical company recently discovered a small secreted protein that blocks apoptosis by binding to and inactivating Fas, a membrane-bound receptor found on the surface of certain cell types. The researcher named her newly discovered 20 kDa "anti-apoptotic" protein Foy (fountain of youth). A *foy* cDNA was isolated by screening an expression cDNA library using an anti-Foy antibody. Deletion mapping of the *foy* coding sequence identified a small region of ~60 amino acids that is absolutely required for Fas inactivation using a bioassay. Moreover, structural biologists at her company are in the process of solving the protein structure of Foy using X-ray crystallography. On the basis of these results, she has decided to obtain additional functional data using oligonucleotide-directed mutagenesis systematically to alter charged residues to alanine at various positions within the 60 amino acid Foy functional domain.

Available Information and Reagents

1. A 650 bp *foy* cDNA, containing the entire 185 amino acid foy open reading frame (ORF), has been cloned into an *E. coli* expression vector and used to produce milligram quantities of functional Foy protein. The addition of 1 ng/ml soluble recombinant Foy to tissue culture media was found to block Fas-mediated apoptosis completely in the bioassay. This same source of bacterially expressed Foy protein is being used for the protein structure determinations and for Foy-Fas in vitro binding studies.
2. The nucleotide sequence of the *foy* functional domain predicts that 10 codons corresponding to charged amino acids (Lys, Arg, Glu and Asp) may be good targets for alanine substitutions. She has designed six oligonucleotide primers that together are sufficient to convert all 10 charged amino acids to alanines.
3. A pET bacterial *foy* expression vector will be used that contains a unique PvuI site in the amp^r gene which can be converted to an XmaIII site by a dinucleotide AT to GC mutation using a 26 nucleotide long "PX" selection primer. A corresponding "XP" reversion primer has been made for a second round of mutagenesis that can be used to mutate the GC dinucleotide of the XmaIII site back to AT and thus re-create the original PvuI site in the amp^r gene.

Basic Strategy
The term "alanine scanning" refers to the systematic substitution of alanine codons at selected positions within a protein coding region. The researcher's plan is to use the six oligonucleotides she designed for site-directed mutagenesis as shown in Figure 3.11. She has extended the flexibility of this technique by designing a selection primer (PX) and reversion primer (XP) to mark the mutated strand by interconversion of a PvuI and XmaIII site (Fig. 3.12). The basis of this enrichment procedure is that under optimal conditions, the same target template molecule is primed by both the selection and target oligos. Because unreplicated dsDNA resulting from incomplete denaturation or renaturation retains the PvuI site, these nonmutated DNA molecules are linearized by PvuI digestion and degraded by *E. coli* nucleases following transformation.

The apoptosis bioassay will be used initially to test the effect of the mutations by determining if the recombinant Foy protein can block Fas-mediated cell death. In addition, in vitro binding assays will be used to determine the effect of alanine substitutions on Foy–Fas interactions. Once the protein structure data are available, she may be able to make more informed structure–function predictions regarding protein–protein interactions necessary for Foy activity.

Comments
Oligonucleotide-directed mutagenesis is a very useful technique for studying nucleic acid and protein function. The original M13 based protocol of Kunkel worked well enough, however, it was hampered by the requirement for sufficient quantities of single-strand template DNA. To circumvent this problem, strategies were devised that used

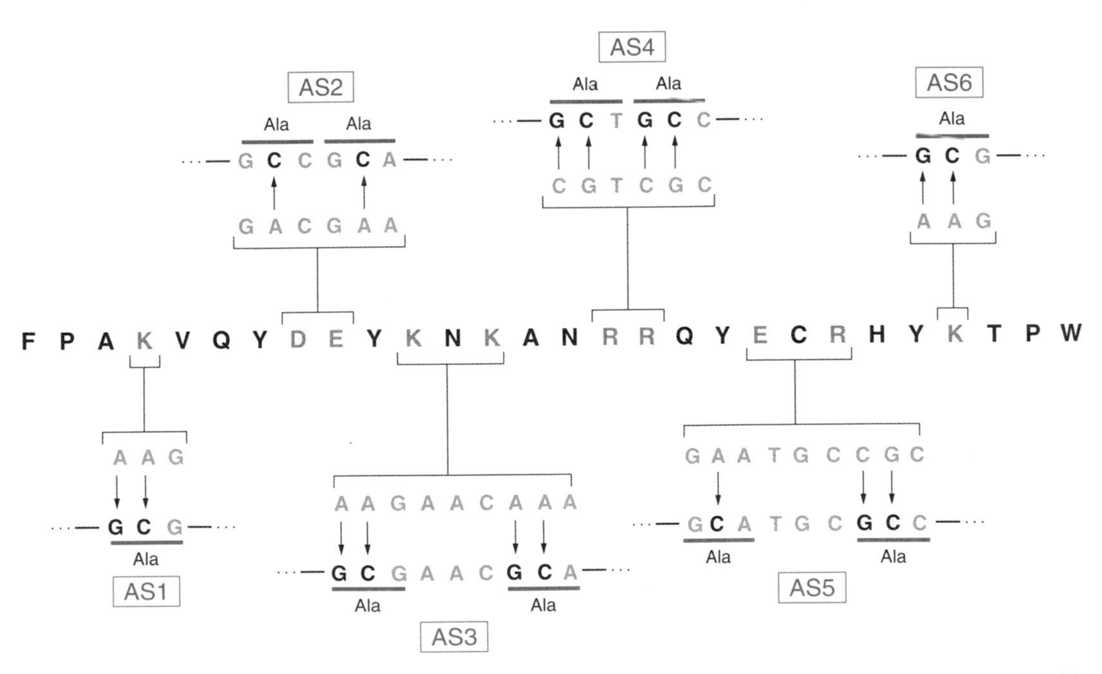

Figure 3.11 Alanine scanning mutagenesis is used systematically to substitute specific amino acids with alanine by altering the codon sequence using standard site-directed in vitro mutagenesis procedures. The amino acid sequence of a hypothetical *foy* coding region is shown with the alanine mutations that could be introduced using six different alanine scan (AS) oligonucleotides (AS1–AS6). Each of the six mutant oligos contain 12–15 complementary nucleotides on either side of the mismatched base pair to increase the thermal stability of the heteroduplex.

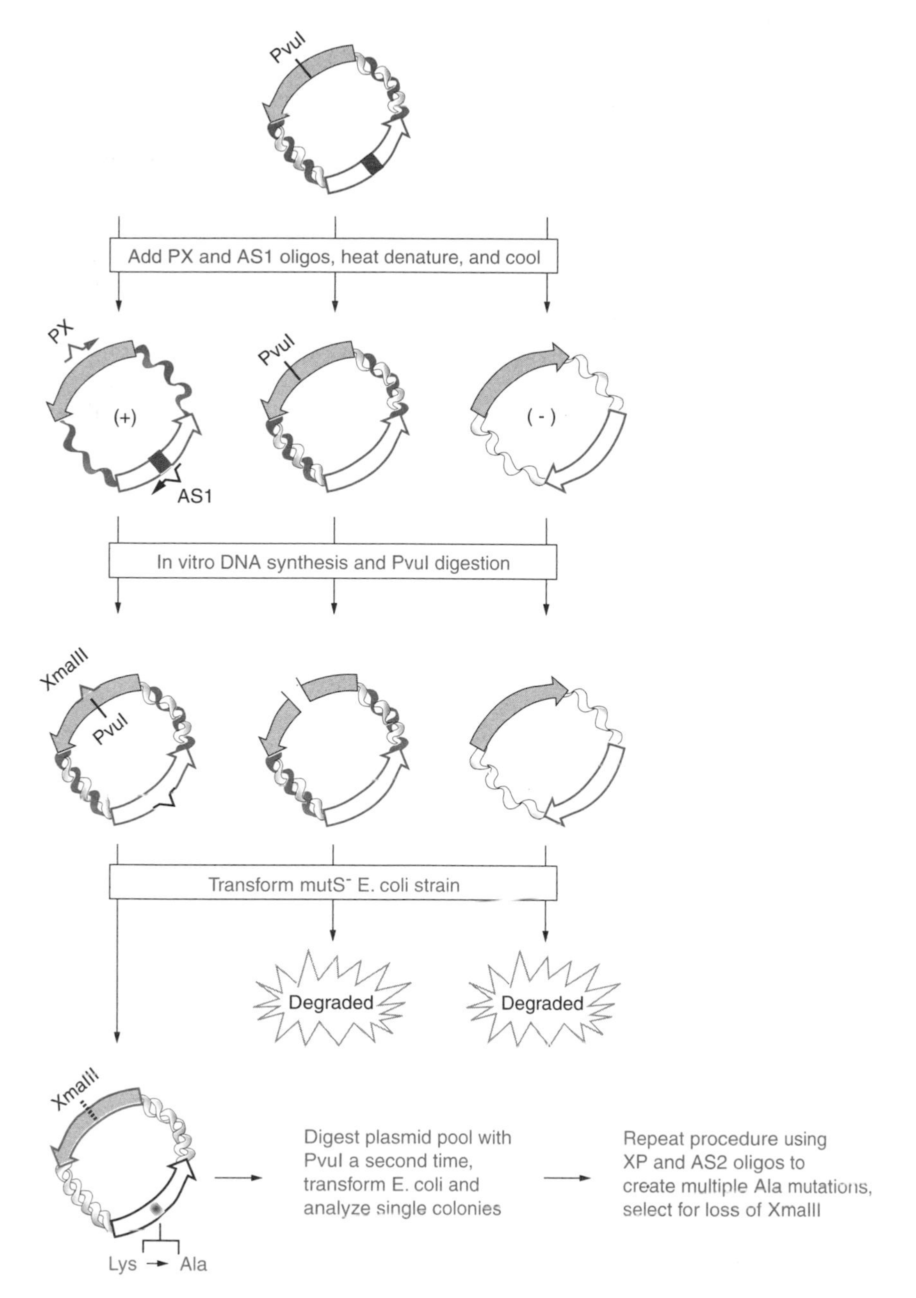

Figure 3.12 The site elimination in vitro mutagenesis strategy uses a "selection" oligo to mark target templates that are primed by the mutant oligo. The mutagenic oligonucleotide changes the amino acid codon(s) within the target insert DNA, and the selection oligo (PX) converts the PvuI restriction enzyme site in the *amp*^{r} gene to an XmaIII site without affecting β-lactamase activity. Following in vitro DNA replication, the reaction mix is treated with PvuI and then used to transform a *mutS* DNA repair-deficient strain of *E. coli*. After a second round of enrichment using PvuI treatment, the DNA is transformed into a standard K-12 strain of *E. coli*, single colonies are picked, and the plasmid DNA is isolated and sequenced to distinguish between mutant (AS1) and wild-type (wt) plasmids. By repeating the in vitro mutagenesis protocol using the recovered AS1 mutant plasmid, the AS2 mutant oligo, and the corresponding "reversion" selection oligo (XP), it is possible to introduce three alanine substitutions into the coding sequence (see Fig. 3.11). Note that it is also possible to use more than one AS mutant oligo during the initial mutagenesis step to create multiple mutations in a single round.

denatured plasmid DNA, but this created new challenges, the most significant of which was eliminating unmutated renatured DNA. The site-elimination technique shown in Figure 3.12 makes it possible to obtain mutation frequencies of >50%. One complication can be that the annealing conditions for both the mutagenic primer and selection primer may not be equivalent owing to differences in sequence composition and local secondary structures near one or both of the priming sites. This problem is usually solved by adjusting primer concentrations or by brute force DNA sequencing to obtain the desired mutation.

Prospective

Aberrant expression of Fas in pancreatic cells can lead to autoimmune disease that results in diabetes due to lack of insulin producing cells. In this disease it is thought that infiltrating thymocytes secrete FasL, which activates Fas on pancreatic cells. The cell biologist is pursuing the idea that the anti-apoptotic function of Foy could somehow be used specifically to block pancreatic cell apoptosis by FasL. By knowing what amino acid residues are required for Foy function, in combination with protein structural data, it might be possible to predict additional Foy modifications that could be used to begin the process of generating candidate therapeutic agents.

REFERENCES

DNA libraries contain a collection of gene sequences

*Blattner, F.R., Williams, B.G., Blechl, A.E., et al. Charon phages: Safer derivatives of bacteriophage lambda for DNA cloning. Science 196: 161–166, 1977.

*Clark, L., Carbon, J. A colony bank containing synthetic ColE1 hybrids representative of the entire *E. coli* genome. Cell 9:91–99, 1976.

Frischauf, A.-M., Lehrach, H., Poustka, A., Murray, N. Lambda replacement vectors carrying polylinker sequences. J. Mol. Biol. 170:827–838, 1983.

*Hohn, B., Murray, K. Packaging recombinant DNA molecules into bacteriophage particles in vitro. Proc. Natl. Acad. Sci. USA 74:3259–3264, 1977.

Huynh, T., Young, R., Davis, R. Construction and screening cDNA libraries in λgt10 and λgt11. In DNA cloning, Vol. 1: A practical approach (D. Glover, ed.). IRL Press, Oxford, pp. 49–78, 1984.

Ish-Horowitz, D., Burke, J.F. Rapid and efficient cosmid vector cloning. Nucl. Acids Res. 9: 2989–2999, 1981.

Kaiser, K., Murray, N.E., Whittaker, P.A. Construction of representative genomic DNA libraries using phage lambda replacement vectors. In DNA cloning 1: Core techniques D.M. Glover, and B.D. Hames, (eds.). IRL Press, Oxford, pp. 37–84, 1995.

Karn, J., Brenner, S., Barnett, L., Cesareni, G. Novel bacteriophage λ cloning vector. Proc. Natl. Acad. Sci. 77:5172–5176, 1980.

Meissner, P.S., Sisk, W.P., Berman, M.L. Bacteriophage λ cloning system for the construction of direction cDNA libraries. Proc. Natl. Acad. Sci. USA 84:4171–4175, 1987.

*Rambach, A., Tiollais, P. Bacteriophage l having EcoRI endonuclease sites only in the nonessential region of the genome. Proc. Natl. Acad. Sci. USA 71:3927–3932, 1974.

*Seed, B., Parker, R.C., Davidson, N. Representation of DNA sequences in recombinant DNA libraries prepared by restriction enzyme partial digestion. Gene 19:201–209, 1982.

Wyman, A.R., Wertman, K.F., Barker, D., Helms, C., Petri, W.H. Factors which equalize the representation of genome segments in recombinant libraries. Gene 49:263–271, 1986.

* Landmark papers in applied molecular genetics.

Screening a λ phage library by DNA hybridization

*Benton, W.D., Davis, R.W. Screening λgt recombinant clones by hybridization to single plaques in situ. Science 196:180–182, 1977.

*Denhardt, D. A membrane filter technique for the detection of complementary DNA. Biochem. Biophys. Res. Commun. 23:641–646, 1966.

*Gillespie, D., Spiegelman, S. A quantitative assay for DNA-RNA hybrids with DNA immobilized on a membrane. J. Mol. Biol. 12:829–842, 1965.

*Grunstein, M., Hogness, D. Colony Hybridization: A method for the isolation of cloned DNA's that contain a specific gene. Proc. Natl. Acad. Sci. USA 72:3961–3965, 1975.

Jacobs, K., Shoemaker, C., Rudersdorf, R., Neill, S.D., Kaufman, R.J., Mufson, A., Seehra, J., Jones, S.S., Hewick, R., Fritsch, E.F., Kawakita, M., Shimizu, T., Miyake, T. Isolation and characterization of genomic and cDNA clones of human erythropoietin. Nature 313:806–810, 1985.

*Maniatis, T., Hardison, R.C., Lacy, E., Lauer, J., O'Connell, C., Quon, D., Sim, G.K., Efstratiadis, A. The isolation of structural genes from libraries of eukaryotic DNA. Cell 15: 687–697, 1978.

*Young, R.A., Davis, R.W. Efficient isolation of genes using antibody probes. Proc. Natl. Acad. Sci. USA 80: 1194–1199, 1983.

Restriction enzyme mapping and subcloning

Ausubel, F., Brent, R., Kingston, R., Moore, D., Seidman, J., Smith, J., Struhl, K. Current protocols in molecular biology. John Wiley & Sons, New York, 1996.

Boseley, P.G., Moss, T., Birnstiel, M.L. 5' labeling and poly(dA) tailing. Methods Enzymol. 65:478–494, 1980.

*Cohen, S.N., Chang, A.C.Y., Boyer, H.W., Helling, R. Construction of biologically functional bacterial plasmid in vitro. Proc. Natl. Acad. Sci. USA 70:3240–3244, 1973.

*Danna, A.J. 1980. Determination of fragment order through partial digests and multiple enzyme digests. Methods Enzymol. 65:449–467, 1980.

*Dugaiczyk, A., Boyer, H.W., Goodman, H.M. Ligation of EcoRI endonuclease-generated DNA fragments into linear and circular structures. J. Mol. Biol. 96:174–184, 1975.

Glover, D.M., Hames, B.D. DNA cloning: A practical approach, Vols. 1–4. Oxford University Press, Oxford, 1995.

Hung, M.C., Wensink, P.C. Different restriction enzyme generated sticky DNA ends that can be joined in vitro. Nucl. Acids Res. 12:1863–1870, 1984.

*Jeffreys, A.J., Flavell, R.J. A physical map of the DNA region flanking the rabbit β globin gene. Cell 12: 429–43, 1977.

Sambrook, J., Fritsch, E.F., Maniatis, T. Molecular cloning: A laboratory manual, 2nd ed. Cold Spring Harbor Laboratory, Cold Spring Harbor, NY, 1989.

*Struhl, K. 1985. A rapid method for creating recombinant DNA molecules. Biotechniques 3:452–453, 1985.

*Ullrich, A., Shine, J., Chirgwin, R., Pictet, R., Tischer, E., Rutter, W.J., Goodman, H.M. Rat insulin genes: Construction of plasmids containing the coding sequences. Science 196: 1313–1316, 1977.

Zimmerman, S.B., Harrison, B. Macromolecular crowding accelerates the cohesion of DNA fragments with complementary termini. Nucl. Acids Res. 13:2241–2249, 1985.

In vitro mutagenesis of cloned DNA sequences

Caldwell, C., Joyce, G.F. Randomization of genes by PCR mutagenesis. PCR Methods Appl. 2:28–33, 1992.

Chamberlain, N.L., Whitacre, D., Miesfeld, R. Delineation of two distinct AF-1 activation functions in the androgen receptor N-terminal domain. J. Biol. Chem. 271:26772–26778, 1996.

Deng, W.P., Nickoloff, J.A. Site-directed mutagenesis of virtually any plasmid by eliminating a unique site. Anal. Biochem. 200:81–88, 1992.

Godowski, P., Rusconi, S., Miesfeld, R., Yamamoto, K. Glucocorticoid receptor mutants that are constituitive activators of transcriptional enhancement. Nature 325:365–368, 1987.

Hill, D.E., Oliphant, A.R., Struhl, K. Mutagenesis with degenerate oligonucleotides: An efficient method for saturating a defined DNA region with base pair substitutions. Methods Enzymol. 155:558–568, 1987.

Kolmar, H., Fritz, H-J. Oligonucleotide-directed mutagenesis with single-stranded cloning vectors. In DNA cloning 1: Core techniques, D.M. Glover, and B.D. Hames, (eds.). IRL Press, Oxford, pp. 193–224, 1995.

*Kunkel, T.A. Rapid and efficient site-specific mutagenesis without phenotypic selection. Proc. Natl. Acad. Sci. USA 82:488–492, 1985.

Lathe, R. Synthetic oligonucleotide probes deduced from amino acid sequence data. Theoretical and practical considerations. J. Mol. Biol. 183:1–12, 1985.

*Leung, D.W., Chen, E., Goeddel, D.V. A method for random mutagenesis of a defined DNA segment using a modified polymerase chain reaction. Techniques 1:11–15, 1989.

Matteucci, M.D., Heyneker, H.L. Targeted random mutagenesis: The use of ambiguously synthesized oligonucleotides to mutagenize sequences immediately 5' of an ATG initiation codon. Nucl. Acids Res. 11:3113–3121, 1983.

*McKnight, S.L., Kingsbury, R. Transcriptional control signals of a eukaryotic protein-coding gene. Science 217:316–324, 1982.

*Myers, R.M., Lerman, L.S., Maniatis, T. A general method for saturation mutagenesis of cloned DNA fragments. Science 229:242–246, 1985.

*Myers, R.M., Maniatis, T., Lerman, L.S. 1987. Detection and localization of single-base changes by denaturing gradient gel electrophoresis. Methods Enzymol. 155:501–527.

Oliphant, A.R., Nussbaum, A.L., Struhl, K. Cloning of random-sequence oligodeoxynucleotides. Gene 44:177–183, 1986.

Piechocki, M.P., Hines, R.N. Oligonucleotide design and optimized protocol for site-directed mutagenesis. BioTechniques 16:702–707, 1994.

*Smith, M. In vitro mutagenesis. Ann. Rev. Genet. 19:423–463, 1985.

Wells, J.A., Vasser, M., Powers, D.B. Cassette mutagenesis: An efficient method for generation of multiple mutations at defined sites. Gene 34:315–323, 1985.

Alanine scanning mutagenesis of a gene coding sequence

Chatellier, J., Mazza, A., Brousseau, R., Vernet, T. Codon-based combinatorial alanine scanning site-directed mutagenesis: Design, implementation and polymerase chain reaction screening. Anal. Biochem. 229:282–290, 1995.

Clackson, T., Wells, J.A. A hot spot of binding energy in a hormone-receptor interface. Science 267:383–386, 1995.

Hassett, D.E., Condit, R.C. Targeted construction of temperature-sensitive mutations in vaccinia virus by replacing clustered charged residues with alanine. Proc. Natl. Acad. Sci. USA 91: 4554–4558, 1994.

Kelley, R.F., O'Connell, M.P. Thermodynamic analysis of an antibody functional epitope. Biochemistry 32:6828–6835, 1993.

Tang, H., Sun, X., Reinberg, D., Ebright, R.H. Protein-protein interactions in eukaryotic transcription initiation: Structure of the preinitiation complex. Proc. Natl. Acad. Sci. USA 93: 1119–1124, 1996.

Wang, L., Liu, L., Berger, S. Critical residues for histone acetylation by Gcn5, functioning in Ada and SAGA complexes, are also required for transcriptional function in vivo. Genes & Dev. 12:640–653, 1998.

SECTION

2

CORE METHODS

CHAPTER

4

CHARACTERIZATION OF GENOMIC DNA

This chapter begins with a brief overview of genomic organization in bacteria, yeast, and humans, to give a perspective of what problems must be overcome by genomic DNA methodologies. This is followed by a presentation of genomic mapping methods and relevant techniques needed to isolate specific genomic DNA segments using artificial chromosome-based libraries. The last section describes how gene regulatory regions can be used to characterize proteins required for control of transcriptional initiation events. Laboratory practicum 4 describes the molecular approaches that are used to identify and isolate candidate human disease genes based on positional cloning techniques.

OVERVIEW OF GENOME ORGANIZATION

The term genome refers to the total DNA content of a cell, or collectively, the DNA-encoded biochemical information of an entire organism. This includes the complete coding and noncoding information contained in the sum total of chromosomes. Cells also contain mitochondrial DNA; however, by convention, "genome" usually refers to nonmitochondrial DNA. To understand better the physical nature of genomic DNA, three representative organismal genomes are described. We first examine genomes of two model laboratory organisms, the bacterium *Escherichia coli* K-12 and the yeast strain *Saccharomyces cerevisiae*, and then the human genome.

Organization of the *E. coli* K-12 Genome

In 1997, Fred Blattner and his group at the University of Wisconsin reported the complete nucleotide sequence of the nonpathogenic laboratory strain of *E. coli* K-12 called MG1655. Their 15 year effort led to the assembly of one continuous DNA strand containing all 4,639,221 bp of the circular *E. coli* genome. A similar feat was accomplished shortly thereafter by a consortium of Japanese research laboratories using the W3110 strain of *E. coli* K-12. Figure 4.1 presents a graphical illustration of the *E. coli* genome

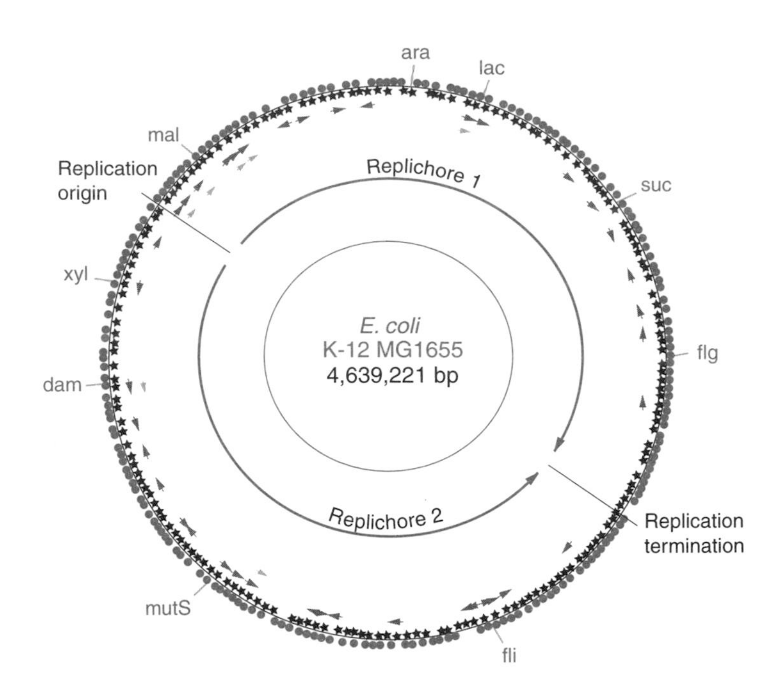

Figure 4.1 Schematic representation of the *E. coli* K-12 genome. The location of gene clusters (~20 or more ORFs) are represented by circles and stars on the two DNA strands. Ribosomal RNA genes are shown as large arrowheads and tRNA genes are represented by small arrowheads. The sites of DNA replication initiation and termination are indicated, as well as the approximate physical position of representative *E. coli* genes and operons.

based on the sequence analysis data of the Blattner research team. It is clear from these data that the *E. coli* genome is very compact with a homogeneous distribution of genes throughout the chromosome. Consistent with earlier genetic data, numerous clusters of genes encoding functionally related proteins were identified as physically linked regulatory operons. This compact arrangement of genes is also observed in the genomes of three other bacteria that have recently been sequenced: 1) the *Bacillus subtilis* genome was found to contain 4100 protein coding sequences distributed across 4,214,810 base pairs, (2) *Mycobacterium tuberculosis,* a pathogenic microorganism with 4000 genes distributed across a 4,411,529 base pair genome, and (3) the gastric pathogen, *Helicobacter pylori*, which is responsible for peptic ulcer disease, has a minimum of 1590 proteins in a genome of 1,667,867 base pairs.

Computer analysis of the *E. coli* DNA sequence identified 4288 actual and proposed gene coding sequences, of which 38 percent have no attributed function based on homology searches with all available databases. It was found that approximately 88% of the genome encodes proteins or RNAs, ~11% appears to be utilized for gene regulatory functions, and <1% consists of repetitive DNA sequences. The average distance between *E. coli* genes is only 120 bp. The coding sequences are continuous and lack the noncoding introns that are commonly found in most eukaryotic genes. *E. coli* K-12 is by far the best characterized free-living organism and research over the past 50 years has provided an abundance of genetic, biochemical, and physiological data. Using this information, Blattner's group was able to assign function to ~3200 genes as summarized in Table 4.1.

Organization of the Yeast *S. cerevisiae* Genome

The yeast genome consists of 16 linear chromosomes, each containing a centromeric region required for chromosome segregation and special repeat sequences at the two

TABLE 4.1 Gene functional groups identified in the DNA sequence of *E. coli* K-12.

Functional class	Number of genes	Percent of total
Transport proteins	427	9.9
Energy metabolism	373	8.7
Replication, transcription, translation	352	8.2
Putative enzymes	251	5.9
Cell structure and membrane proteins	237	5.5
Nucleotide, amino acid, fatty acid metabolism	237	5.5
Regulatory functions	178	4.2
Miscellaneous gene functions	601	14.0
Hypothetical, unclassified, unknown	1632	38.1
Total	4288	100

ends, called telomeres, which provide important functions during DNA replication and in chromosome maintenance. Figure 4.2 shows the approximate size and organization of yeast chromosomes.

Unlike the *E. coli* chromosome, which contains a single origin of replication (Fig. 4.1), eukaryotic chromosomes contain multiple initiation sites for bidirectional DNA replication. The nucleotide sequence of the entire *S. cerevisiae* genome has been determined and found to contain 12,068 kb of DNA. Sequence analysis has identified 5885 potential protein coding genes and another 455 RNA coding genes (rRNA, snRNA, and tRNA genes). Almost 70% of the yeast genome is devoted to protein coding sequences, with the average distance between genes being ~2 kb. Interestingly, unlike most other eukaryotic genes, only 4% of the ~6000 yeast genes have introns, and even then, most of these genes contain only a single intron near the 5′ end of the coding sequence.

The yeast genome sequencing effort was accomplished through a "divide and conquer" strategy with different research groups around the world taking responsibility for specific regions of the genome. The data collected by these researchers are most easily accessed through the Internet using linked WWW sites. In the example shown in Figure 4.3, a DNA segment near the centromere of yeast chromosome III is shown to illustrate how relevant physical, genetic, and nucleotide sequence data can be accessed through a graphical interface.

Organization of the Human Genome

Having the complete DNA sequence of the *E. coli* and yeast genomes is certainly of great interest to basic science researchers, but there is even more enthusiasm for obtaining similar information for the human genome. Unfortunately, cloning, mapping and sequencing of the human genome are much more complex for several reasons. First, the human genome contains at least 3 billion bp and is therefore 250 times larger than the yeast genome. This increased size has meant that it will probably take until the year 2005 to complete the human genome sequencing project. Second, whereas the *E. coli* and yeast genomes are very compact and contain a high density of uniformly distributed genes, the human genome contains large amounts of repetitive DNA sequences and non-

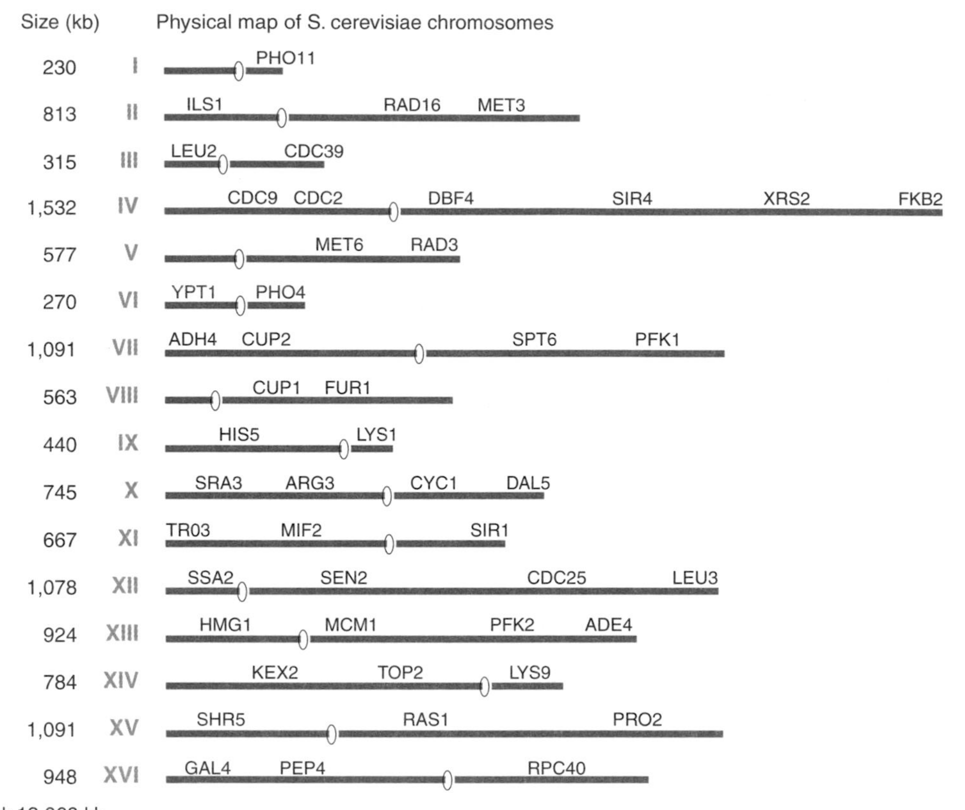

Figure 4.2 The genome of the yeast *S. cerevisiae* is contained on 16 linear chromosomes. Schematic representation of the physical map of the yeast genome is shown. The centromeres are drawn as open circles and the locations of representative genes on each chromosome are indicated. The total length in kilobases of each yeast chromosome is known from DNA sequence information.

coding intronic DNA, as described in Chapter 1. In fact, it is estimated that the protein coding sequence of 100,000 human genes could be contained within <10% of the human genome (~3×10^8 bp is enough to encode 10^5 genes of 1000 amino acids each). Therefore, when taking into account that most human genes contain 10–20 introns, a typical human gene "hunt" requires both clever detective work and a bit of luck. Third, human genetic mapping cannot be done by experimental manipulation. In other words, a human geneticist cannot arrange a standard backcross or F_1 x F_1 mating using a laboratory strain of *Homo Sapiens*!

Figure 4.4 shows a karyotype of a typical mitotic spread of a human cell that was being analyzed for the purpose of prenatal diagnosis. The expected finding of 22 different autosomal and two sex chromosomes per human mitotic cell can be seen. The paired chromosomes represent the maternal and paternal homologues. Karyotyping is a rather crude method, however, it is sufficiently sensitive to identify a number of genetic abnormalities that are characteristic of certain fetal syndromes. The banding pattern of each chromosome results from the fixation and staining procedure used for the cytological preparation of metaphase cells. A more sensitive method of chromosome analysis is based on sequence-specific DNA–DNA hybridization directly to metaphase spreads and is called fluorescent in situ hybridization (FISH).

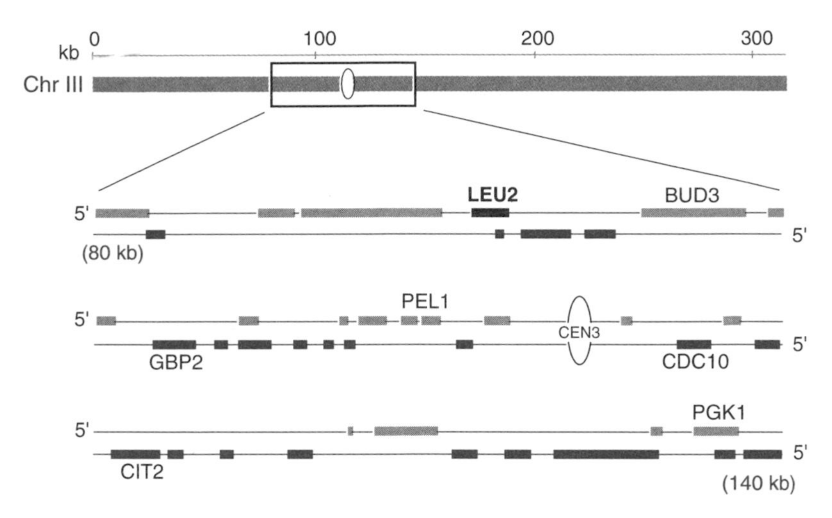

Figure 4.3 Genomic organization of a DNA segment on yeast chromosome III in the region of the L*eu* 2 gene. A 60 kb segment of chromosome III is shown with all known genes and open reading frames that have been identified by DNA sequencing. The centromeric region of chromosome III is called *CEN3*. The corresponding nucleotide sequence for this region of chromosome III is linked to this graphical interface and can be directly accessed though WWW sites on the Internet.

Similar to yeast chromosomes, each human chromosome has a single centromere, telomeres at each end, and multiple initiation sites for DNA replication. Based on the detailed analyses of hundreds of human genes we know that most protein-coding genes are contained on a segment of DNA about 10–100 kb long. Only 1–5 kb of this DNA corresponds to exon coding sequence, with the rest being intronic regions and transcriptional regulatory sequences. cDNA sequences representing expressed sequence

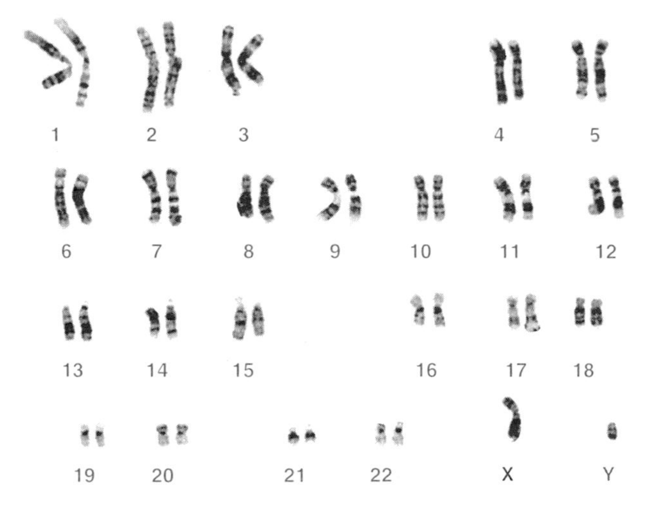

Figure 4.4 Analysis of a mitotic spread showing a normal 46, XY human karyotype. The banding pattern on each chromosome is used to identify the 22 pairs of autosomes and the X and Y chromosomes. Each of the chromosomal homologues is derived from the maternal or paternal gamete. The X chromosome was donated by the mother, whereas the Y chromosome came from the father.

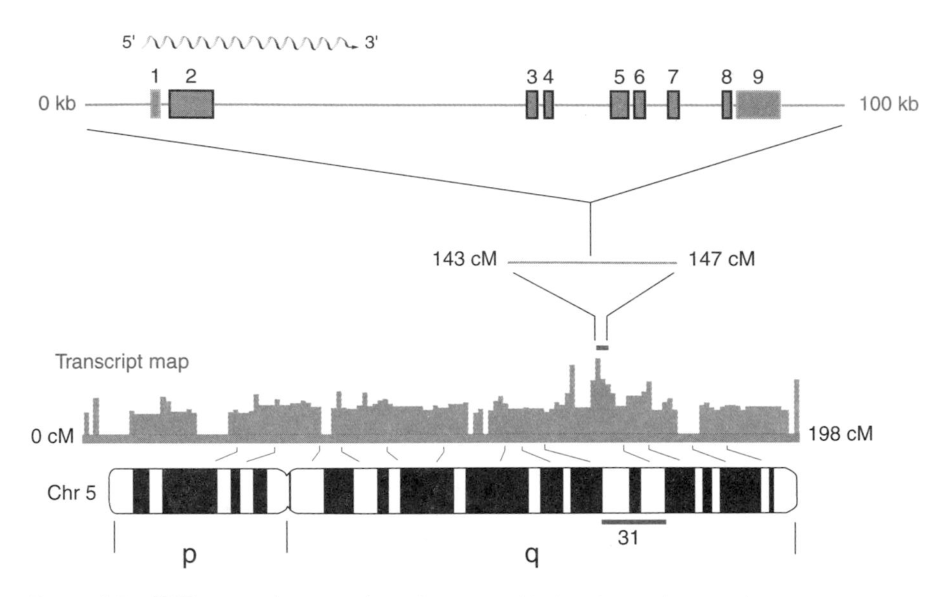

Figure 4.5 EST transcript maps have been used to locate markers on human chromosomes. Schematic representation showing the distribution and frequency of ESTs that have been mapped to human chromosome 5. The chromosomal region in the vicinity of 5q31 is expanded to show the genomic structure of the glucocorticoid receptor gene, which represents just one of ~50 transcripts that have been mapped within a segment located near the end of chromosome 5. A centimorgan (cM) is a genetic unit that corresponds to ~1 Mb (10^6 bp) of human DNA. The glucocorticoid receptor coding sequence is contained in nine exons that together span almost 100 kb. The bold rectangles denote exons encoding the majority of the 777 amino acids of the receptor protein. The human gene transcript map can be accessed through the Internet at http://www.ncbi.nlm.nih.gov/SCIENCE96/.

tags (ESTs) have been cloned and sequenced and used to map genomic transcription units in the human genome using hybridization analyses. A human gene map has been constructed based on these EST data; an example is shown in Figure 4.5 for human chromosome 5 in the region encoding the human glucocorticoid receptor. As of early 1997, ~50,000 transcription units had been mapped to the human genome using ESTs, and about 4500 of these transcription units were found to correspond to known genes already present in the database.

GENOMIC MAPPING

The ultimate goal of many genome researchers is to construct a physical and functional map of an entire genome to be used as a research tool to study biological processes. In the Human Genome Project, for example, researchers are currently focused on constructing a functional genomic map that, in combination with a physical map and catalogue of cloned DNA fragments, can be used to determine the nucleotide sequence of all 24 human chromosomes (22 autosomes, X, and Y). In this section we describe the two basic approaches to genomic mapping that have been used to determine where genes are located on chromosomes. One is a functional mapping strategy in which genetic phenotypes are scored to determine the frequency of genetic recombination during meiosis

mal recombination in *Drosophila* has been used as a method to isolate genes based on alterations in the genetic and physical maps of specific chromosomes. This approach has also been used to produce subchromosomal fragments of human DNA in the genetic background of a human–rodent hybrid somatic cell line. David Cox and his colleagues used radiation-induced recombination in Chinese hamster–human hybrid cell lines to generate a radiation hybrid (RH) linkage map of human chromosome 21. Similar radiation-induced subchromosomal hamster–human hybrid cell lines are available for all 24 human chromosomes. These hybrid cell lines have been used for physical mapping, gene isolation, and linkage mapping. Figure 4.7 illustrates how radiation-induced subchromosomal hamster–human hybrid cell lines are generated.

Genomic Mapping Using DNA Sequence Polymorphisms as Genetic Markers

Classic genetic linkage mapping provides functional information about genetic phenotypes. Molecular genetic methods have been developed to monitor genetic inheritance using biochemical assays that identify DNA sequences using restriction enzyme mapping and PCR. These techniques are independent of gene function and therefore can be used to map genomic regions in human populations. This type of human genomic mapping takes advantage of the fact that nucleotide sequences can be different between two individuals without causing overt or deleterious mutations. A nucleotide difference found at the same physical location in the genome is called a DNA sequence polymorphism if it varies in at least 1% of the population. As shown in Figure 4.8, two types of DNA polymorphisms are those represented by altered restriction enzyme digestion patterns or a change in the number of repeat units in a tandem array. Single nucleotide polymorphisms (SNPs) have also been used as genome markers. SNPs, pronounced "snips," can be identified by PCR (Chapter 6), or by DNA microchip assays (Chapter 9). DNA polymorphisms serve as genetic markers and can therefore be used to follow the inheritance of specific chromosomal regions. If a disease phenotype is tightly linked to the inheritance of a specific set of DNA markers, then it provides a starting point for gene identification using DNA segments that contain these same DNA markers.

Using Genetic Data, DNA Markers, and Cloned Segments to Construct a Contig Map

How can information generated from a genetic linkage map be aligned with genomic DNA sequences from the same chromosomal region? The answer is that genes used for linkage analysis are physically localized to cloned genomic segments through a shared set of DNA sequence markers. If enough of a chromosome has been cloned and characterized as overlapping genomic segments called contigs, then it is possible to superimpose a linkage map measured in centimorgans, with a physical genomic map measured in kilobases. For both *E. coli* and *S. cerevisiae*, these two types of maps were used ultimately to determine the complete nucleotide sequence of the genome.

The strategy being used to construct a human genomic contig map is based on localizing chromosome-specific DNA markers, called sequence tagged sites (STSs), to cloned genomic segments contained within YAC (yeast artificial chromosomes), BAC (bacterial artificial chromosomes) or phage P1 cloning vectors. An STS marker is any DNA sequence that can be assayed by PCR using specific primer pairs, most commonly

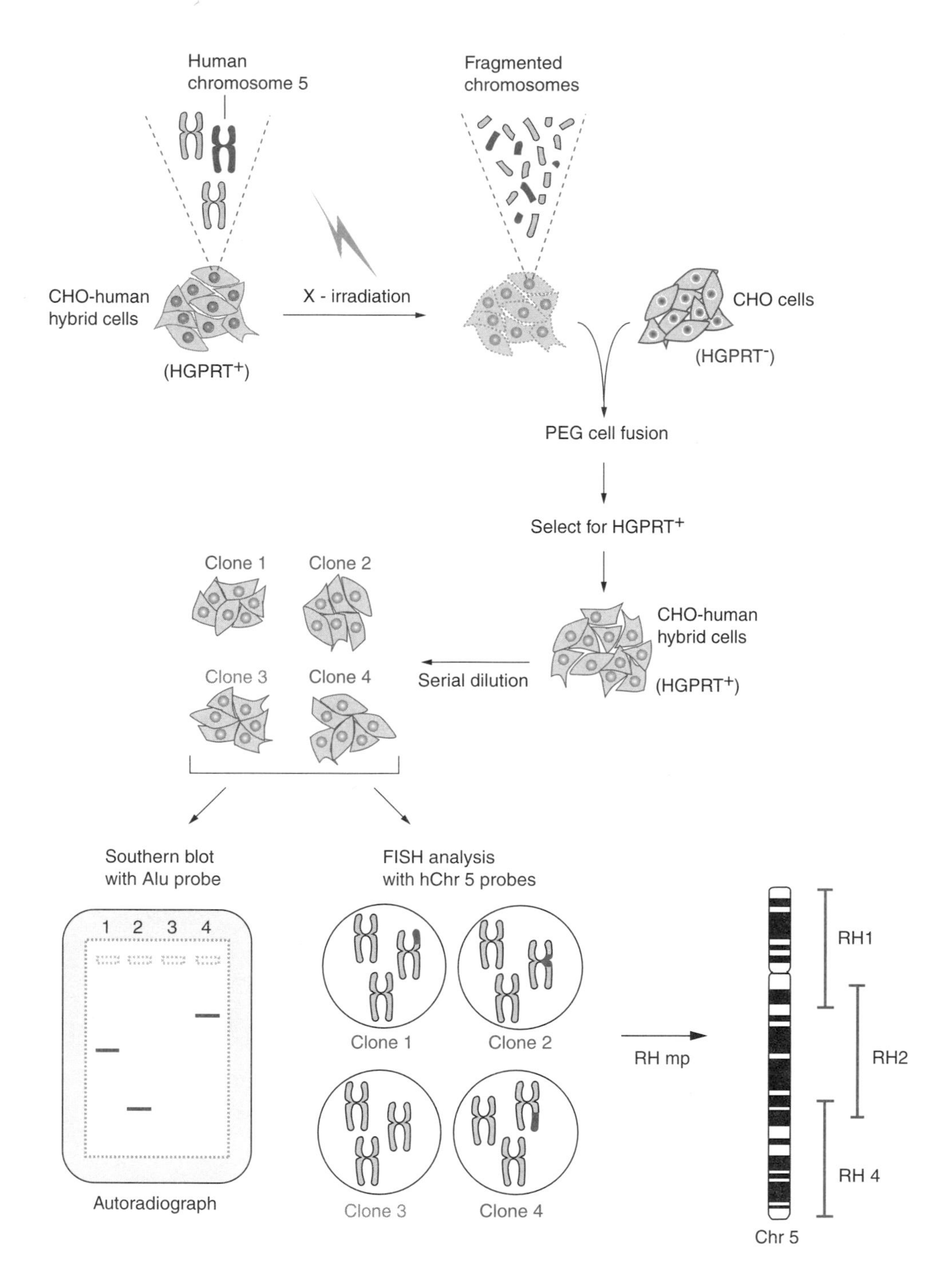

Figure 4.7 Creation of radiation hybrid panels using hamster–human cell lines. Flow scheme showing how subfragments of human chromosome 5 (hChr 5) can be generated and maintained in hybrid somatic cell lines. Irradiation of a Chinese hamster ovary (CHO) cell line, containing an intact hChr 5, produces fragments of both hamster and human chromosomes that can be fused to intact CHO cells that contain no human chromosomes. Because the hChr5 CHO cell line has a wild-type hypoxanthine guanosine phosphoribosyl transferase (HGPRT$^+$) gene, and the donor CHO cell does not (HGPRT$^-$), fused cells can be isolated by selecting for growth in HAT (hypoxanthine/aminopterin/thymidine) media. Radiation hybrid (RH) cell lines are characterized for the presence of human chromosome subfragments by Southern blotting and fluorescent in situ hybridization (FISH). In this example the RH lines 1, 2, and 4 contain portions of hChr 5.

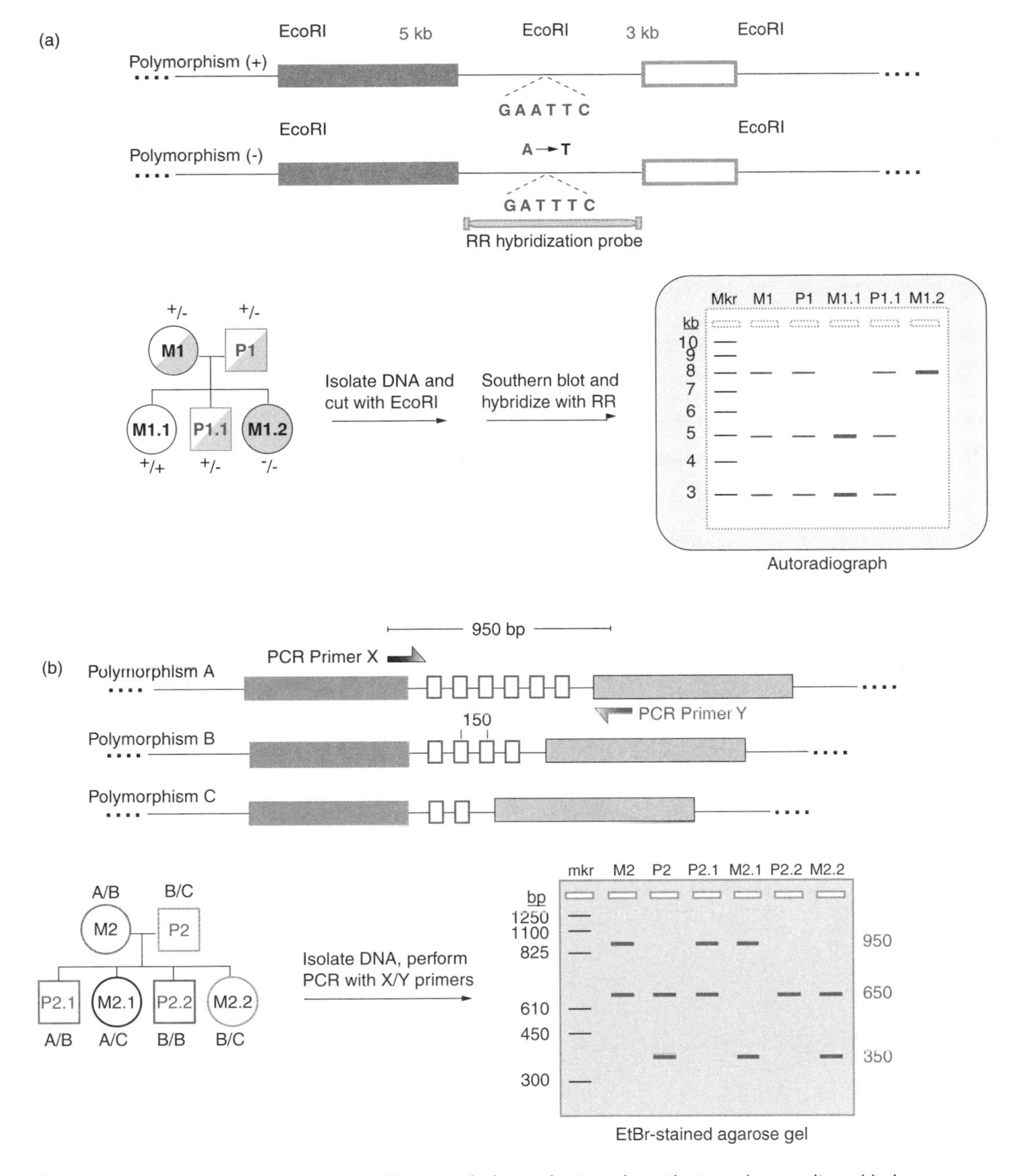

Figure 4.8 DNA polymorphisms provide gene-independent markers that can be monitored in human samples using standard biochemical assays. (*a*) Restriction fragment length polymorphisms (RFLPs) can be monitored by Southern blotting. DNA extracted from related individuals is analyzed using a restriction enzyme (EcoRI) that distinguishes between two alleles ($EcoRI^+$ and $EcoRI^-$). Heterozygous individuals contain one copy of each allele (+/–), and homozygous individuals contain only one allele type (+/+ or –/–). (*b*) Short tandem repeat polymorphisms (STRPs) can easily be identified by PCR using sequence specific primers that flank the polymorphic region. Because the number of tandem repeats can vary by a single repeat length (e.g., 150 bp), multiple polymorphisms can be detected with the same PCR primers (A, B, C).

STRPs or SNPs. A large number of STSs have so far been mapped to the human genome (>15,000), making it possible to align genetic linkage maps with physical maps. A genomic contig is defined as a region of continuous DNA that has been assembled from multiple overlapping cloned segments. Because artificial chromosome based vectors can contain artifacts from spurious ligations between two genetically unlinked genomic DNA segments, a high confidence contig map requires agreement between at least two overlapping clones. Alignment with genetic loci that have been physically mapped to the same chromosomal region is also necessary. Figure 4.9 shows an example of how genetic maps and STRPs were used to identify the defective gene responsible for an inherited premature aging disease called Werner's syndrome.

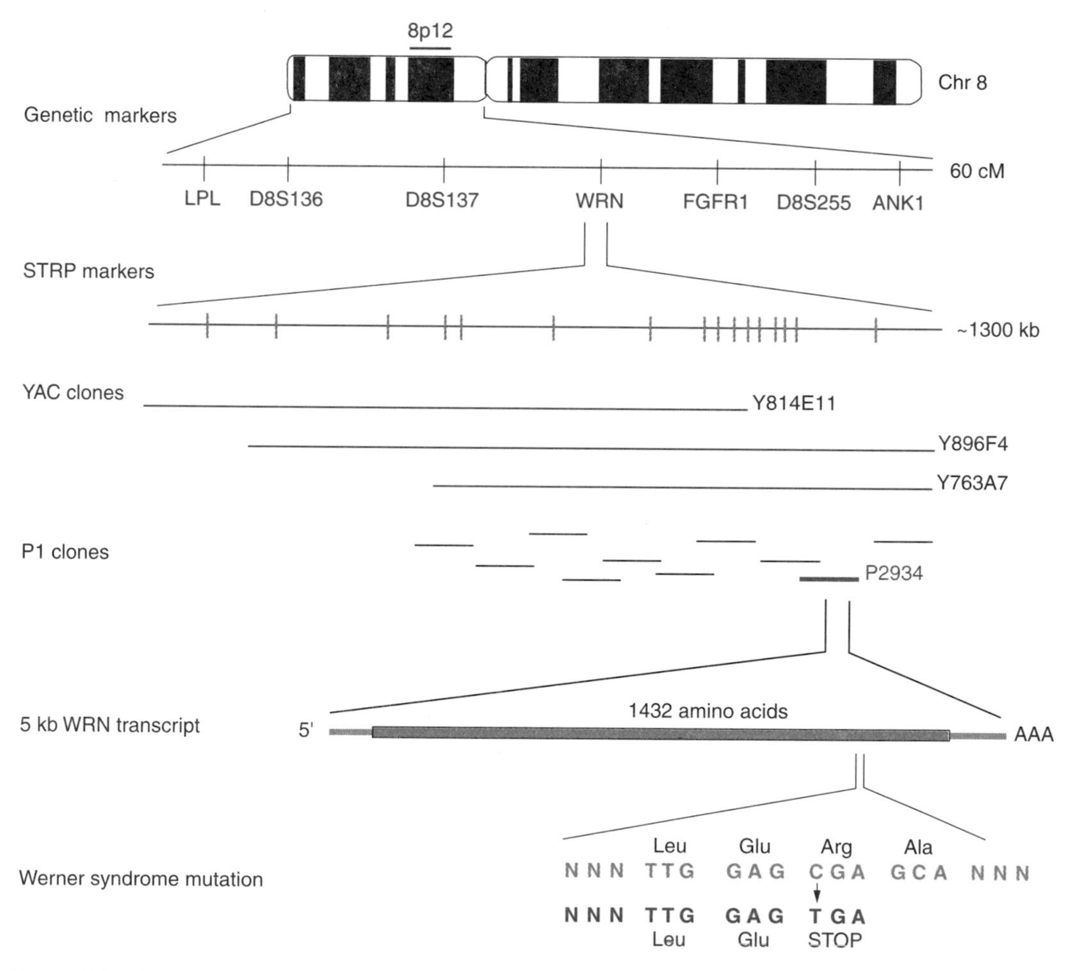

Figure 4.9 Chromosomal contigs can be used to align DNA segments with genetic linkage maps. Schematic representation of the human chromosome 8p12 region illustrating alignment of the genetic and physical maps using STRP markers, YAC, and P1 clones. These chromosome 8 reagents were used to identify the Werner's syndrome gene (WRN) by positional cloning. One of the Werner syndrome alleles was found to be a C to T mutation resulting in the insertion of a termination codon in the open reading frame.

MANIPULATION OF LARGE GENOMIC DNA SEGMENTS

This section describes the biochemical techniques that have been developed to manipulate very large DNA fragments. These biochemical methods are essential for the initial construction of representative genomic libraries and for preparing DNA inserts from individual clones once they have been identified by genomic library screening.

FACS Sorting of Intact Chromosomes

The largest single molecule of DNA in a cell is the chromosome. As shown in Figure 4.4, mitotic cells can be trapped in metaphase by the addition of colcemid, a methylated derivative of colchicine. This plant alkaloid binds to tubulin and prevents the formation of microtubules, a process required for chromosome segregation at anaphase. By staining mitotic cells with a fluorescent DNA-binding dye such as ethidium bromide, it is possible to separate individual mitotic chromosomes physically using a technique called fluorescent activated cell sorting (FACS). This FACS-based approach has been used to purify enough material to construct chromosome-specific genomic libraries. Subgenomic libraries are easier to store and screen than a representative whole genomic library because fewer independent isolates are required.

Chromosome Microdissection

Direct microdissection of a metaphase chromosome using hand-drawn glass pipette microneedles is a second method that has been developed for constructing chromosome-specific subgenomic libraries. This method has been used for subcloning regions of *Drosophila* polytene chromosomes, and as a direct approach to generating human chromosome-specific DNA sequences for FISH analysis. In this technique, metaphase or polytene chromosomes are spread on a microscope slide and the chromosomal region of interest is isolated by literally scraping the DNA off the slide with a microneedle. All these micromanipulation steps are carried out in an oil chamber using high power light microscopy. The tiny amount of DNA on the tip of the needle is transferred to a microdrop of collection buffer located inside the chamber. About 20 mitotic chromosomes, or one polytene chromosome, are required to obtain enough material for processing. The DNA collected in the microdrop is purified by adding an extraction buffer to the droplet containing proteinase K and sodium dodecylsulfate, which digests chromatin-associated proteins. Following a short incubation period at room temperature, the DNA is extracted with phenol and chloroform to remove the undigested proteins and proteinase K. The DNA remaining in the droplet is then digested with a frequent cutting restriction enzyme (e.g., an enzyme with a 4 bp recognition site such as Sau3A), primer adaptors are ligated to the termini, and the isolated genomic DNA is amplified by PCR and cloned into a plasmid vector. This highly specialized technique has been used to isolate new STS sequences corresponding to chromosomal regions lacking suitable markers and to isolate specific chromosomal segments associated with human cancers.

Pulsed-Field Gel Electrophoresis

Agarose gel electrophoresis is the standard technique for separating nucleic acids, however, conventional agarose gels cannot resolve fragments >20 kb. This is because large

DNA fragments are unable to migrate through the gel matrix by the same sieving mechanism that smaller fragments do. In fact, the only way that large DNA fragments can even enter an agarose gel is by "squirming" through the gel like a snake in a parallel direction relative to the electric field. The technique of pulsed-field gel electrophoresis (PFGE) takes advantage of this electrophoretic property of large DNA fragments. PFGE resolves DNA molecules of 100–1000 kb by intermittently changing the direction of the electric field in a way that causes large DNA fragments to re-align more slowly with the new field direction than do smaller molecules. The principle of PFGE is illustrated in Figure 4.10. PFGE was developed as a way to separate yeast chromosomes and to facilitate long-range mapping of human genomic DNA that had been digested with rare-cutting enzymes such as NotI (8 bp recognition site). Preparative PFGE is used to purify very large genomic DNA fragments for insert cloning into artificial chromosome vectors.

Two basic types of PFGE systems have been developed. The simplest is called field inversion PFGE, which works by periodically reversing the direction of the electric field using a standard gel box apparatus. Each time the direction of the field is inverted, larger molecules take longer to reverse direction than do smaller molecules. Net movement in one direction is accomplished by keeping the time of field inversion in the forward direction greater than in the reverse orientation. A second type of PFGE is based on alternating the angle of the electric field. The original PFGE design for alternating-angle electric fields was developed by David Schwartz and Charles Cantor and was based on a 90 degree field alteration, however, this created an uneven DNA migration path through the gel. Gilbert Chu and Ron Davis improved the basic PFGE design by changing the shape of the gel box so that multiple electric fields could be produced by placing electrodes around the perimeter of a hexagonal box (Fig. 4.10). This type of PFGE gel box uses alternating angles of 120 degrees and produces a contoured clamped homogeneous electric field, or CHEF. Commercially available CHEF gel box systems have been designed that control the angle of the alternating electric fields, and the pulsed time in each direction, to optimize the resolution capacity for any range of DNA sizes.

DNA sample preparation for PFGE gels cannot rely on standard biochemical extraction procedures because very large DNA fragments are randomly sheared by physical forces. To solve this problem, PFGE DNA samples are prepared using immobilized cells in agar blocks that are soaked in lysis buffer containing detergent and proteinase K. This method gently purifies the cellular DNA without subjecting it to the shearing forces produced by pipetting. Following cell lysis, a protease inhibitor is soaked into the agar block to inactivate the proteinase K and the DNA is then digested with restriction enzymes prior to placing the agar block into the well of the PFGE gel. Once electrophoresis is initiated, the DNA fragments migrate out of the agar block and directly into the gel.

GENOMIC DNA CLONING VECTORS

Chapter 3 describes λ cloning vectors that are commonly used to isolate DNA inserts of ~1–18 kb. Although this range of insert size is ideal for cDNA libraries, it is not necessarily the best choice for constructing representative genomic DNA libraries. Several specialized genomic cloning vector systems have been developed that can accommodate large DNA inserts with an average size of 35–50 kb (cosmid vectors), 50–300 kb (BAC and P1 vectors), and 200–1000 kb (YAC vectors). Genomic DNA cloning vectors are utilized in a hierarchical manner based on insert size. For example, a researcher may

(a)

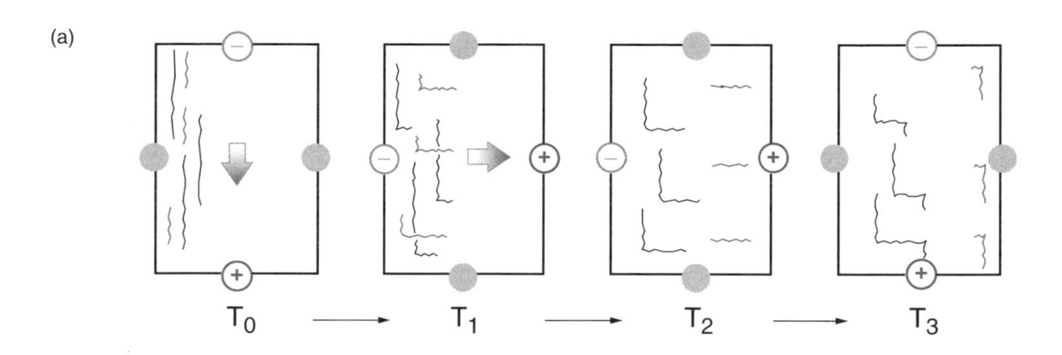

(b)

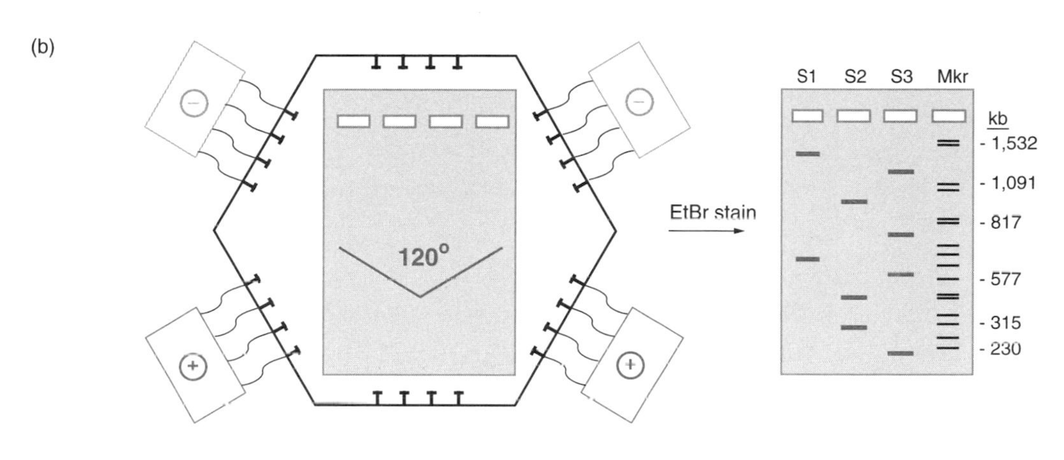

Figure 4.10 Pulsed-field gel electrophoresis (PFGE) is based on the principle that large DNA molecules migrate at different rates through an alternating-angle electric field. (*a*) Very large DNA molecules that are migrating parallel to the electric field in one direction take more time (*T*) to reorient in an alternating electric field than do smaller molecules. (*b*) CHEF pulsed-field gel systems use a hexagonal gel box that alters the angle of the fields relative to the agarose gel. PFGE gels are loaded with DNA samples that are embedded within agar blocks to minimize random breakage of large molecules. *S. cerevisiae* chromosomes are a convenient source of molecular weight markers (Mkr) for PFGE gels (see Fig. 4.2).

identify a specific YAC or BAC clone with an insert of ~500 kb, and then subclone smaller fragments (20–50 kb) into a λ or cosmid cloning vector. These smaller inserts would then be characterized and eventually subcloned further as 0.5–5 kb fragments into a plasmid vector for fine structure analysis and DNA sequencing.

Cosmid Vectors

The development of cosmid vectors was based on the observation that a ~200 bp DNA sequence in the λ genome called *cos* is required for λ DNA packaging into phage particles during lytic infection (Chapter 3). Cleavage at *cos* by the λ terminase protein during phage packaging produces a 12 nucleotide cohesive end at the termini of the linear

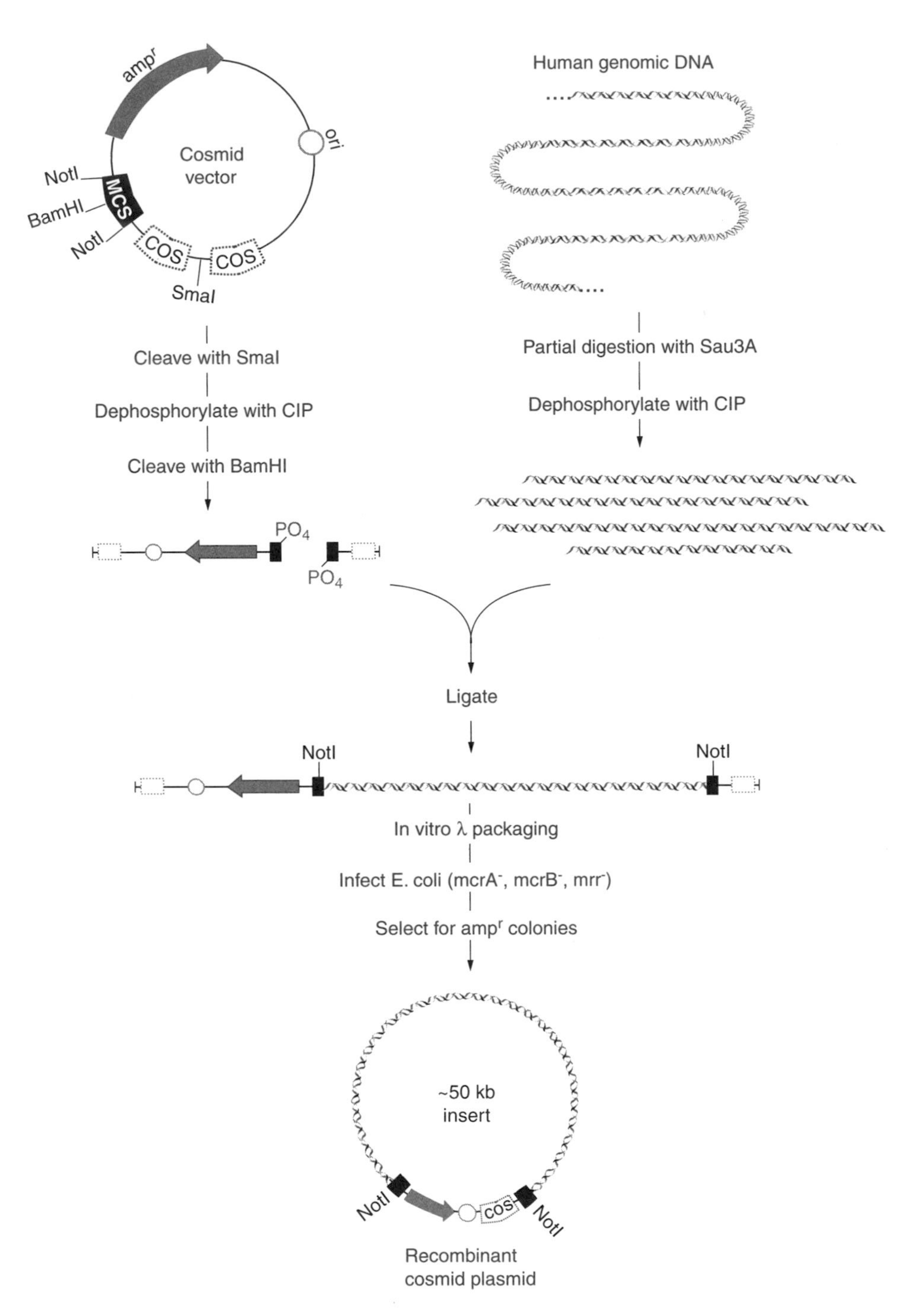

Figure 4.11 Cosmid cloning vectors contain *cos* packaging sequences, a plasmid origin of replication and antibiotic resistance gene. Cosmid DNA libraries are constructed by ligating linearized cosmid vector DNA with genomic DNA that has been digested with a compatible restriction enzyme. Sucrose gradient centrifugation can be used to purify partially digested genomic DNA of the appropriate length (30–50 kb), or alternatively, the biological size constraints of λ packaging can be exploited as a selection strategy. In this example, a double *cos* site vector is shown that can be used with dephosphorylated genomic DNA to enrich for primary recombinants based on insert size. Because many types of eukaryotic genomic DNA contain methylated cytosines, the λ packaged cosmids are used to infect a strain of *E. coli* that lacks the *mcr* methylation-restriction system (Chapter 2). Recombinant cosmid plasmids are stored as pooled bacterial cultures or as individual colony isolates in microtiter plates. CIP is calf intestinal phosphatase.

λ genome. Recircularization of the λ genome after bacterial infection is facilitated by base pairing between the complementary cohesive ends. Cosmid cloning vectors with DNA inserts of 30–45 kb can be packaged in vitro into λ phage particles, provided that the ligated double-strand DNA contains λ *cos* sequences on either side of the insert DNA (Fig. 4.11). Because the λ phage head can hold up to 45 kb of DNA, the optimal ligation reaction in cosmid cloning produces recombinant molecules with *cos* sequences flanking DNA segments of ~40 kb. *E. coli* cells infected with cosmid-packaged phage are able to support autonomous replication of circular cosmid plasmids because a ColE1 origin of replication and the β lactamase gene (*amp*^{r}) are linked to the *cos* sequence (Fig. 4.11). Cosmid libraries are maintained in *E. coli* and are screened by standard filter hybridization methods using radioactive DNA probes.

Yeast Artificial Chromosome (YAC) Vectors

Maynard Olsen and colleagues described the first generation of yeast cloning vectors that were based on artificial chromosomes. YAC vectors use DNA sequences that are required for chromosome maintenance in *S. cerevisiae*. Plasmids containing these yeast sequences are digested with appropriate restriction enzymes and ligated to PFGE size-fractionated genomic DNA. Figure 4.12 shows how the pYAC4 vector can be used to construct a representative human YAC library that can be stored as individual transformants in a series of microtiter plates. pYAC4 is an example of a YAC cloning vector that contains an autonomously replicating sequence (*ARS1*) to function as an origin of replication, centromere elements (*CEN4*) for chromosome segregation during cell division, telomeric sequences (*TEL*) for chromosome stability, and a growth selectable marker (*URA3*) to select positively for chromosome maintenance. The plasmid vector itself can be grown in bacteria because it contains a ColE1 plasmid origin and an antibiotic-resistance gene (*amp*^{r}). Yeast transformants containing recombinant YAC molecules can be identified by red/white color selection using a yeast strain that contains the *ADE2-1* ochre mutation, which is suppressed by the *SUP4* gene product. Inactivation of *SUP4* by DNA insertion into the EcoRI site results in the formation of a red colony.

Bacterial Artificial Chromosome (BAC) Vectors

BAC cloning vectors were developed by Mel Simon and his colleagues as a an alternative to YAC vectors. BAC vectors are maintained in *E. coli* as large single copy plasmids and contain inserts of 50–300 kb. BAC libraries are easy to work with because most procedures utilize standard methods developed for bacterial plasmid cloning. This is in contrast to P1 phage libraries, which have a maximum insert size of 100 kb and require the production of specially prepared P1 packaging extracts and purified phage arms. BAC vectors contain the F plasmid origin of replication (*oriS*), F plasmid genes that control plasmid replication (*repE*) and plasmid copy number (*parA, B, C*), and the bacterial chloramphenicol acetyltransferase gene (*CM*^{r}) for plasmid selection. The methods used to construct a BAC library are essentially the same as a standard plasmid library, except that the insert DNA must be prepared by preparative PFGE. Figure 4.13 illustrates how the pBeloBAC 11 vector from the Simon lab is used to construct a human BAC library.

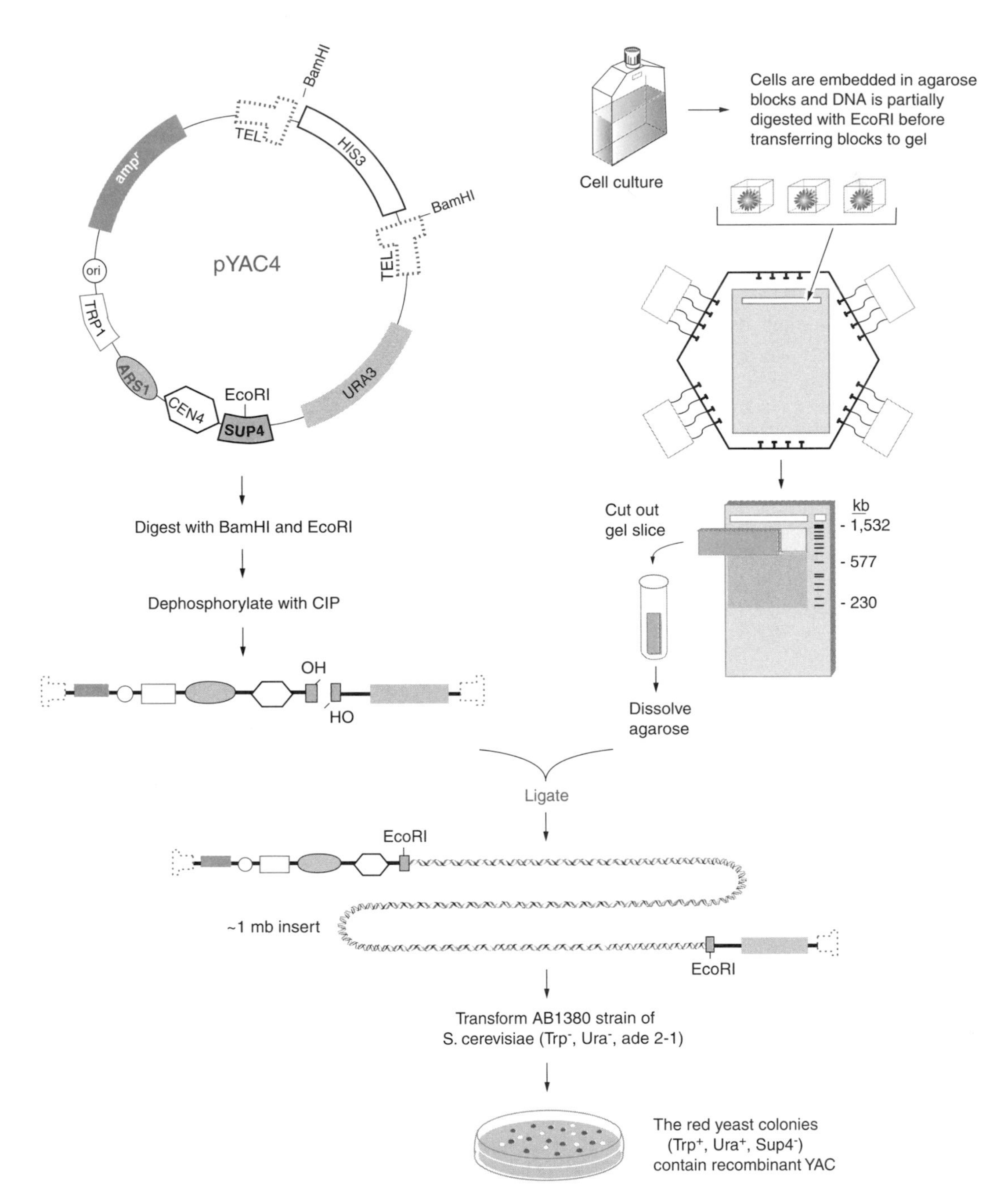

Figure 4.12 YAC cloning vectors contain functional elements for chromosome maintenance in *S. cerevisiae*. To make a human YAC library, human genomic DNA is partially digested with EcoRI and size-fractionated by preparative PFGE. This material is then ligated to dephosphorylated pYAC4 vector DNA that has been digested with BamHI to liberate the telomeric ends and with EcoRI to create the insert cloning site. An appropriate yeast strain is transformed with the ligated material, and Trp+, Ura+ red colonies (disrupted *SUP4* tRNA gene in YAC vector) are isolated and characterized.

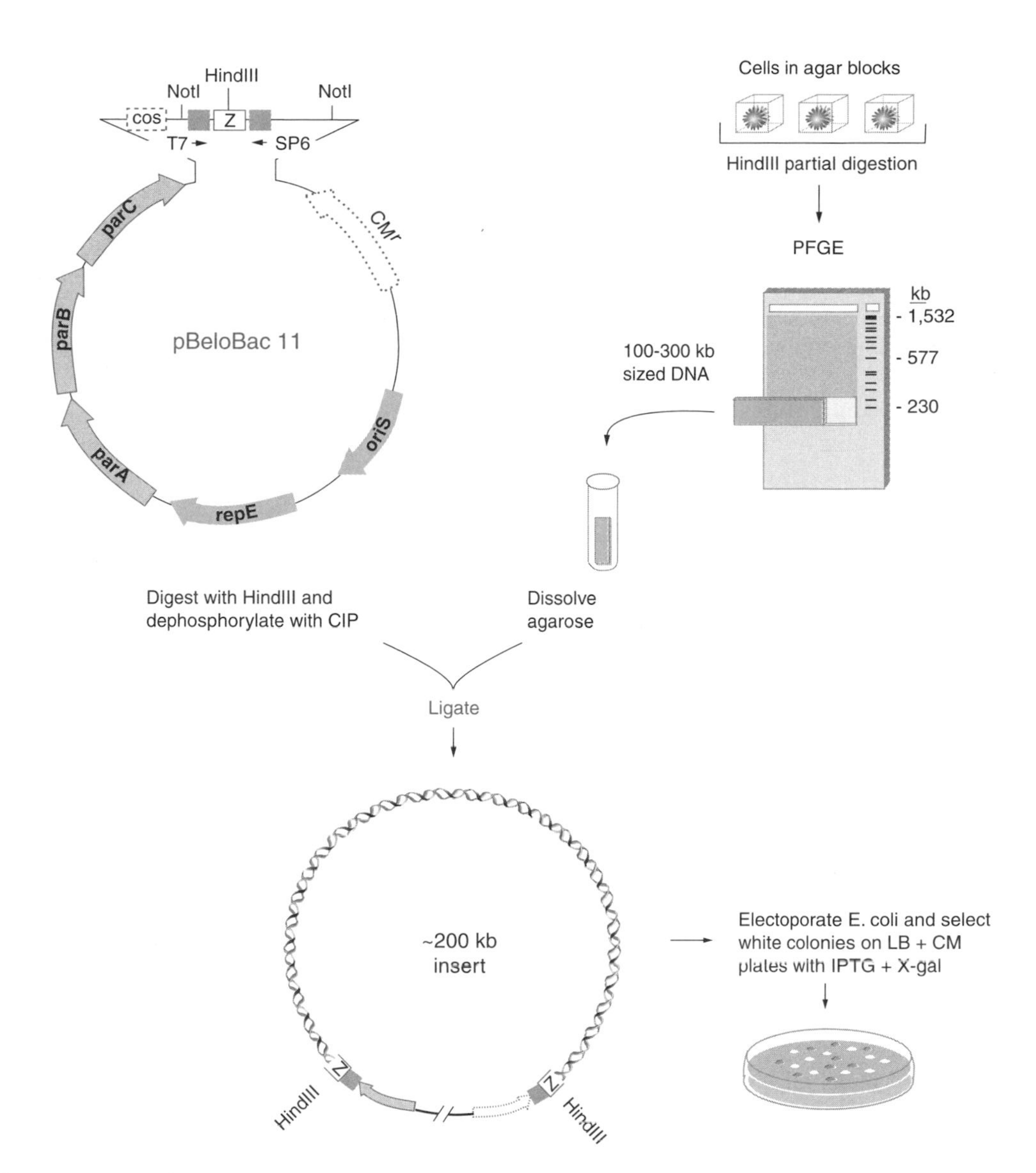

Figure 4.13 BAC vectors contain the F plasmid origin of replication and are maintained in *E. coli* as stable single-copy plasmids. pBeloBAC 11 is an example of a BAC vector developed in the Simon laboratory. High molecular weight DNA is partially digested with the enzyme HindIII, size-fractionated by PFGE, and ligated to the HindIII-digested and phosphatase-treated pBeloBAC 11 vector. *E. coli* cells are electroporated with the ligated material and white colonies (DNA insertion into the *lacZ* α subunit gene) are isolated on the basis of chloramphenicol-resistance (CM^r). The insert DNA can be excised with the rare-cutting enzymes NotI (8 bp recognition sites). The plasmid can also be linearized at the *cosN* site by λ terminase to facilitate end-labeling for restriction enzyme mapping. The bacteriophage T7 and SP6 transcriptional promoter sequences flanking the HindIII cloning site provide a means to generate end-specific RNA probes for library screening.

Screening Genomic Libraries Using DNA Pools

High-quality human and mouse genomic libraries in YAC, BAC, and P1 phage vectors are commercially available from genome research companies. Therefore, the task of cloning a particular gene is actually initiated by screening, rather than constructing, one of these libraries. For example, a gene mapping study based on positional cloning may have identified a region of the human genome that has been mapped with several STS markers covering a 1 Mb span of DNA. To identify the YAC clones that contain the chromosomal region of interest, a YAC library would be screened by PCR using specific STS primers and pools of yeast transformants. By systematically testing different YAC pools, it is possible to identify individual YACs that contain insert DNA with the desired STS marker. The use of pooled transformants for library screening is feasible because of the sensitivity of PCR assays (see Chapter 6). Using the screening strategy outlined in Figure 4.14, only 45 PCR reactions would be required to confirm one positive YAC clone out of 38,400 independent yeast transformants.

MAPPING GENE REGULATORY SEQUENCES

Genomic DNA contains important gene regulatory sequences that function as binding sites for sequence-specific DNA binding proteins called transcription factors. For any given gene, the exact start site of transcription, as well as the control of transcriptional initiation rates, is determined by transcription factor binding to DNA sequences located upstream, and sometimes downstream, of gene coding sequences. One of the objectives of genomic DNA studies can sometimes be to map gene regulatory sequences functionally and physically using in vivo and in vitro approaches. In this section, three general strategies are presented that can be used to characterize genomic DNA sequences required for controlling the temporal and spatial expression of eukaryotic genes.

Mapping Regulatory Sequences by in vivo Expression Assays

Because a large number of proteins in a cell extract are capable of nonspecific, or even nonphysiological, binding to DNA sequences, it is important to identify bona fide gene regulatory elements using an in vivo functional assay. For example, the glucocorticoid receptor zinc finger protein can bind in vitro as a dimer to essentially any DNA fragment containing the consensus sequence 5′-nGnACAnnnTGTnC-3′. However, the high frequency random occurrence of this sequence in genomic DNA indicates that in vitro glucocorticoid receptor binding to purified DNA is not necessarily a predictor of in vivo relevance. One type of functional gene regulation assay is transient DNA transfection of cultured cells (Chapter 7) using a plasmid construct called a heterologous "reporter gene." Reporter genes contain transcriptional regulatory sequences that are linked to the coding sequence of a gene encoding a protein function that can be quantitatively measured. Figure 4.15 shows how a firefly luciferase reporter gene can be used to map gene regulatory sequences required for the transcription of a liver-specific gene. Deletion mapping can be used to identify two types of sequence elements based on reporter gene assays: genomic DNA sequences required for basal promoter function and sequences necessary for liver-specific expression. These functionally defined sequences provide biochemical reagents to characterize sequence-specific DNA binding proteins using in vitro assays.

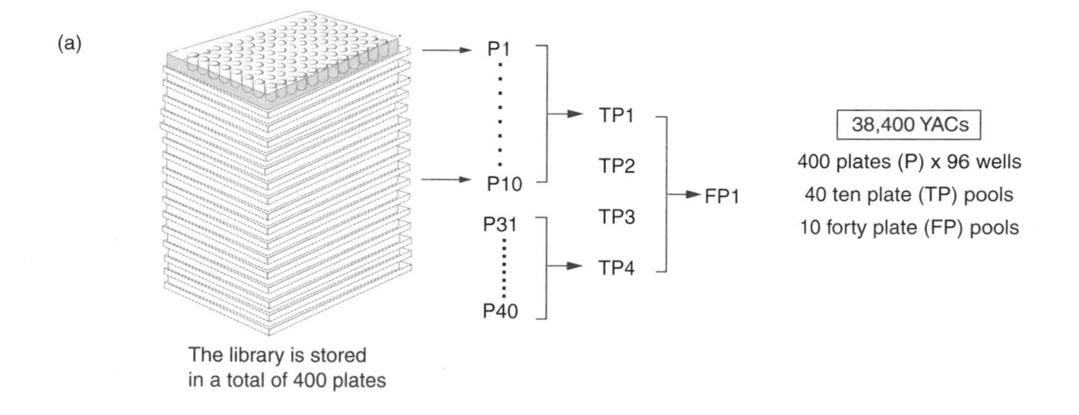

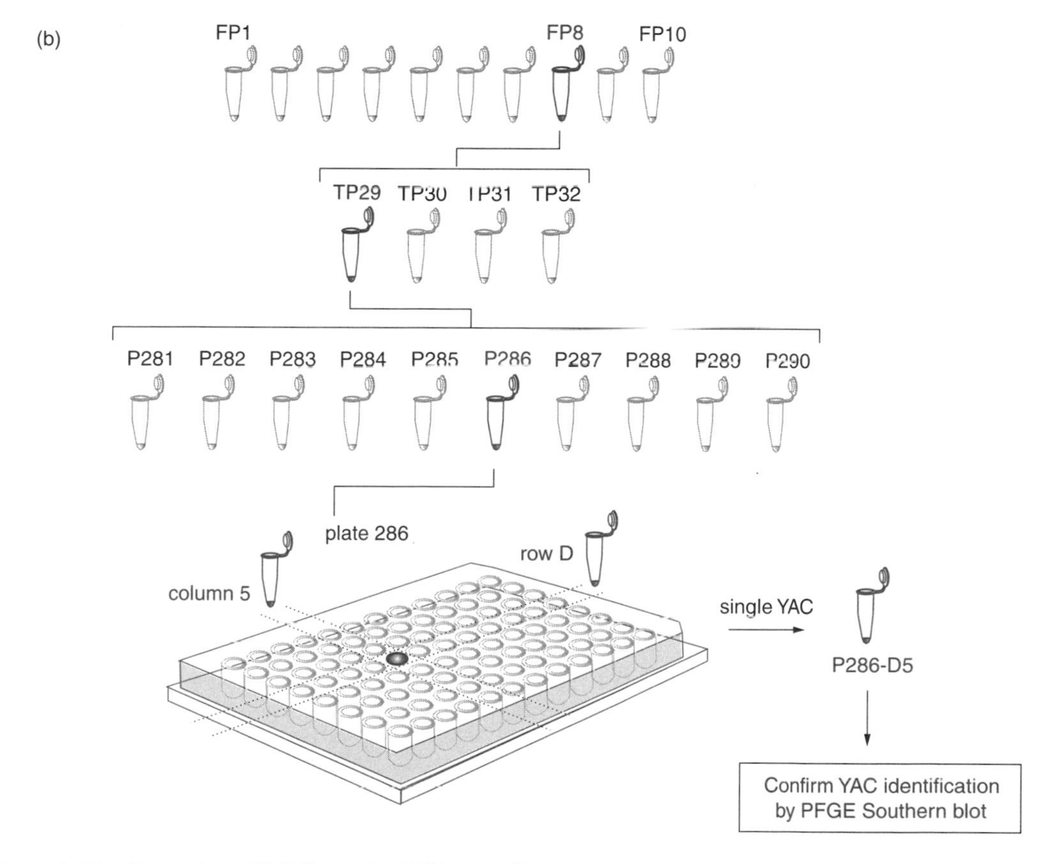

Figure 4.14 Screening a YAC library by PCR can efficiently be done using DNA pools. (*a*) Three levels of yeast pools are generated: P1-P400, each containing 96 strains from one plate; TP1-TP40, each containing 10 pooled plates (960 strains per pool); and FP1-FP10, each containing four sets of TP pools (40 plate equivalents which is 3840 strains per pool). (*b*) Beginning with the FP1-10 pools, successive subpools are screened by PCR until a single strain is identified by the intersection of a column and row pool on one microtiter plate. Using this strategy, a total of only 45 PCR reactions is required to identify positively one YAC strain from the original FP pools.

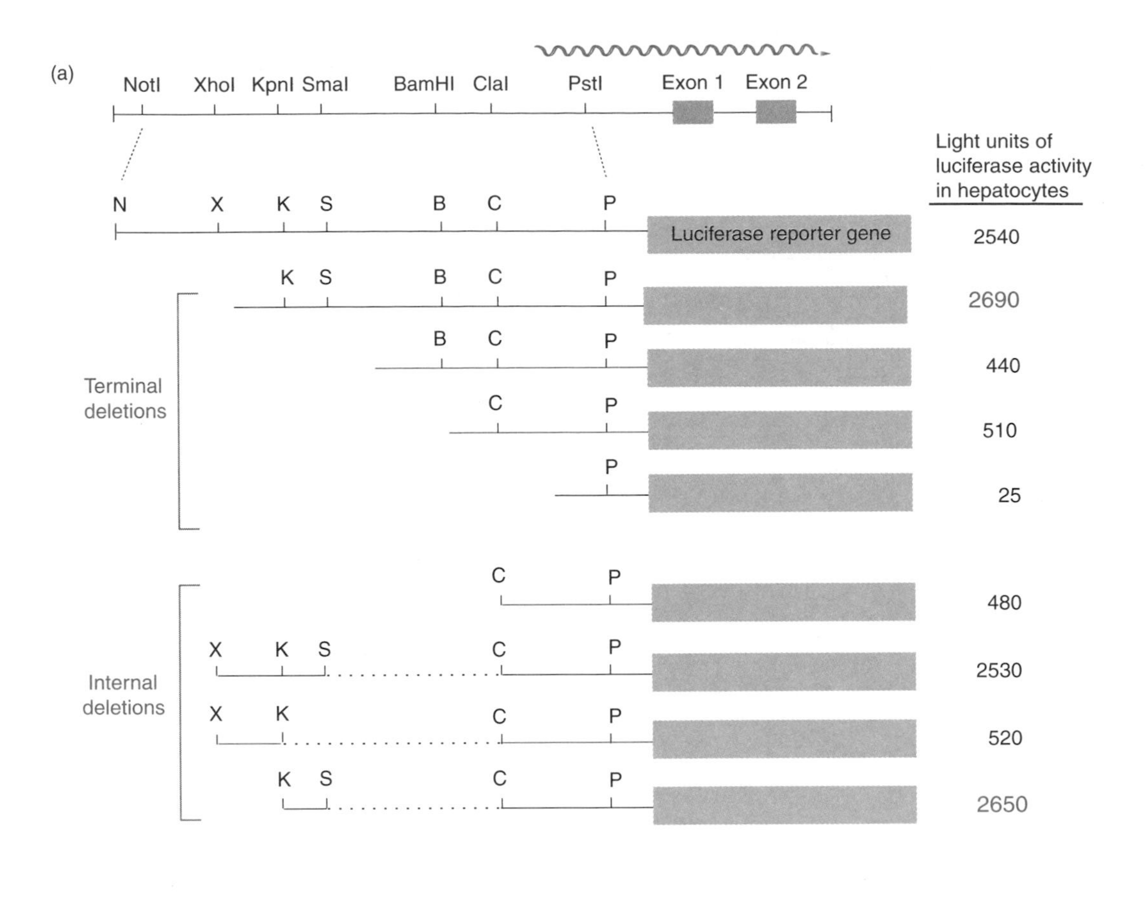

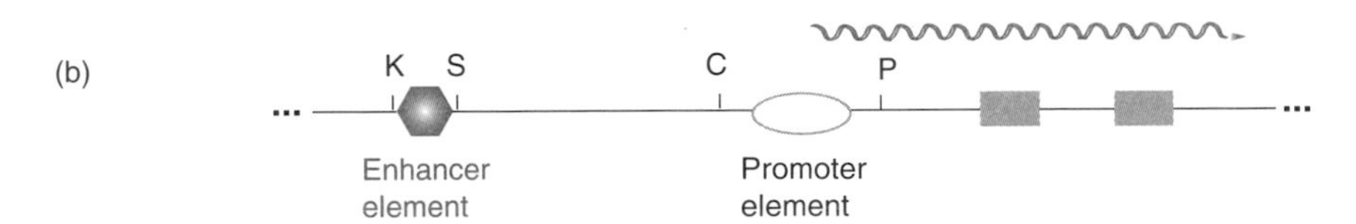

Figure 4.15 Reporter genes can be used for in vivo mapping of gene regulatory sequences. Genomic segments containing putative gene regulatory sequences can be inserted into reporter gene plasmids encoding enzymes that are readily measured by biochemical assays such as the firefly luciferase protein. (*a*) Deletion mutagenesis of the genomic segment can be performed to map gene regulatory sequences required for basal promoter activity and cell-specific gene expression. (*b*) Analysis of these data identifies the location of an upstream enhancer element and basal promoter sequences. A description of reporter gene assays is given in Chapter 7.

Mapping Protein Binding Sites by DNaseI Protection

The endonuclease DNaseI cleaves the phosphodiester backbone of DNA randomly and can be used to map protein binding sites on DNA. Two basic DNaseI mapping techniques are commonly used to map protein interaction sites; DNaseI footprinting and DNaseI hypersensitivity. The principle of DNaseI mapping is that when sequence-specific DNA binding proteins interact with DNA, they alter the DNaseI cleavage pattern in the region surrounding the binding site. Protein binding sites on DNA can be

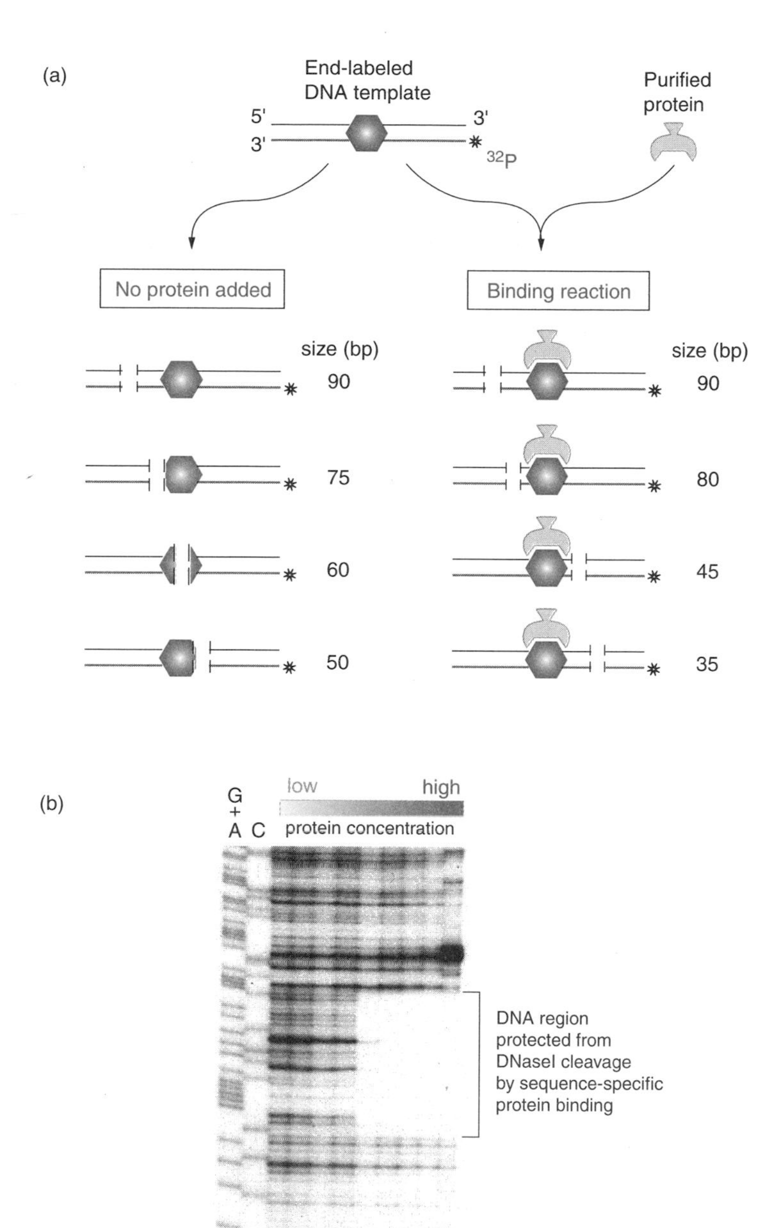

Figure 4.16 In vitro DNaseI footprinting is used to map high affinity protein binding sites on naked DNA. (*a*) An end-labeled double-strand DNA fragment is incubated in binding buffer with and without a purified DNA binding protein. Limited digestion with DNaseI (~1 cleavage per molecule) produces a population of DNA fragments of various sizes that can be resolved by denaturing acrylamide gel electrophoresis. By comparing the DNA fragment sizes from binding reactions containing no protein to that from reactions with protein, it is possible to identify a region of protection from DNaseI cleavage. (*b*) An autoradiogram of a DNaseI in vitro footprinting gel showing specific binding of purified glucocorticoid receptor DNA binding domain protein to a hormone response element sequence located in the 5′ regulatory region of the rat prostatein gene. The molecular weight markers (G+A and C) correspond to DNA sequencing reactions performed with the same end-labeled template. Crude protein preparations can also be used for DNA footprinting, however, nonspecific DNA binding proteins must be sequestered by preincubating the protein preparation with a polydeoxyribonucleotide such as poly dI-dC.

mapped relative to known restriction enzyme sites using gel electrophoresis and autoradiography. For both types of DNaseI mapping strategies to work, it is critical that each protein-bound DNA molecule in the reaction be cleaved only once by DNaseI to produce a pool of digested fragments.

Figure 4.16 illustrates how in vitro DNaseI footprinting can be used to define the 5′ and 3′ boundaries of a protein binding site using "naked" DNA (nucleic acid that has been stripped of proteins by phenol extraction) and a purified protein preparation. By comparing the pattern of DNaseI digested products produced by partial digestion of unbound and bound DNA, it is possible to identify a protected region that corresponds to the sequence specific binding site of the protein. The ability to detect the protein "footprint" depends on the relative affinity of the protein for specific and nonspecific DNA sequences. At very high protein concentrations, the footprint is obscured by the nonspecific DNA binding properties of the protein. The location of the protein binding site is determined by using a DNA sequencing ladder of the same end-labeled template DNA as a molecular weight marker.

DNaseI hypersensitivity is a technique used to map protein binding sites on chromatin using isolated nuclei that have been incubated in a buffered solution containing small amounts of DNaseI. In contrast to protein binding sites that are *protected* from DNaseI cleavage in in vitro footprinting reactions, DNaseI cleavage at "hypersensitivity" sites in chromatin results from transcription factor binding and nucleosome perturbation, which leads to the DNA being *more susceptible* to DNaseI cleavage. This interpretation comes from the observed correlation between mapped DNaseI hypersensitive sites in chromatin, in vitro sequence-specific transcription factor binding sites, and deletion analysis of gene regulatory sequences using an in vivo assay. DNaseI hypersensitivity mapping is done by using information from a restriction enzyme map within the genomic region of interest to plan a Southern blotting strategy that can reveal regions of chromatin that are highly susceptible to DNaseI cleavage (Fig. 4.17).

Electrophoretic Mobility Gel Shift Assays

The electrophoretic mobility gel shift assay, or EMSA, is an in vitro DNA binding assay used to map transcription factor binding sites in gene regulatory regions, and to monitor sequence-specific DNA binding activities in nuclear extracts. This assay is based on the reduced electrophoretic mobility of a DNA–protein complex, compared to unbound DNA; EMSA is also called a gel retardation assay. The bound and unbound forms of DNA are easily separated in a low percentage nondenaturing acrylamide gel because of large differences in their migration rates. The DNA template, or probe, is usually a short (~30 bp) double-strand oligonucleotide that has been radioactively labeled at the termini by the fill-in reaction of Klenow polymerase. The sequence of the DNA probe contains a known (or putative) recognition site for a sequence-specific DNA binding protein. The oligonucleotide probe is incubated with purified protein or nuclear extracts in a binding buffer for a short period prior to electrophoresis. The amount of DNA in the shifted and unshifted complexes is quantified by phosphorimaging as shown in Figure 4.18. The favorable buffer conditions and caging effect of the gel matrix contribute to the stability of the protein–DNA complex during electrophoresis. By including unlabeled specific and non-specific DNA templates in the binding reaction, EMSA can be used to measure binding affinities by competition analyses. The protein in the shifted DNA complex can be identified by adding a monoclonal antibody to the binding reaction, which can lead to the appearance of a "super shifted" band.

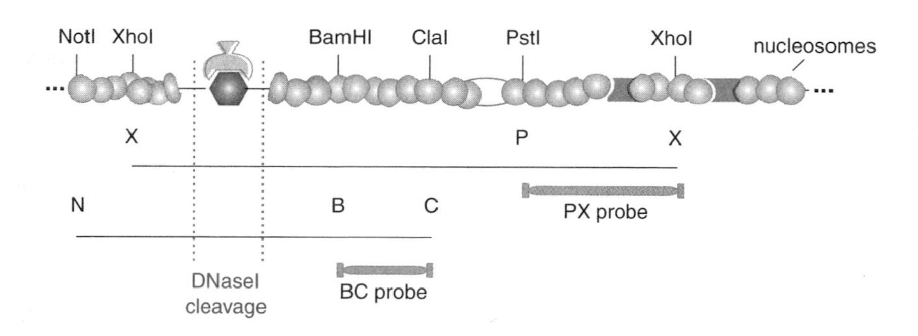

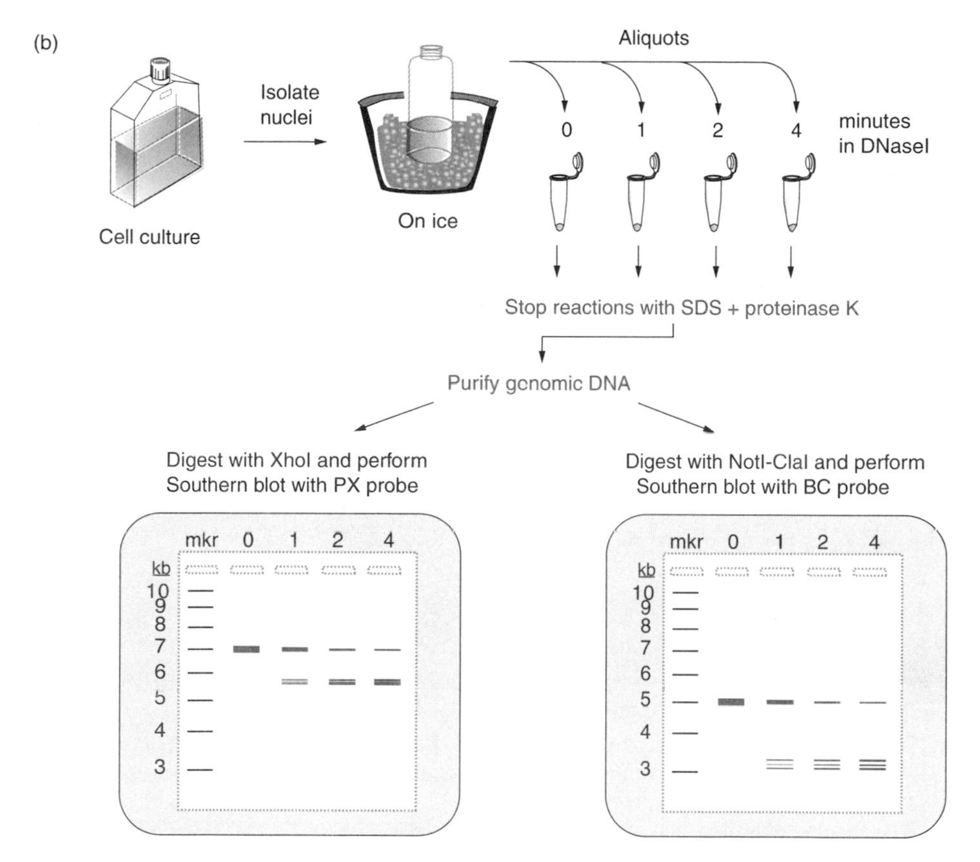

Figure 4.17 DNaseI hypersensitivity of chromatin in isolated nuclei is a technique that can be used to map gene regulatory regions. (*a*) DNaseI hypersensitive site mapping strategy. Enzyme digests and probes are designed to localize sites of DNaseI cleavage in isolated nuclei, based on a genomic DNA restriction map in the vicinity of the putative in vivo binding site. (*b*) Nuclei from mouse liver cells were isolated and treated for various amounts of time (0, 1, 2, or 4 minutes) with DNaseI under conditions that permit limited digestion of chromatin. The reaction is stopped and genomic DNA is extracted and digested with restriction enzymes (XhoI or NotI + ClaI) that release DNA fragments encompassing the region to be mapped. The digested DNA is separated by agarose gel electrophoresis and then Southern blotted with DNA probes (PX or BC) that serve as reference points for mapping based on molecular weight markers (mkr). The appearance of DNA fragments shorter than the expected restriction enzyme fragment indicates a DNaseI hypersensitive site.

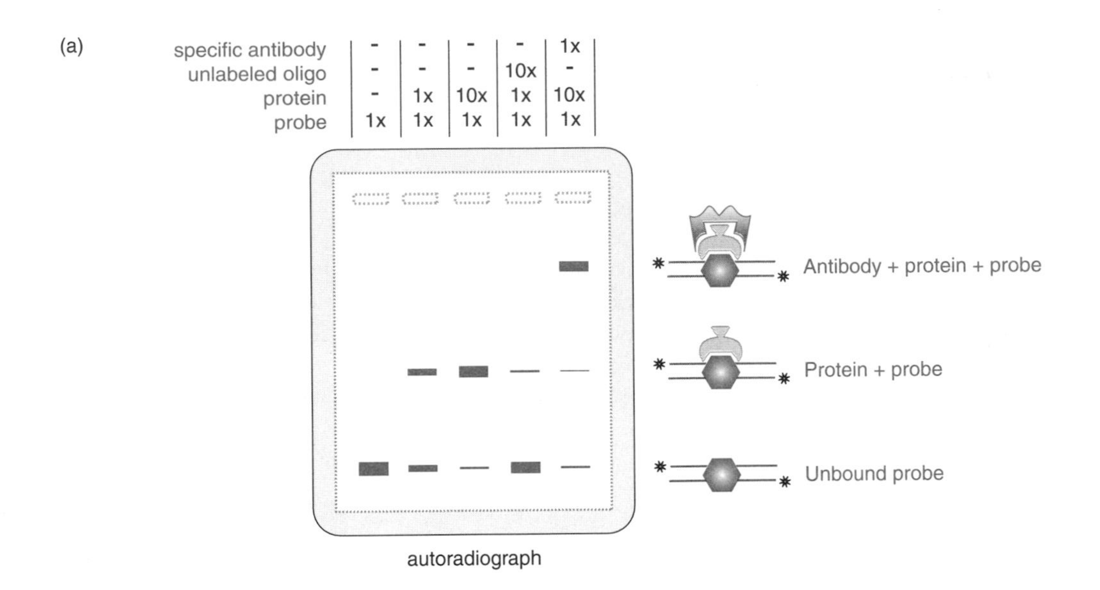

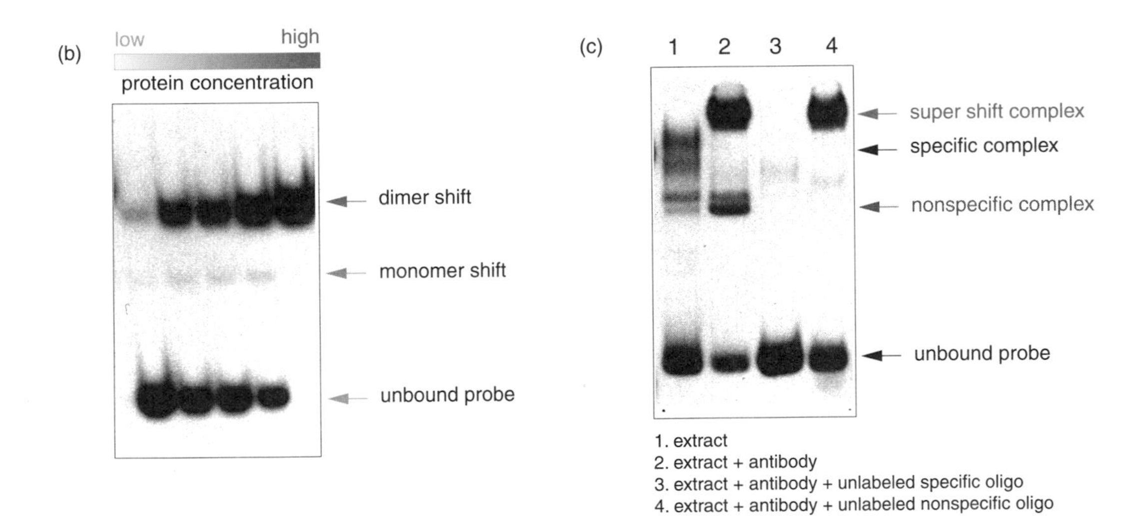

Figure 4.18 The binding of transcription factors to gene regulatory sequences can be characterized using EMSA. (*a*) The principle of EMSA is that protein–DNA complexes migrate more slowly in an electrophoretic field than unbound DNA. Depending on the DNA binding affinity of the protein being tested, and its relative concentration in the binding reaction, some proportion of the ^{32}P-labeled double strand oligo is shifted during electrophoresis. Controls for binding specificity in EMSA assays include the addition of excess unlabeled oligo and protein-specific antibodies to the binding reaction. (*b*) An autoradiogram of an EMSA assay performed with increasing amounts (1–20 ng) of purified androgen receptor DNA binding domain protein and with a ^{32}P-labeled oligonucleotide (27-mer) encoding a high affinity androgen receptor binding site (5′-N_8GGTACAN_3TGTTCTN_4 -3′). Androgen receptor dimers form the most stable DNA–protein complexes in this assay. (*c*) Cell extracts can also be used for EMSA analysis. The autoradiogram shows that specific androgen receptor binding activity can be identified in extracts from stably transfected rat hepatocytes based on control binding reactions using an androgen receptor antibody (lane 2) and unlabeled specific (lane 3) and nonspecific (lane 4) oligos. As is common with cell extract EMSAs, nonspecific complexes can sometimes be observed.

Laboratory Practicum 4. *Isolation of a human disease gene by positional cloning*

Research Objective

An autosomal recessive mutation has been mapped by linkage analysis to the human chromosomal region 7q31.3. The team of genomic researchers working on this project plans to identify the candidate disease gene by systematically comparing the sequence of genes in normal individuals (homozygous wild type), carriers (heterozygous), and patients with the disease (homozygous mutant). The assumption is that homozygous individuals afflicted with the disease will have nucleotide alterations in both copies of the gene, whereas carriers and normal individuals will have one or two copies of the wild-type gene, respectively. Biochemical evidence suggests that the disease results from a complete loss of a key enzyme required for glucose metabolism.

Available Information and Reagents

1. A physical map of human chromosome 7 has recently been constructed using 5400 YACs that are linked together into 22 contigs covering 98% of the chromosome. A total of 2150 chromosome 7-specific STS markers has been mapped with an average marker distance of 80 kb. This information can be readily accessed through the Internet.
2. Initially, RFLP analysis was performed using three generations of individuals from four different families that have a high incidence rate of the disease. A low resolution map of 7q31.3 indicated that the disease gene was likely contained within a ~1 Mb region. The physical map of chromosome 7 indicates that eight STS markers and four ESTs map to this same region of 7q31.3 that is covered by a yeast contig of 850 kb.
3. Fibroblast cell lines were established from some of the individuals in the study and these can be used as a source of freshly prepared RNA. All of the relevant STS primer pairs, EST clones, and YAC-containing yeast strains can be obtained commercially.

Basic Strategy

The DNA samples are subjected to PCR analysis using the STS primer pairs to delineate further the region most likely to contain the disease gene. Based on the results of this linkage analysis, EST clones from the region are used to recover full-length cDNAs from a human cDNA library. The DNA sequences of these cDNAs and of the genomic regions encoding the transcribed sequences are determined. PCR primers flanking each of the predicted exons of the candidate genes in the region are synthesized and used to obtain sequence information from normal and diseased individuals. Northern blots are performed using cDNA probes and freshly prepared RNA from the fibroblasts cell lines to identify differences in the expression pattern of these genes in normal and diseased individuals.

Comments

A direct correlation between the presence of the disease (phenotype) and a specific nucleotide(s) alteration in a gene (genotype) would be taken as strong evidence that the metabolic disease is due to a functional defect in the encoded gene product. Confirmation requires DNA analysis of samples from other diseased individuals not represented in the initial set of 45 related family members. Owing to the nature of this particular disease, it may even be possible to obtain biochemical evidence that the encoded protein is required for a key step in the affected metabolic pathway. The fibroblast cell lines would be expected to have a defect in glucose metabolism because the Northern blot showed

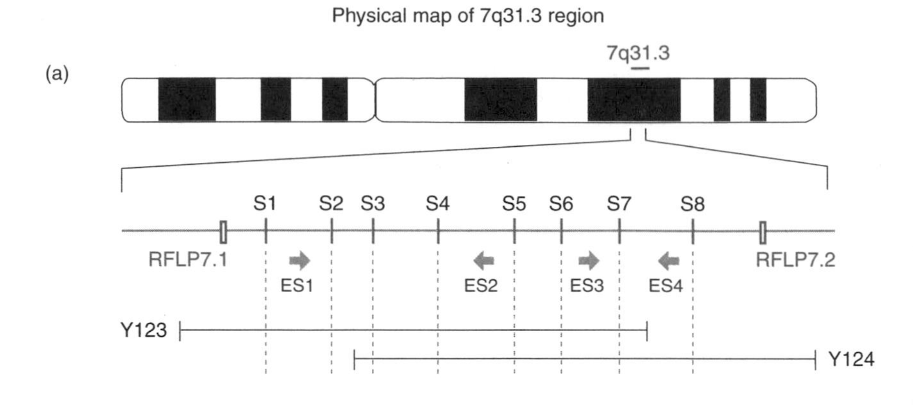

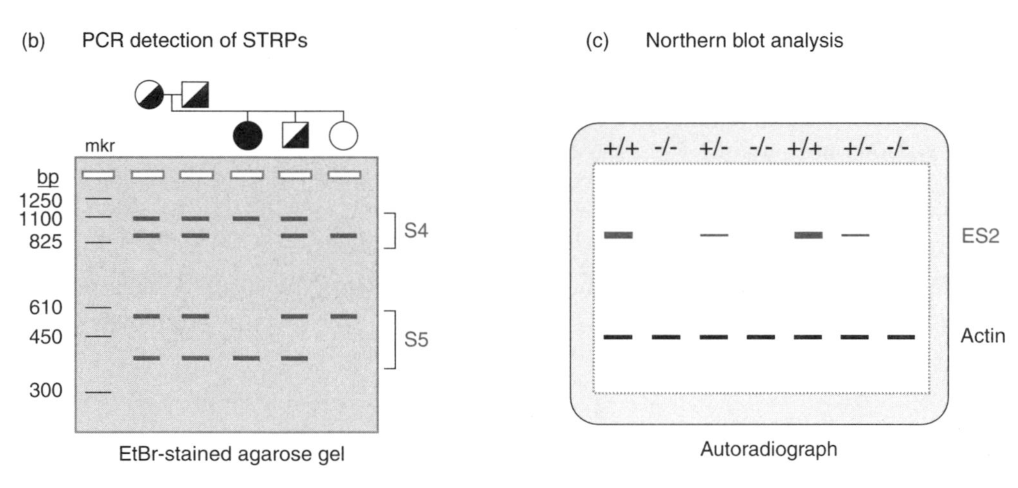

Figure 4.19 Physical maps provide a framework for positional cloning strategies. (*a*) Hypothetical physical map in the 7q31.3 region where the disease gene has been mapped by RFLP linkage analysis using the RFLP7.1 and RFLP7.2 markers. The relative position of STS primer sets (S1–S8), ESTs (E1–E4), and YAC clones (Y123 and Y124) in this 850 kb region are indicated. (*b*) Results of PCR reactions using primer pairs for STS4 and STS5 that detect a polymorphic short tandem repeat. The inferred genotype of normal individuals (open symbol), carriers (half-filled symbol), and patients with the disease (filled symbol) is shown above each lane. (*c*) Northern blot results using the ES2 cDNA as a probe and RNA isolated from patient-derived fibroblast cell lines. The actin cDNA probe is used as an internal control to monitor RNA sample loading.

that diseased individuals lacked the expression of the putative ES2 disease gene. Biochemical experiments could be performed using extracts from the fibroblast cell lines to determine if addition of the ES2 encoded protein restores functional glucose metabolism.

Prospective

The identification of a disease gene provides the opportunity to develop molecular genetic tests for diagnostic purposes. The required information comes directly from the DNA sequence of various genetic alleles identified during the initial gene characterization. In some cases, the elucidation of a disease gene coding sequence provides new insights into possible biochemical mechanisms that would explain the physiological defect. For example, if the ES2 gene product encodes a putative kinase, based on sequence homology comparisons to other kinases in the GenBank database, then it might indicate that altered phosphorylation signaling causes a defect in glucose metabolism. Another research direction could be to develop a mouse model of the human metabolic disease as a means to begin testing potential therapeutic agents. One way to develop such an animal model would be to clone the homologous mouse ES2 gene and then use it as a reagent to create a gene knockout mouse (Chapter 8).

REFERENCES

Overview of genome organization

Bevan, M., Bancroft, I., Bent, E. et al. Analysis of 1.9 Mb of contiguous sequence from chromosome 4 of *Arabidopsis thaliana*. Nature 391:485–488, 1998.

*Blattner, F. R., Plunkett, G., Bloch, C.A., et al. The complete genome sequence of *Escherichia coli* K-12. Science 277:1453–1462,1997.

Cole, S., Brosch, R., Parkhill, J. et al. Deciphering the biology of *Mycobacterium tuberculosis* from the complete genome sequence. Nature 393:537–44, 1998.

Dib, C., Faure, S. Fizames, C., et al. A comprehensive genetic map of the human genome based on 5,264 microsatellites. Nature 380:152–154, 1996.

Doolittle, R.F. Microbial genomes opened up. Nature 392:339–342, 1998.

Fraser, C., Casjens, S., Huang, W., et al. Genomic sequence of a Lyme disease spirochaete, *Borrelia burgdorferi*. Nature 390:580–586, 1997.

*Goffeau, A., Barrell, B.G., Bussey, H., et al. Life with 6000 genes. Science 274:546–567, 1996.

Klenk, H P., Clayton, R., Tomb, J-F. et al. The complete genome sequence of the hyperthermophilic, sulphate-reducing archeon *Archaeoglobus fulgidus*. Nature 390:363–370, 1997.

*Kunst, F., Ogasawara, N., Moszer, I. et al. The complete genome sequence of the Gram-positive bacterium *Bacillus subtilis*. Nature 390:249–256, 1997.

Miklos, G. and Rubin, G. The role of the genome project in determining gene function: Insights from model organisms. Cell 86:521–529, 1996.

Rowen, L., Mahairas, G., Hood, L. Sequencing the human genome. Science 278:605–607, 1997.

*Schrock, E., du Manoir,S., Veldman, T., et al. Multicolor spectral karyotyping of human chromosomes. Science 273:494–497, 1996.

Schuler, G.D., Buguski, M.S., Stewart, E.A., et al. A gene map of the human genome. Science 274: 540–546, 1996.

Strauss, E.J., Falkow, S. Microbial pathogenesis: Genomics and beyond. Science 276:707–712, 1997.

Tomb, J-F., White, O., Kerlavage, A., et al. The complete genome sequence of the gastric pathogen *Helicobacter pylori*. Nature 388:539–547, 1997.

Genomic mapping

*Botstein, D., White, R.L., Skolnick, M., Davis, R.W. Construction of a genetic linkage map in man using restriction fragment length polymorphisms. Am. J. Human Genet. 32:314–331, 1980.

*Chumakov, I., Rigault, P., Guillou, S., et al. Continuum of overlapping clones spanning the entire chromosome 21q. Nature 359:380–387, 1992.

*Cox, D. R., Burmeister, M., Price, E.R., Kim, S., Myers, R.M. Radiation hybrid mapping: A somatic cell genetic method for constructing high resolution maps of mammalian chromosomes. Science 250:245–250, 1990.

* Landmark papers in applied molecular genetics.

*Edwards, A., Hammond, Chakraborty, R., Caskey, C.T. DNA typing and genetic mapping with trimeric and tetrameric tandem repeats. Am. J. Human Genet. 49:746–756, 1991.

Encio, I., Detera-Wadleigh, S. The genomic structure of the human glucocorticoid receptor. J. Biol. Chem. 266:7182–7188, 1991.

*Foote, S., Vollrath, D., Hilton, A., Page, D.C. The human Y chromosome: Overlapping DNA clones spanning the euchromatic region. Science 258:60–66, 1992.

*Hudson, T.J., Stein, L.D., Gerety, S.S., et al. An STS-based map of the human genome. Science 270:1945–1954, 1995.

NIH/CEPH Collaborative Mapping Group. A comprehensive genetic linkage map of the human genome. Science 258:67–86, 1992.

*Olson, M., Hood, L., Cantor, C., Botstein, D. A common language for physical mapping of the human genome. Science 245:1434–1435, 1989.

Sprecher, C. J., Puers, C., Lins, A.M., Schumm, J.W. General approach to analysis of polymorphic short tandem repeat loci. Biotechniques 20:266–276, 1996.

White, R., Lalouel, J-M. Sets of linked genetic markers for human chromosomes. Ann. Rev. Genet. 22:259–279, 1988.

Yu, C-E., Oshima, J., Fu, Y-H. et al. Positional cloning of the Werner's syndrome gene. Science 272:258–262, 1996.

Yuan, B., Vaske, D., Weber, J.L., Beck, J., Sheffield, V.C. Improved set of short-tandem-repeat polymorphisms (STRPs) for screening the human genome. Am. J. Human Genet. 60: 459–460, 1997.

Manipulation of large genomic DNA segments

*Chu, G., Volrath, D., Davis, R.W. Separation of large DNA molecules by contour-clamped homogeneous electric fields. Science 234:1582–1585, 1986.

Lai, E., Birren, B.W., Clark, S.M., Simon, M.I., Hood, L. Pulsed field gel electrophoresis. Biotechniques 7:34–41, 1989.

*Ludecke, H-J., Senger, G., Claussen, U., Horsthemke, B. Cloning defined regions of the human genome by microdissection of banded chromosomes and enzymatic amplification. Nature 338:348–350, 1989.

Nizetic, D., Zehetner, G., Monaco, A.P., Gellen, L., Young, B.D., Lehrach, H. Construction, arraying and high-density screening of large insert libraries of human chromosomes X and 21: Their potential use as reference libraries. Proc. Natl. Acad. Sci. USA 88:3233–3237, 1991.

*Schwartz, D. C., Cantor, C.R. Separation of yeast chromosome-sized DNAs by pulsed field gradient gel electrophoresis. Cell 37:67–75, 1984.

Willard, H.F. Chromosome manipulation: A systematic approach toward understanding human chromosome structure and function. Proc. Natl. Acad. Sci. USA 93:6847–6850, 1996.

Zhang, J., Trent, J.M., Meltzer, P.S. Rapid isolation and characterization of amplified DNA by chromosome microdissection: Identification of IGF1R amplification in malignant melanoma, Oncogene, 8:2827–2831, 1993.

Genomic DNA cloning vectors

Ahand, R. Cloning into yeast artificial chromosomes. In DNA cloning 3: A practical approach (D. Glover and B. Hames, eds.). IRL Press, Oxford, pp. 103–127, 1995.

*Burke, D.T., Carle, G.F., Olson, M.V. 1987. Cloning of large segments of exogenous DNA into yeast by means of artificial chromosome vectors. Science 236:806–812.

Cai, W., Jing, J., Irvin, B. et al. High-resolution restriction maps of bacterial artificial chromosomes constructed by optical mapping. Proc. Natl. Acad. Sci. USA 95:3390–3395, 1998.

*Collins, J., Hohn, B. Cosmids: A type of plasmid gene-cloning vector that is packageable in vitro in bacteriophage λ heads. Proc. Natl. Acad. Sci. USA 75:4242–4246, 1978.

Cross, S.H., Little, P.F. A cosmid vector for systematic chromosome walking. Gene 49:9–16, 1986.

Green, E.D., Riethman, H.C., Dutchik, J.E., Olson, M.V. 1991. Detection and characterization of chimeric yeast artificial-chromosome clones. Genomics 11:658–669.

Hamaguchi, M., O'Connor, E., Chen, T., Parnell, L., McCombie, R., Wigler, M.H. Rapid isolation of cDNA by hybridization. Proc. Natl. Acad. Sci. USA 95:3764–3769, 1998.

Kim, U-J, Birren, B.W., Slepak, T., et al. Construction and characterization of a human bacterial artificial chromosome library. Genomics 34:213–218, 1996.

*Shizuya, H., Birren, B., Kim, U-J., Mancino, V., Slepak, T., Tachiiri, Y., Simon, M.I. Cloning and stable maintenance of 300-kilobase-pair fragments of human DNA in *Escherichia coli* using an F-factor-based vector. Proc. Natl. Acad. Sci. USA 89:8749–8797, 1992.

Sternberg, N. Library construction in P1 phage vectors. In DNA cloning 3: A practical approach (D. Glover and B. Hames, eds.). IRL Press, Oxford, pp. 81–101, 1995.

Mapping gene regulatory sequences

Alam, J., Cook, J.L. Reporter genes: Application to the study of mammalian gene transcription. Anal. Biochem. 188:245–254, 1990.

Brenowitz, M., Senear, D.F., Shea, M.A., Ackers, G.K. Footprint titrations yield valid thermodynamic isotherms. Proc. Natl. Acad. Sci. USA 83:8462–8466, 1986.

*De Wet, J.R., Wood, K.V., DeLuca, M., Helinski, D.R., Subramani, S. Firefly luciferase gene: Structure and expression in mammalian cells. Mol. Cell. Biol. 7:725–737, 1987.

*Fried, M., Crothers, D.M. Equilibria and kinetics of lac repressor-operator interactions by polyacrylamide gel electrophoresis. Nucl. Acids Res. 9:6505–6525, 1981.

*Galas, D., Schmitz, A. DNase footprinting: A simple method for the detection of protein-DNA binding specificity. Nucl. Acids Res. 5:3157–3170, 1978.

Gordon, D., Chamberlain, N., Flomerfelt, F., Miesfeld, R. A cell-specific and selective effect on transactivation by the androgen receptor. Exp. Cell Res. 217:368–377, 1995.

*Gorman, C.M., Moffat, L.F., Howard. B.H. Recombinant genomes which express chloramphenicol acetyltransferase in mammalian cells. Mol. Cell. Biol. 2:1044–1051, 1982.

*Hendrickson, W., Schleif, R.F. Regulation of the Escherichia coli L-arabinose operon studied by gel electrophoresis DNA binding assay. J. Mol. Biol. 174:611–628, 1984.

*Jackson, P.D., Felsenfeld, G. A method for mapping intranuclear protein-DNA interactions and its application to a nuclease hypersensitive site. Proc. Natl. Acad. Sci. USA 82:2296–2300, 1985.

Jantzen, H., Strahle, U, Gloss, B., Stewart, F., Schmid, W., Boshart, M., Miksicek, R., Schutz, G. Cooperativity of glucocorticoid response elements located far upstream of the tyrosine aminotransferase gene. Cell 49:29–38, 1987.

*Johnson, A.D., Meyer, B.J., Ptashne, M. Interactions between DNA-bound repressors govern regulation by the lambda phage repressor. Proc. Natl. Acad. Sci. USA 76:5061–5065, 1979.

Liu, J-K., Bergman, Y., Zaret, K.S. The mouse albumin promoter and a distal upstream site are simultaneously DNaseI hypersensitive in liver chromatin and bind similar liver-abundant factors in vitro. Genes Dev. 2:528–541, 1988.

Rundlett, S.E., Miesfeld, R.L. Quantitative differences in androgen and glucocorticoid receptor DNA binding properties contribute to receptor-selective transcriptional regulation. Mol. Cell. Endocrinol. 109:1–10, 1995.

Singh, H., Sen, R., Baltimore, D., Sharp, P.A. 1986. A nuclear factor that binds to a conserved sequence motif in transcriptional control elements of immunoglobulin genes. Nature 319: 154–158.

Tullius, T.D., Dombroski, B.A., Churchill, M.E.A., Kam, L. Hydroxyl radical footprinting: A high resolution method for mapping protein-DNA contacts. Methods Enzymol. 155: 537–558, 1987.

Wenger, R.H., Moreau, H., and Nielson, P.J. A comparison of different promoter, enhancer, and cell type combinations in transient transfections. Anal. Biochem. 221:416–418, 1994.

Isolation of a human disease gene by positional cloning

*Bouffard, G.G., Idol, J.R., Braden, V.V., et al. A physical map of human chromosome 7: An integrated YAC contig map with average STS spacing of 79 kb. Genome Res. 7:673–692, 1997.

Campuzano, V., Montermini, L., Molto, M.D., et al. Friedreich's ataxia: Autosomal recessive disease caused by an intronic GAA triplet repeat expansion. Science 271:1423–1427, 1996.

Chandrasekharappa, S.C., Guru, S.C., Manickam, P., et al. Positional cloning of the gene for multiple endocrine neoplasia-type 1. Science 276:404–407, 1997.

Collins, F. Positional cloning moves from perditional to traditional. Nature Gen. 9:347–350, 1995.

Ghosh, S., Collins, F.S. The geneticist's approach to complex disease. Ann. Rev. Med. 47: 333–353, 1996.

Mushegian, A.R., Bassett, D.E., Boguski, M.S., Bork, P., Koonin, E.V. Positionally cloned human disease genes: Patterns of evolutionary conservation and functional motifs. Proc. Natl. Acad. Sci. USA 94:5831–5836, 1997.

Pennacchio, L.A., Lehesjoki, A-E., Stone, N.E., et al. Mutations in the gene encoding cystatin B in progressive myoclonus epilepsy (EPM1). Science 271:1731–1734, 1996.

Schafer, A., Hawkins, J. DNA variation and the future of human genetics. Nature Biotech. 16: 33–39, 1998.

CHAPTER

5

ISOLATION AND CHARACTERIZATION OF GENE TRANSCRIPTS

One of the most useful applied molecular genetic methods is the cloning of complementary DNA (cDNA) sequences derived from the coding sequences of specific genes. The ability to "capture" mRNA as cDNA was made possible by the independent discovery of a viral RNA-dependent DNA polymerase called reverse transcriptase (RTase) by Howard Temin and David Baltimore. These two molecular virologists shared the 1975 Nobel Prize in Physiology and Medicine for their discovery of RTase with another virologist, Renato Dulbecco. RTase is a retroviral enzyme required to convert the single strand viral RNA genome into a double-strand linear DNA molecule that integrates into the host genome. The first part of this chapter describes biochemical reactions used to synthesize double-strand cDNA from mRNA and general methods for constructing representative cDNA libraries. This introduction is followed by a discussion of various library screening strategies that can be used to isolate specific cDNA sequences. The last section describes molecular genetic methods utilizing cloned cDNA as a reagent to characterize gene expression. Laboratory practicum 5 presents a cDNA screening strategy based on in vitro expression of pooled gene sequences as a means to identify cDNAs that encode defined biochemical functions.

CONVERTING mRNA TRANSCRIPTS INTO cDNA

The first consideration in planning a cDNA cloning strategy is to determine if a commercially available cDNA library can be obtained or if it will be necessary to construct a specialized cDNA library. Oftentimes both types of cDNA libraries are needed. For example, after screening a commercial cDNA library to obtain several overlapping clones corresponding to a large portion of the encoded transcript, it still may be necessary to construct a specialized library that is enriched for the cDNA sequences being sought. In this section we describe the three basic steps required to construct a useful cDNA library: (1) isolation of high quality mRNA, (2) synthesis of double-strand cDNA, and (3) efficient cDNA vector ligation.

mRNA Purification

RNA is not as biochemically stable as DNA and care must be taken to purify intact mRNA molecules from cells. Because RNA is single-strand, *any* hydrolysis event that breaks the phosphate backbone will result in cleavage of the molecule into subfragments. Two factors contribute to the biochemical instability of RNA. First, endoribonucleases (RNases) are very stable enzymes that cannot be easily inactivated. In fact, human hands are a rich source of RNase and it is therefore necessary to wear clean latex gloves during RNA isolation procedures and to use RNase-free labware. Second, RNA is thermodynamically less stable than DNA because of the 2′ hydroxyl group on the ribose ring that promotes hydrophilic attack on the 5′-3′ phosphodiester bond to form a 2′-3′ cyclic phosphate. This cyclic phosphate intermediate is stabilized by Mg^{2+}, a component of many biochemical reactions. Therefore, even if all RNases are eliminated or inhibited during RNA purification, RNA spontaneously degrades while in solution. To circumvent this biological decay of RNA, purified samples are stored at -20°C as ethanol precipitates.

The purification of mRNA involves two basic steps; 1) biochemical separation of total cellular RNA from DNA and protein using a strong protein denaturant to inhibit cellular RNases, and 2) isolation of poly A+ mRNA using an oligo dT affinity matrix. Unopened sterile plastic ware is essentially RNase-free and glassware can be treated with a 0.1% solution of diethyl pyrocarbonate (DEPC) to inactivate RNases by covalent modification. Unreacted DEPC must be removed from the glassware by autoclaving prior to RNA isolation because DEPC can carboxymethylate RNA and inhibit subsequent enzymatic reactions. There are also two types of specific RNase inhibitors that can be added directly to RNA-containing solutions to protect against RNA degradation during the cDNA synthesis reaction. One inhibitor is a fortuitously discovered angiogenin binding protein isolated from human placenta called Rnasin® (Promega Corporation, Madison, Wisconsin), and the other is a mixture of vanadyl-ribonucleoside complexes that function as inhibitory transition state analogues of RNase activity.

A common method used to isolate mRNA from tissue culture cells is outlined in Figure 5.1. Guanidinium thiocyanate is a protein denaturant that lyses the cells and inhibits cellular RNases. This purification method is based on the finding that RNA preferentially partitions to the aqueous phase in a solution containing 4*M* guanidine isothyocyanate at pH 4 in the presence of phenol and chloroform. Under these conditions, proteins partition to the organic phase and most of the large DNA fragments are trapped in the interphase. Poly A+ mRNA is purified from total cellular RNA using oligo dT as an affinity ligand. Approximately 1–5% of the total RNA in a cell is polyadenylated, and therefore, oligo dT affinity purification of poly A+ mRNA results in a ~100-fold enrichment.

cDNA Synthesis

First strand cDNA synthesis is the most critical reaction because it determines the maximum length of cDNA products. Libraries made with oligo dT primers (a poly dT strand 12–18 nucleotides long) have the best chance of containing long cDNAs because the priming event is targeted to the 3′ poly A tail of the mRNA. However, many mouse and human genes contain long stretches of 3′ noncoding transcript sequences, and therefore oligo dT primed cDNA libraries may lack complete ORF sequences. Short random hexanucleotide primers can also be used to initiate cDNA synthesis. These nonspecific

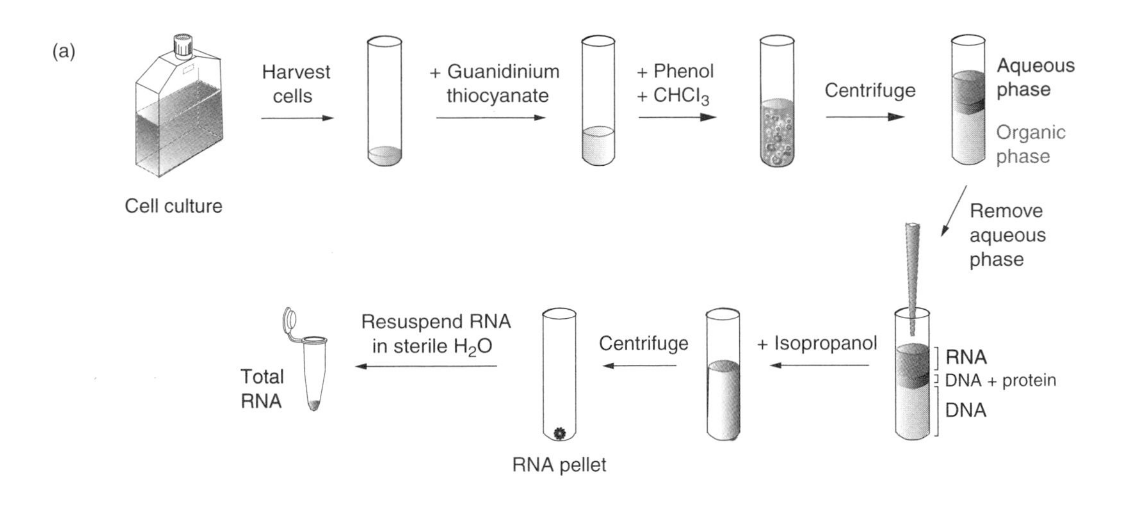

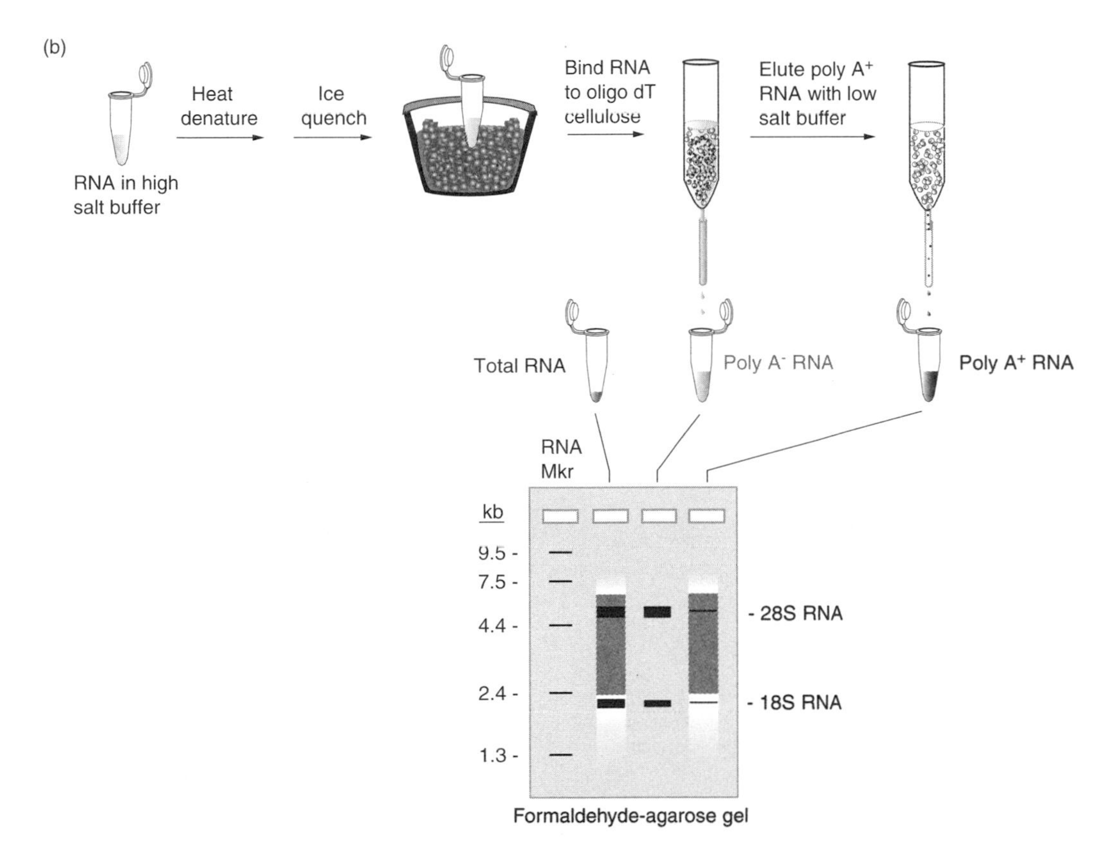

Figure 5.1 Purification of mRNA from tissue culture cells using guanidinium thiocyanate and oligo dT cellulose. Total RNA remains in the aqueous phase under acidic pH conditions following phenol:chloroform extraction. After precipitation with isopropanol, the RNA solution is adjusted to high salt (0.5M NaCl) and loaded onto an oligo-dT cellulose column. RNA integrity can be analyzed by EtBr-staining using formaldehyde–agarose gel electrophoresis under denaturing conditions.

template primers bind to multiple sites along the mRNA and can provide a better distribution of coding and noncoding sequences. It is sometimes possible to use gene-specific primers to initiate first strand synthesis in cases where a portion of the 3′ cDNA sequence is known. Figure 5.2 summarizes the advantages and disadvantages of using oligo dT, random primers, and gene-specific primers for first strand cDNA synthesis.

In addition to template primers and dNTPs, the first strand cDNA synthesis reaction requires the enzyme RTase which uniquely synthesizes DNA products from primed RNA templates. Avian myelobastosis virus (AMV) reverse transcriptase was the first RTase specifically purified for use in first strand cDNA reactions. The active enzyme

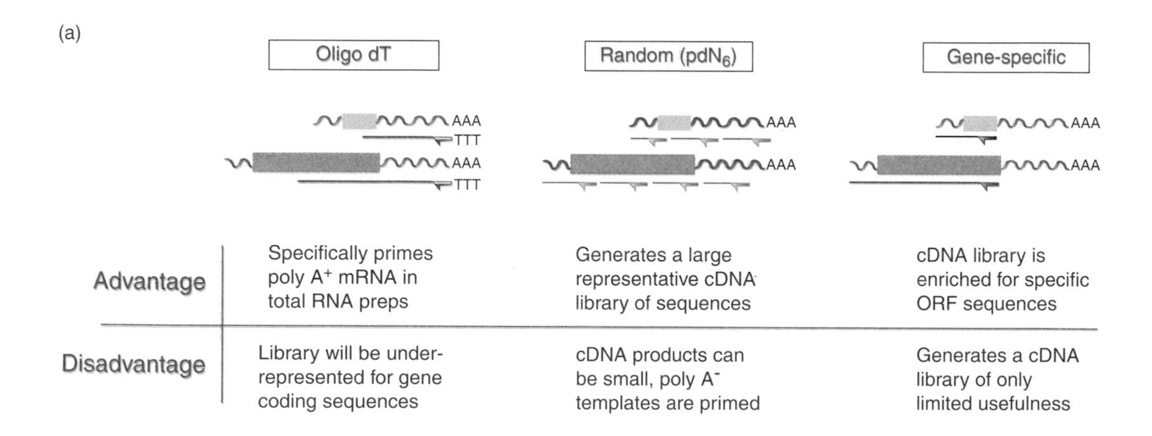

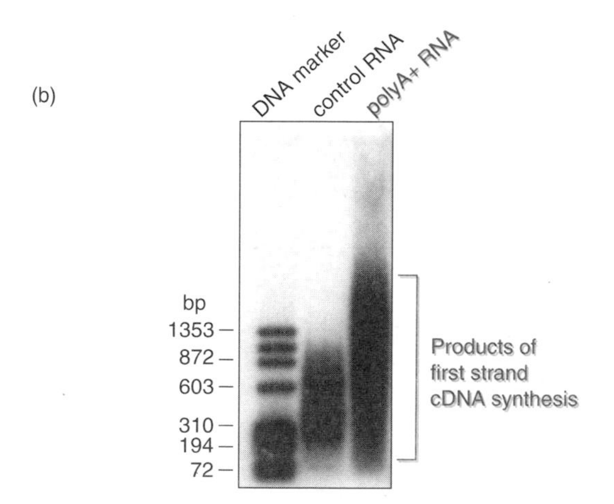

Figure 5.2 First strand cDNA synthesis can be initiated with oligo dT, random primers, or gene-specific primers to achieve different outcomes from the same template RNA pool. (*a*) There are advantages and disadvantages to using each of these three priming strategies. (*b*) The size distribution of first strand cDNA products can be visualized with autoradiography by adding trace amounts of ^{32}P-dNTPs to the first strand reaction. An autoradiograph is shown of radioactive cDNA generated by oligo dT priming of an in vitro transcribed control poly A+ RNA, or by priming cellular poly A+ RNA.

consists of two subunits that together encode the DNA polymerizing activity and an RNase activity called RNase H which degrades RNA in RNA:DNA heteroduplexes. RNase H activity is normally required for second strand DNA synthesis of the retroviral genome during genome replication. This RNase H activity, however, is a potential problem during in vitro first strand cDNA synthesis because it can degrade the RNA template near the 3′ end of the elongating cDNA chain. Because of this interference by the endogenous RNase H activity of AMV RTase, a second retroviral RTase is now more commonly used for first strand cDNA synthesis. The RTase of Moloney murine leukemia virus (MMLV) is a single polypeptide chain that encodes all the required RTase functions. The MMLV RTase has been cloned and re-engineered to have negligible levels of RNase H activity, without compromising its first strand cDNA polymerizing function. Side-by-side comparisons of first cDNA synthesis reactions using either the purified AMV RTase or the RNase H-deficient MMLV RTase have shown that the MMLV RTase produces cDNA products that are on average 25% longer than AMV RTase products.

Second strand cDNA synthesis presents some challenges that aren't encountered during replication of the retroviral genome in infected cells. The big problem is how to prime second strand synthesis without creating DNA termini that block vector ligation. In vivo retroviral replication involves a complicated series of priming events that are initiated by host cell tRNA molecules and require complementary sequence elements within the long terminal repeats (LTR) of the retroviral genome. Because cDNA copies of mRNA molecules do not contain these special priming sequences, an artificial priming event is required to initiate second strand synthesis. Several different priming strategies have been developed, one of which was based on the observation that RTase will occasionally create first strand 3′ termini that serve as primers for second strand synthesis through the formation of "hairpin loops." These double strand RNA structures are created when the 3′ end of the cDNA folds back and re-anneals to the nascent cDNA strand, usually through stabilizing G-C base pairs. Although this hairpin structure provides the necessary free 3′ OH for priming second strand synthesis, it occurs only on <10% of the cDNA products. Moreover, the looped structure must be cleaved by S1 nuclease in order to create a double-strand DNA terminus for vector ligation. Because the S1 nuclease cleavage reaction is difficult to control, and the number of cDNA molecules containing hairpin structures is low, this second strand priming method results in <1% yield of cDNA from mRNA.

To improve the efficiency of second strand cDNA synthesis, researchers reexamined the retroviral RTase reaction and discovered that by adding small amounts of *E. coli* RNase H to the second strand synthesis reaction, they were able to find conditions that resulted in the production of short RNA primers. In the presence of *E. coli* DNA polymerase I, these RNA primers promote DNA synthesis at multiple sites along the cDNA template as shown in Figure 5.3. The "replacement synthesis" reaction for second strand cDNA synthesis was first described by Hiroto Okayama and Paul Berg. A short time later, Ueli Gubler and Beth Hoffman developed a modified version of the RNA replacement reaction. The Gubler and Hoffman protocol is the method of choice for commercial cDNA synthesis kits, because both the first and second strand reactions can be performed in a single eppendorf tube.

Modification of cDNA termini is the final step of cDNA synthesis that must be performed prior to vector ligation. The most straightforward, but least efficient, method is to ligate blunt-ended cDNA molecules to a vector that has been digested with an enzyme that produces blunt (flush) ends. One way to increase the efficiency of cDNA vector lig-

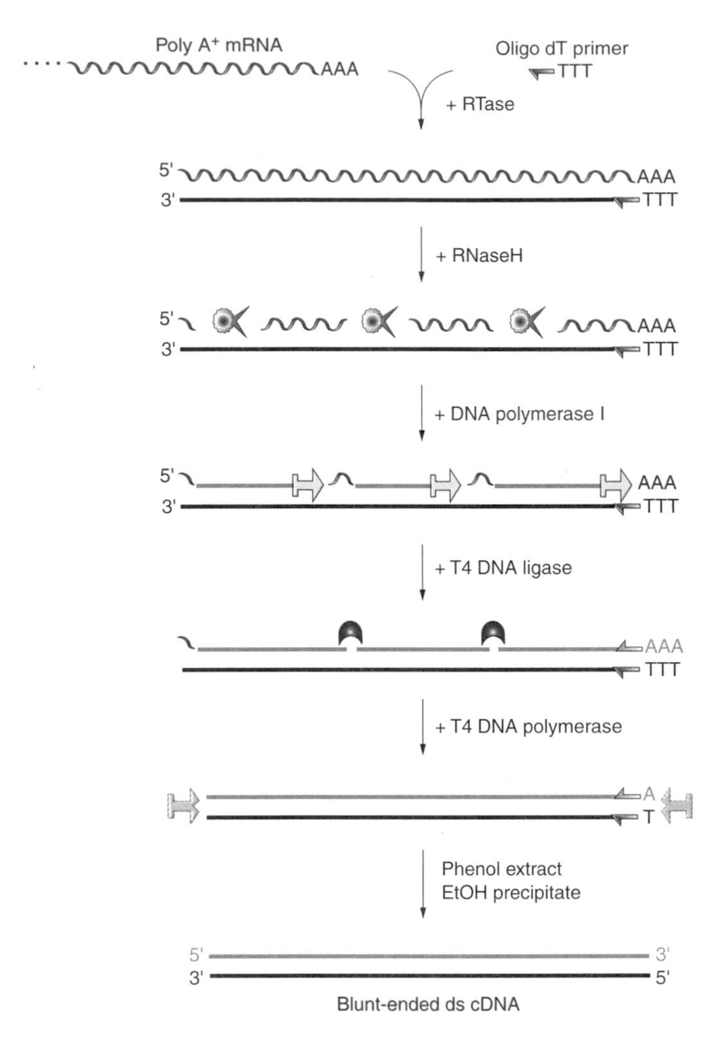

Figure 5.3 The replacement reaction for second strand cDNA synthesis using *E. coli* RNase H and DNA pol I. The RNase H activity creates short RNA fragments in the RNA:DNA heteroduplex that function as primers for second strand cDNA synthesis by *E. coli* DNA pol I. The addition of bacteriophage T4 DNA polymerase to the reaction creates blunt double-strand DNA termini that are suitable substrates for adaptor ligation.

ation is to add complementary polynucleotidyl "tails" to the 3′ end of the cDNA molecules and the plasmid vector using terminal deoxynucleotidlytransferase (TdT). A major disadvantage to TdT tailing, however, is that this method incorporates long stretches of CpG or ApT dinucleotides between the vector and the cDNA that cannot be removed by restriction enzyme digestion. Because of these drawbacks, the preferred cloning method is to ligate double-strand oligonucleotide adaptor molecules to the cDNA termini. This converts blunt-end cDNA molecules to more conventional DNA fragments possessing cohesive ends. An outline of the adaptor ligation method is shown in Figure 5.4. Note that even though covalent attachment of the adaptor to the cDNA molecule requires blunt-end ligation, the large molar excess of adaptors to cDNA termini greatly facilitates bimolecular collision and ligation.

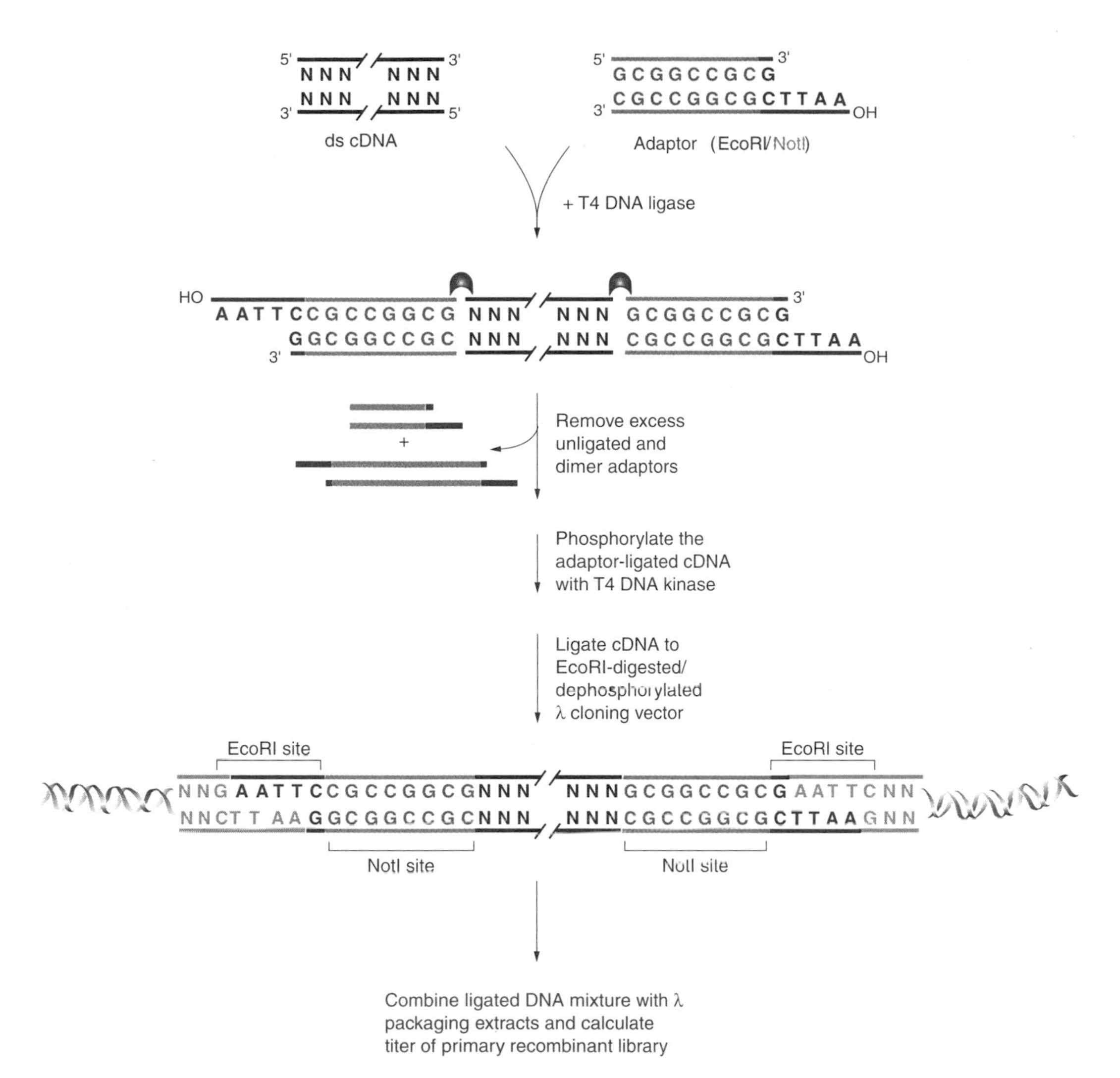

Figure 5.4 Oligonucleotide adaptors are used to convert blunted cDNA termini into cohesive ends to increase the efficiency of cDNA cloning. EcoRI adaptors can be ligated onto blunted cDNA termini as a first step toward constructing a cDNA library in an EcoRI-digested λ cloning vector. Importantly, adaptor dimers and unligated adaptors must be removed after the first ligation reaction, usually by spin column chromatography, to prevent insertion of adaptor dimers into the cloning vector. Most adaptors contain internal recognition sites for rare-cutting enzymes, such as NotI, to facilitate subcloning of insert fragments that contain internal EcoRI sites.

cDNA Cloning Vectors

The choice of using a plasmid vector or λ phage vector for constructing a cDNA library depends on the screening method to be used. Plasmid vectors are commonly used for cDNA expression libraries that will be screened by a functional assay. Plasmid vectors are also convenient vectors to use when the cDNA library has been enriched by a pre-screening technique, such as subtraction hybridization or RACE PCR. If the cDNA library is going to be screened by conventional DNA hybridization or by antibody

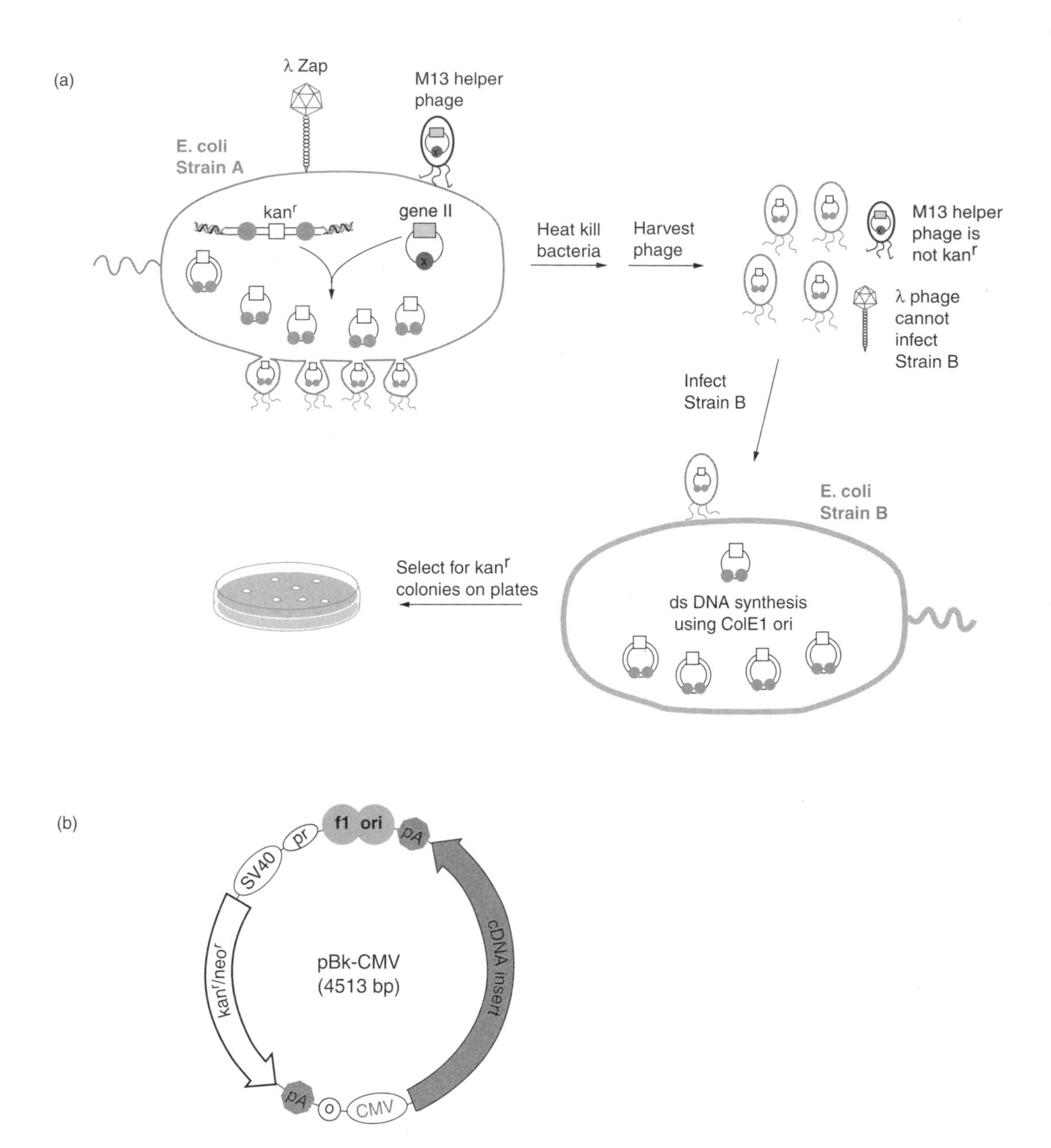

Figure 5.5 The linear λZap Express™ cDNA cloning vector can be converted in vivo to a circular eukaryotic expression plasmid by an excision process called phagemid rescue. (*a*) Sequential infection of two different *E. coli* strains, which permit or restrict phage production, respectively, results in the recovery of antibiotic-resistant (*kan*^{r}) bacterial colonies containing multiple copies of double-strand plasmid. (*b*) The excised λZap Express™ cDNA plasmid contains the same functional elements as the Stratagene pBluescript vector (Fig. 4.6) but, in addition, includes eukaryotic gene regulatory elements that allow high level expression of the cDNA insert in mammalian cells and selection for stable DNA integration (*neo*^{r} phenotype).

screening, then λ phage vectors are a better choice because large numbers of clones can be screened using the plaque forming assay.

Some of the most convenient λ phage cDNA cloning vectors are those that incorporate an f1 bacteriophage origin of replication into the vector backbone to permit in vivo excision of insert cDNA. The λ Zap cDNA cloning vectors from Stratagene Cloning Systems are examples of λ phagemid vectors (Fig. 5.5). The primary advantage of λ phagemid vectors is that they combine the high transfection efficiency of recombinant λ phage vectors with the rapid in vivo excision capability of phagemids. The cDNA cloning vector λZap Express™ contains a eukaryotic promoter and polyadenylation signals flanking the cDNA multiple cloning site (MCS). In vivo excision of λZap Express™ phagemid plasmid from the λ vector backbone produces a eukaryotic expression plasmid that can be directly transfected into mammalian cells and analyzed for cDNA-encoded functional activities.

Most types of cDNA libraries are constructed as representative libraries in which the sequence complexity of the cDNA inserts reflects the sequence complexity of the mRNA sample. In a representative cDNA library, cDNAs corresponding to rare transcripts are represented infrequently among the primary recombinants, whereas cDNAs synthesized from abundant mRNAs constitute a large fraction of the library. One way to increase the relative fraction of cDNAs from rare transcripts is to "normalize" the primary recombinant cDNA products prior to vector ligation. Ideally, a normalized cDNA library contains only one copy of each *different* transcript sequence present in the original mRNA sample. The most common way to construct a normalized cDNA library is to use several rounds of cDNA denaturation and annealing under conditions that permit removal of duplex DNA molecules that anneal at low C_0t values (Chapter 1). This strategy is also used to produce subtraction hybridization probes to screen cDNA libraries as described below. Normalized cDNA libraries are useful for expression cloning approaches that depend on plasmid sib selection strategies, because each plasmid in the pool represents a different RNA sequence.

SCREENING REPRESENTATIVE cDNA LIBRARIES

A high quality representative cDNA library should have a large number of primary recombinants (>10^6), and the majority of clones should have inserts with an average size of ~1.5 kb or greater. The determining factor for screening cDNA libraries is what type of probe will be used. Unlike genomic library screening, which utilizes a unique or homologous DNA sequence probe, cDNA library screening is often the first step taken to obtain a novel gene sequence. Therefore, straightforward nucleic acid hybridization screening is not possible. This section describes three strategies used to screen a representative λ phage cDNA library with the most commonly used probes: (1) degenerate oligonucleotides corresponding to predicted amino acid codons, (2) antibodies directed against a purified protein, and (3) subtracted cDNA probes representing differentially expressed genes.

Degenerate Oligonucleotide Probes

Many types of proteins can be purified to homogeneity using a series of chromatographic steps based on the size of the protein, its charge, and its hydrophobicity. The amino acid sequence of the *N*-terminus and internal peptide fragments of a highly purified protein can usually be determined by automated protein microsequencing using the Edman degradation procedure. Although it is possible to obtain peptide sequence with as little as 1–5 pmol of protein, more commonly ~20 pmol of protein (1–10 μg of protein, depending on the molecular weight) is required for a peptide sequence of 10–20 amino acids. Peptide sequences can be used to predict a nucleic acid sequence based on triplet codons in the genetic code. Custom oligonucleotide synthesis can then be used to generate a nucleic acid probe for library screening. However, the genetic code is degenerate (multiple codons specify the same amino acid), and therefore, the oligonucleotide must contain all possible codon sequences to ensure that at least one sequence is 100% complementary to the protein-coding sequence of the gene.

Figure 5.6 illustrates how two polypeptide sequences from the same protein can be used to generate separate mixed oligonucleotide pools for use as probes for cDNA library screening. By initially screening the cDNA library with one oligonucleotide pool from a specific region of the protein, and then rescreening all candidate clones with a second oligonucleotide pool, it is possible to identify a subset of double-positive clones that warrant further characterization. It has been empirically determined that oligonucleotides 17 nucleotides long (17-mer) are optimal for use as mixed degenerate probes. To minimize inherent differences in the duplex melting temperature (T_m) between the various 17-mers within the mixed probe pool, the hybridization and wash solutions contain 3*M* tetramethylammonium chloride (TMAC). This molecule contains a high density of positive charges that effectively stabilize hybrid molecules, independent of sequence composition.

Antibody Probes

It is not always possible to obtain amino acid sequence information from purified protein because of covalent protein modifications that block Edman degradation reactions or because of limitations in instrument accessibility. A ~90% pure protein preparation, however, can provide a source of antigen for producing monoclonal or polyclonal antibodies. Protein-specific antibodies can be used as "probes" immunologically to detect bacterial fusion proteins encoding the protein antigen. Antibody screening of cDNA libraries is facilitated by special λ phage expression vectors that produce fusion proteins that can be detected on membrane filters. The most commonly used cDNA cloning vectors for antibody screening are the λgt11 and λZap phage vectors, which contain an *E. coli* transcriptional promoter linked to the β-galactosidase (*lacZ*) coding sequences. By inserting cDNA fragments within the lacZ coding sequence of the vector, it is possible to produce recombinant phages that express LacZ–cDNA fusion proteins in infected bacteria. Figure 5.7 illustrates the key steps involved in antibody screening of a λgt11 cDNA library.

Antibody screening of cDNA libraries can present problems not encountered with nucleic acid probes. First, nonbacterial sequences can be growth inhibitory to *E. coli* when expressed at high levels as LacZ fusion proteins; therefore lytic infection could be compromised, resulting in the underrepresentation of some recombinants. This potential limitation can be circumvented by controlling the expression of the *lacZ*–cDNA fusion gene using the *lac* promoter system. Once the initial phase of infection has occurred, IPTG can be added to inhibit lac repressor function and thus activate fusion gene expression during the burst phase of lytic infection. A second problem with antibody screening is that the antibody "probe" is only an indirect measurement of protein function, and

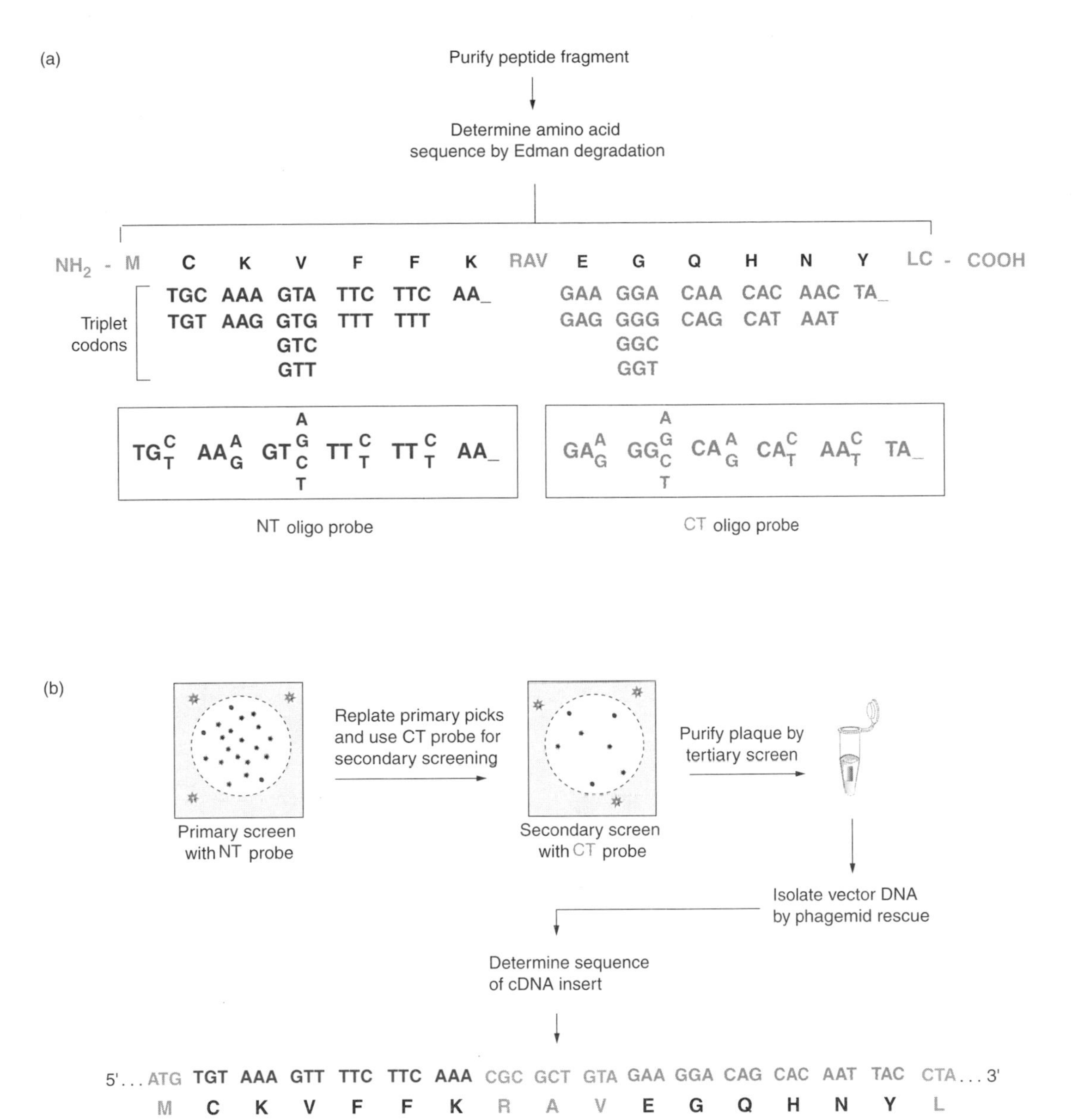

Figure 5.6 A mixed pool of oligonucleotides can be used to screen a cDNA library for the corresponding protein coding sequence. (*a*) A total of 64 oligonucleotides are required to account for the codon redundancy needed to specify the six amino acids in each polypeptide. (*b*) Sequential library screening with the NH_2-terminal (NT) oligo probe pool, and then the COOH-terminal (CT) oligo probe, can be used to identify double-positive cDNA clones. DNA sequencing of cDNA inserts is required to confirm that the predicted ORF and known protein sequence are an exact match.

therefore candidate cDNAs must be verified using a functional assay that is independent of the antibody–antigen interaction. One way to do this is to use the cDNA in a coupled in vitro transcription and translation system to determine if it directly encodes the protein of interest. Alternatively, the cDNA can be used to block in vitro translation of the bona fide protein by pre-incubating single strand cDNA with total poly A+ prior to translation. This translational interference strategy is called hybrid arrest translation.

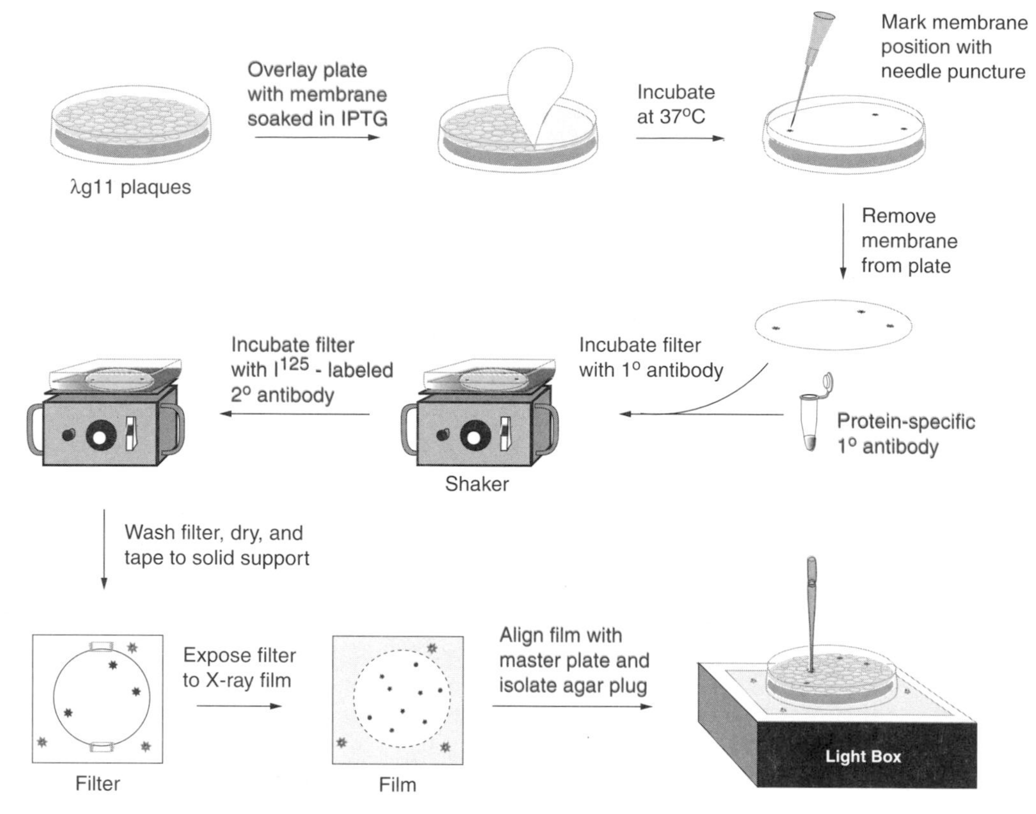

Figure 5.7 Expression cDNA libraries can be screened with a protein-specific antibody to identify recombinant clones expressing reactive fusion proteins. Libraries made in the cDNA expression vector λgt11 can be screened with monoclonal or purified polyclonal antibodies that have been shown to be antigen-specific by Western blotting. Nitrocellulose filters soaked in IPTG are used to transiently induce fusion gene expression during late phase infection. Antigen-containing fusion proteins are detected by ^{125}I-labeled secondary antibodies, and the double-positive plaques are isolated from the master plate using a pipette.

Subtracted cDNA Probes

A third cDNA screening strategy has been devised that exploits differences in mRNA complexity between two cell samples to identify gene sequences that are differentially expressed. In its simplest form this approach is called "differential" cDNA screening. Subtraction hybridization is a variation of differential screening that provides a more sensitive method for detecting low abundance transcripts. In this screening strategy, mRNA is isolated from cells that express the phenotype (or protein) of interest (+), and from cells that do not (–). By converting the (+)mRNA to radioactive (+)cDNA using RTase and ^{32}P-dNTPs, and then hybridizing this probe to a mass excess of (–)mRNA, followed by removal of the double-strand cDNA:mRNA heteroduplexes, it is possible to obtain an enriched probe containing (+)cDNA sequences. This is done by separating the single strand cDNA (unique transcript sequences) from the cDNA:mRNA heteroduplexes (common transcript sequences).

Subtraction hybridization cDNA screening was used by Steve Hedrick and Mark Davis to isolate cDNA sequences encoding the T cell receptor (TCR) gene. The two mRNA populations used in the subtraction were derived from T cells (TCR$^+$) and B cells (TCR-). Figure 5.8 shows how a subtraction hybridization probe could be used to screen a cDNA library that had been constructed with mRNA from salt-stressed plants. Although subtraction hybridization screening has been useful in the isolation of a number of important cDNAs, this method has largely been superseded by PCR-dependent strategies (Chapter 6), and by high-throughput screening using cDNA microarrays (Chapter 9).

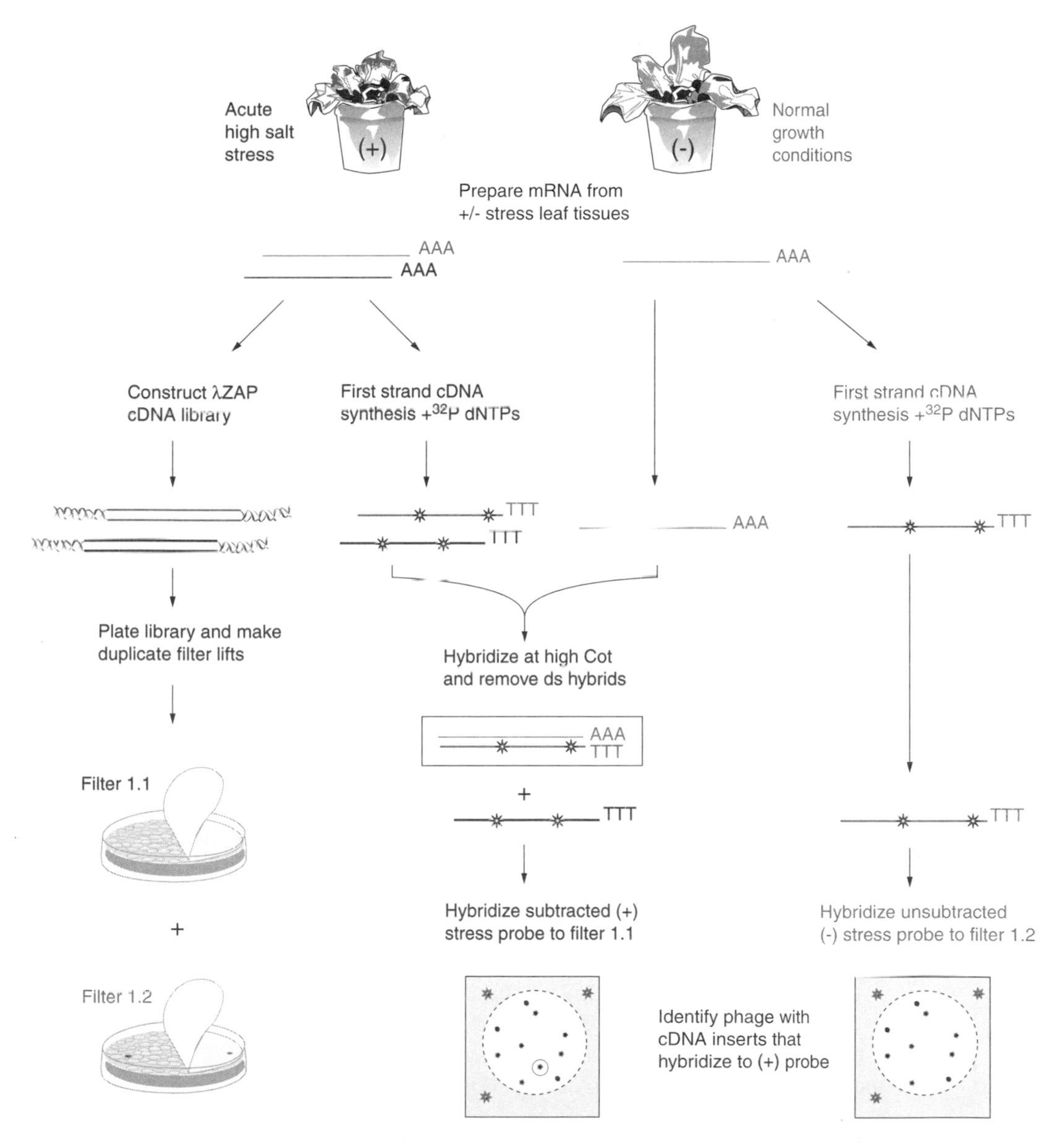

Figure 5.8 Subtraction hybridization is a method to enrich a pool of cDNA probes for sequences that are differentially expressed. This example illustrates how salt stress in plants can alter the expression of genes required for osmotic regulation. High specific activity cDNA probes are hybridized to duplicate filter lifts to identify plaques that contain cDNA inserts corresponding to differentially expressed genes. Because the subtraction step is an enrichment, rather than a purification, very abundant transcripts are still represented in the subtracted probe. Candidate cDNAs must be tested by analyzing the expression pattern of encoded transcripts using Northern blots.

FUNCTIONAL SCREENING OF cDNA EXPRESSION LIBRARIES

Cloned cDNA sequences have the potential to encode active proteins. The technique of cDNA "expression cloning" uses a functional assay to identify positive recombinants without prior knowledge of the gene product. Antibody screening of cDNA expression libraries is a type of functional screen; however, it is in principle different from expression cloning because sufficient information about the protein antigen is already available. This section describes several types of functional screening strategies that have been designed to clone genes solely on the basis of a positive signal in an in vitro or in vivo assay. Ideally, it would be best to use normalized cDNA libraries for these types of screens to reduce the number of recombinants that have to be analyzed.

Protein Activity Assays

There are three basic protein activity assays commonly used for functional screening of cDNA expression libraries. The first approach uses a eukaryotic expression plasmid library that can be expressed at high levels in transiently transfected monkey kidney cells (COS-7 cells). In this method, functional assays are used to identify COS-7 cells that express a protein activity with the desired attributes. A second strategy is to screen λ phage expression libraries with functional probes in much the same way that antibodies are used to screen cDNA expression libraries for antigen binding activity. For example, fusion proteins can be screened for sequence-specific DNA binding activity using DNA probes. The third strategy is to use pools of directionally cloned cDNAs as a means to produce in vitro transcribed RNAs that can be injected into *Xenopus* oocytes and assayed for protein function in vivo. Representative examples of these three types of protein activity screens are listed in Table 5.1.

Yeast Two-Hybrid System

Biological processes require protein–protein interactions that can be defined by noncovalent molecular interactions. The idea of using a cloned protein to screen cDNA expression libraries for interacting proteins follows from the work of Stan Fields and colleagues, who initially used a transcriptional activation assay in yeast to identify mutations that disrupt the in vivo protein–protein interactions between two known proteins. This in vivo protein interaction assay became known as the yeast two-hybrid screen. A similar approach was developed by Roger Brent and colleagues and is called the interaction trap assay.

The yeast two-hybrid strategy was based on an earlier finding by Mark Ptashne and co-workers demonstrating that the yeast GAL4 transcription factor is a modular protein. The Ptashne lab showed that the transcriptional regulatory activity of GAL4 requires a sequence specific DNA binding domain (DBD) and a transcriptional activation domain (AD). Several key experiments revealed that the GAL4 AD could function as a transcriptional activator when fused to the heterologous DNA binding domain of the *E. coli* LexA repressor protein. Similarly, the GAL4 DBD could activate the transcription of GAL4 target genes when fused to a heterologous AD from the herpes simplex virus VP16 protein. Based on this observed modularity of GAL4 protein domains, Fields and others devised a clever three component system to identify protein–protein interactions in yeast cells. The three required elements are (1) a fusion protein consisting of the GAL4

TABLE 5.1. Examples of protein activity assays used to screen cDNA libraries.

Gene product	cDNA screening strategy	Reference
GMCSF	Supernatant from transiently transfected COS cells was collected and assayed for GMCSF activity using a bioassay	Lee et al. (1985)
CD2 antigen	Transiently transfected COS cells were selected for cell adherence using anti-CD2 antibody-coated dishes	Seed and Aruffo (1987)
Oct-2	Double-strand oligonucleotide probes were used to screen λgt11 filters for sequence-specific DNA binding activity	Staudt et al. (1988)
γ interferon receptor	Transiently transfected COS cells were identified that expressed γ interferon receptor binding activity using radiolabeled γ interferon ligand	Munro and Maniatis (1989)
Xwnt-8	*Xenopus* embryos were assayed for dorsalizing activity after injection with in vitro transcribed RNA from pooled directionally cloned cDNAs	Smith and Harland (1991)
XPCC	Ultraviolet-resistant cells were selected from a DNA repair-deficient cell line that was transfected with an episomal vector expression library	Legerski and Peterson (1992)
RBAP-1	GST-Rb fusion protein was labeled in vitro with ^{32}P using protein kinase A and used to screen a λgt11 library for protein-binding activity	Kaelin et al. (1992)
Angiopoietin-1	Transfected COS cells were formaldehyde-fixed and assayed in situ for the localized secretion of ligands that bound to soluble TIE2 receptor	Davis et al. (1996)
MNK1	λ phage filter lifts were screened for expression of GST fusion proteins that can be labeled by soluble ERK1 MAP kinase and [γ-^{32}P]ATP	Fukunaga and Hunter (1997)
Capsaicin receptor	Transfected HEK293 cells were monitored for capsaicin-dependent calcium release; sib selections were used to identify a single cDNA clone	Caterina et al. (1997)

DBD and a cloned protein coding region which serves as the "bait," (2) a cDNA expression library that inserted into a *GAL4* AD gene fusion vector, and (3) a selectable GAL4 reporter gene that can be activated only by a reconstituted GAL4 DBD-AD protein complex. The strategy of yeast two hybrid library screening is illustrated in Figure 5.9.

Numerous variations of the yeast two-hybrid system have been developed to exploit the sensitivity of the assay. For example, the "one-hybrid" screen can be used to identify DNA binding proteins that recognize a specific DNA sequence cloned upstream of the yeast reporter gene. In the one-hybrid screen, cDNA sequences are cloned into the AD plasmid and yeast cells are identified that contain functional DBD-AD proteins that activate the test binding site reporter gene. Another variation is the "reverse" two-hybrid system, which is used to identify point mutants that disrupt functional interactions between two known proteins. The basis of the reverse two-hybrid assay is to identify positively yeast strains expressing AD fusion proteins that fail to induce the expression of a toxic gene, but which retain the ability to activate a second reporter gene weakly. Most two-hybrid screens suffer from a high rate of identifying false-positive protein interactions. These false positives arise from target proteins that are capable of interacting with functionally unrelated bait proteins, and therefore lack the specificity that would be expected of novel biologically significant interactions. Empirical data have

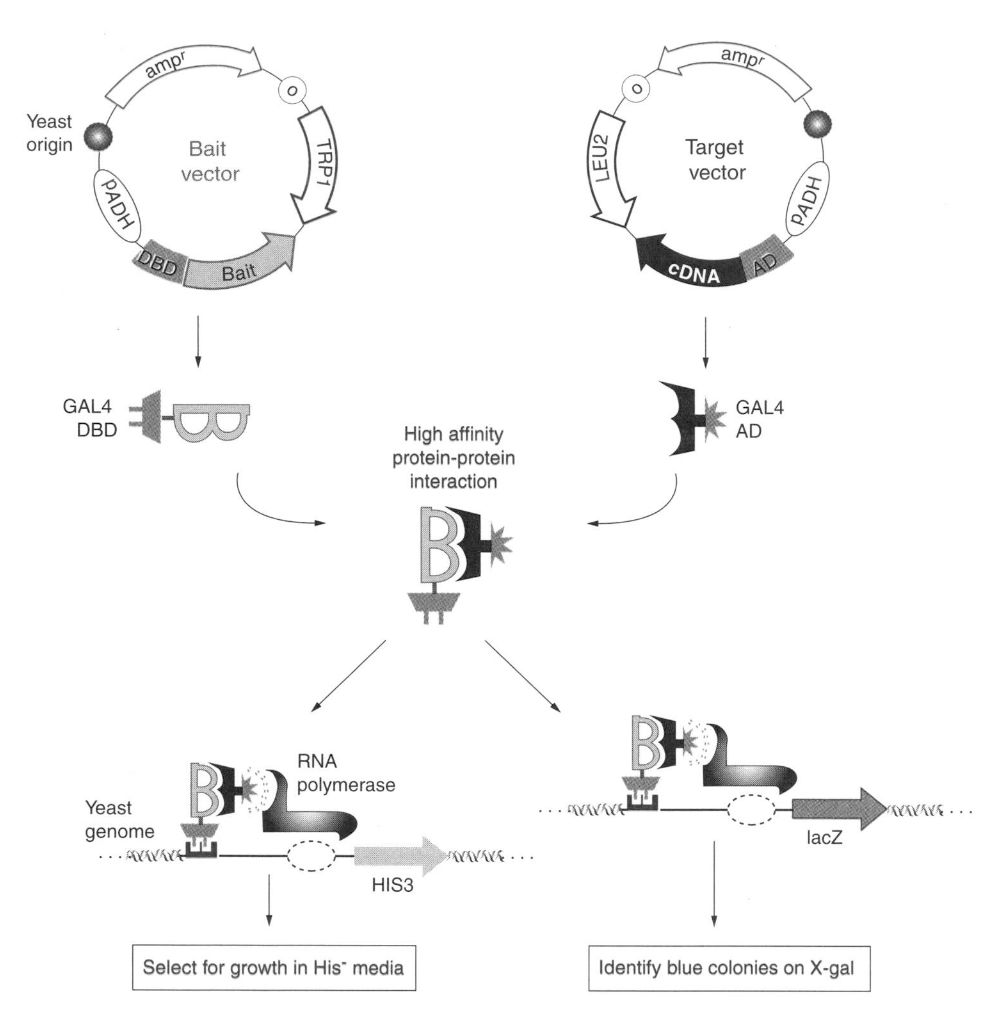

Figure 5.9 The yeast two-hybrid system is used to identify cDNA sequences encoding protein domains capable of interacting with a "bait" protein. The DBD-bait and AD-cDNA plasmids replicate episomally in yeast, using the 2 μ origin of replication, and are maintained in the transformed yeast by growth selection (*TRP1* and *LEU2*). Two *GAL4* reporter genes are integrated into the host yeast genome and encode (1) a growth selectable marker such as *HIS3* and (2) the *lacZ* coding sequence that provides a color marker in the presence of X-gal. The plasmid AD-cDNA library pool is initially constructed and amplified in *E. coli* as *amp*^{r} colonies.

shown that the most common false positives are heat shock proteins, ribosomal proteins, cytochrome oxidase, and proteosome subunit proteins.

cDNA Phage Display

The yeast two-hybrid system is well-suited for identifying protein interacting domains, but it is limited to bimolecular interactions involving two DNA encoded protein products. There are a number of important biological interactions that occur between one protein product, usually a cell surface receptor, and a small diffusible molecule, such as a ligand or antigen. To screen cDNA libraries for these types of receptor proteins,

George Smith and colleagues developed a strategy using M13 filamentous phage vectors that express phage coat fusion proteins that could interact with a molecular bait covalently attached to the bottom of a plastic dish. Because the fusion protein is expressed on the outside of the phage, it is possible physically to separate phages that bind to the bait from those that do not. This in vitro binding technique is called "panning," and it has long been used by immunologists to enrich for B cell populations that express antigen-specific antibodies. Figure 5.10 illustrates how filamentous phage display cDNA screening is performed.

A limitation of phage display cDNA libraries is that many types of fusion proteins render the phage biologically inert by preventing correct phage assembly or by inhibiting phage functions required for *E. coli* infection. For example, fusion proteins with foreign polypeptides longer than 80 amino acids can often be too large to accommodate phage head packaging. Because of this problem, phage display cDNA screening approaches have largely been replaced by phage display peptide libraries containing randomized oligonucleotides in place of cDNAs. This phage display peptide screening approach played a central role in stimulating the development of high-throughput combinatorial chemistry.

USING CLONED cDNA AS A REAGENT

In most cases, cDNA sequence isolation and characterization are only the first step in a more broadly defined research objective ultimately aimed at determining the biological function of an encoded gene product. For example, once a cDNA clone is isolated, DNA sequence analyses would be performed to predict protein coding sequences and possible functional domains (Chapter 9). If the full-length ORF had been obtained, then it is likely that biochemical studies would be initiated utilizing protein expression systems in cultured cells (Chapter 7) and eventually gene targeting and transgenesis in whole organisms (Chapter 8). However, even without the complete gene transcript in hand, it is possible to perform a variety of gene expression studies using cDNA fragments. Knowing where and when a gene is transcribed can be important information that may help in understanding the function of the protein product. This section describes three of the most common gene expression assays that use cDNA as a reagent: (1) Northern blotting, (2) RNase protection, and (3) nuclear run-ons.

Northern Blotting

Often the very first assays to be performed with a newly isolated cDNA fragment is to radiolabel the cloned insert for use as a probe to determine the approximate size of the corresponding full-length gene transcript by a filter hybridization technique called Northern blotting. Analogous to its predecessor the Southern blot, which is used to characterize DNA fragments (Fig. 1.14), the Northern blot is made by transferring size-fractionated denatured RNA onto a nylon or nitrocellulose filter membrane using capillary action. The filter membrane is hybridized with a denatured cDNA probe under conditions that promote the formation of cDNA:RNA heteroduplexes. Following a series of washes in low ionic strength buffer to remove unhybridized probe, the filter membrane is dried and exposed to film or a phosphorimager screen to determine the size and abundance of the corresponding gene transcript. By using a control cDNA probe that detects transcripts from a "housekeeping" gene, such as actin, cyclophilin, or glyceraldehyde-

(a)

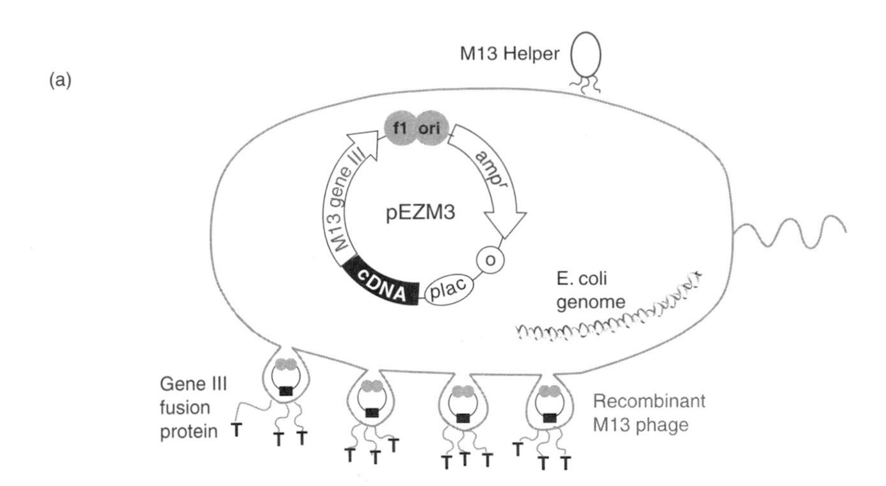

(b)

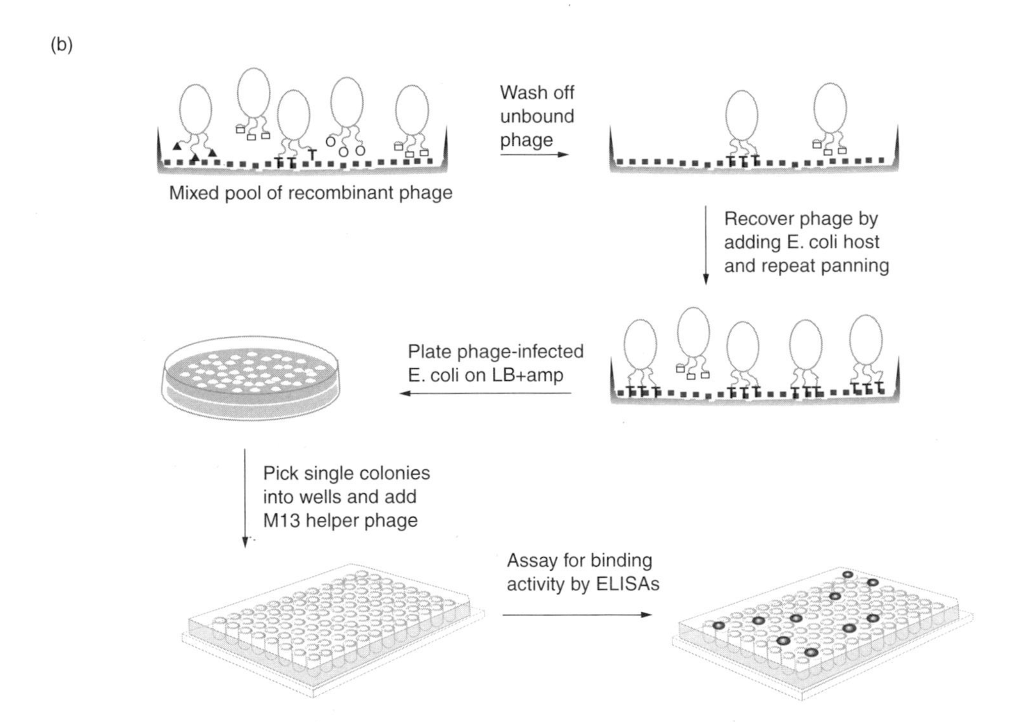

Figure 5.10 Filamentous phage cDNA expression libraries can be screened for fusion proteins that bind to a solid support using the panning method. (*a*) The fusion protein phagemid vector pEZM3 (Clontech) is used to insert cDNA sequences into a multiple cloning site (MCS) within the M13 gene III protein coding sequence. In-frame coding sequences have the potential to produce a fusion protein that is expressed on the phage surface. (*b*) Sequential panning and infection cycles are carried out to enrich for phage that bind to the "bait" attached to the solid support. The phagemids are rescued in *E. coli* and individual picks can be assayed by superinfection with M13 helper phage to produce phage for a 96-well ELISA (enzyme-linked immunosorbent assay). In addition to the filamentous phage vectors, several efficient λ and T7 bacteriophage display vector systems have also been developed.

3-phosphate dehydrogenase (*GAPDH*), it is possible to determine the expression level of the experimental gene, relative to that of the control gene. Northern blots are often used to characterize RNA that has been isolated from different cell types, and if appropriate, RNA from the same cell type under different conditions. The analysis of total RNA preparations is sufficient in most cases, however, some very low abundance transcripts can be detected only with blots that contain the same amount (~10 μg) of poly A+ RNA. RNA splicing variants that differ in molecular weight can also be characterized by Northern blot analysis.

RNase Protection Assay

Northern blotting provides a good approximation of gene transcript size in kilobases; however, the technique is not very sensitive and low abundance mRNAs can be difficult to detect. An alternative to Northern blotting is a method called RNase protection, which can provide a more accurate measurement of RNA expression levels owing to the enhanced sensitivity and reproducibility of the assay. In this procedure, an antisense RNA "riboprobe" is synthesized in vitro in the presence of ^{32}P-α-rCTP using a cloned DNA as the template and purified bacteriophage (T7 or SP6) RNA polymerase. Riboprobes are hybridized in solution to total RNA to form specific RNA:RNA homoduplexes. Excess unhybridized probe and single strand regions of noncomplementary RNA are digested using a mixture of RNases that cleave single-strand RNA. Double-strand homoduplex RNA molecules are not cleaved by these RNases. The resulting mixture is then treated with proteinase K to degrade the RNases and the labeled RNA products are size-fractionated on a polyacrylamide/urea gel. The increased sensitivity of RNase protection assays, as compared to Northern blot analysis, is due primarily to the higher specific activity of the RNA probe and the efficiency of solution hybridization. Figure 1.15 shows an example of how the RNase protection assay can be used to map the start site of transcription using an antisense riboprobe made from cloned genomic DNA. Besides being used as a quantitative assay to measure steady-state mRNA levels, and to map transcript termini (both 5′ and 3′ ends can be determined with appropriate overlapping riboprobes), RNase protection assays are also a convenient method for localizing splice site junctions.

In situ hybridization is another technique that uses in vitro synthesized high specific activity (isotopic or nonisotopic) antisense RNA probes to characterize gene expression. In this method, the RNA probe is hybridized to histological preparations, or whole mount specimens, that have been fixed in paraformaldehyde. The processed microscope slides are then analyzed for the presence of positive hybridizing signals that co-localize with morphological structures or cell types. A modification of this technique, called in situ PCR, is described in Chapter 6.

Nuclear Run-on Assay

The steady-state levels of gene transcripts can be controlled by changes in the rate of transcriptional initiation, RNA processing and nuclear export, or cytoplasmic RNA turnover. RNase protection assays are a good method to quantify changes in the steady state levels of RNA that result from altered RNA processing or turnover rates. However, to determine if observed changes in steady-state levels of RNA are due to increased or decreased rates of transcriptional initiation, assays are required that can distinguish between transcriptional and post-transcriptional mechanisms. The most straightforward way to investigate transcriptional control of gene expression is to analyze reporter genes that contain only the nontranscribed regions of a gene linked to a heterologous transcript coding sequence (see Chapters 4 and 7). Reporter gene assays, however, may not always be feasible. A second approach is to measure the number of nuclear "run-on" transcripts produced in isolated nuclei that have been obtained from cells treated in vivo under various conditions. These assays were originally called nuclear "run-offs," but a more accurate name is nuclear run-ons because the in vitro synthesized transcripts do not actually run off the template DNA.

The principle of nuclear run-on assays is that isolated nuclei contain transcription complexes that are stalled on the DNA template owing to an acute loss of ribonucleotide substrates. By providing ^{32}P-α-UTPs to isolated nuclei in vitro, it is possible to reactivate the stalled RNA polymerase complexes on the DNA templates. The amount of gene-specific radiolabeled RNA synthesized in one nuclei preparation, as compared to another, should reflect the number of transcriptional initiation events present at the time the nuclei were isolated. Figure 5.11 shows how the nuclear run-on assay can be used to determine if elevated steady-state mRNA levels are the result of increased transcriptional initiation events. Although the nuclear run-on assay can be informative, it does not work well for the detection of low abundance transcripts owing to the high sequence complexity and low specific activity of the radiolabeled RNA probe. A more sensitive, albeit less direct, measurement of nascent mRNA synthesis can be obtained using an RTase-PCR assay that quantifies the amount of heterogeneous RNA (hnRNA) produced under one condition as compared to another.

Laboratory Practicum 5. *In vitro expression cloning of genes encoding intracellular signaling proteins*

Research Objective

A graduate student has isolated a mutant neuronal cell line that contains a genetic defect in a receptor-stimulated intracellular signaling pathway required for post-translational activation of a well-characterized cysteine protease. He plans to use in vitro expression cloning (IVEC) to identify cDNA clones that encode protease-activating functions in extracts prepared from receptor-stimulated mutant cells. The biochemical complementation assay he has developed utilizes a high-throughput screen based on detecting in vitro cleavage of a cysteine protease-specific fluorogenic substrate using a 96 well-formatted plate fluorimeter.

Available Information and Reagents

1. Extracts from stimulated mutant cells contain unactivated protease and therefore lack substrate cleaving activity under normal assay conditions. It is possible to stimulate substrate cleavage in the mutant extracts by adding a small amount of extract prepared from unstimulated wild-type cells.
2. A directional cDNA library has been constructed using mRNA that was isolated from the wild-type cell line. This library is cloned into a plasmid vector containing a T7 bacteriophage promoter that can be used to direct the synthesis of "sense" RNA that is suitable for in vitro translation using a reticulocyte protein synthesis reaction mix.
3. The cDNA plasmid library has been normalized and subdivided into 384 pools (four 96-well plates) of independent *E. coli* strains, with each pool containing ~50 members. This normalized library is estimated to consist of approximately 20,000 different cDNA recombinants.

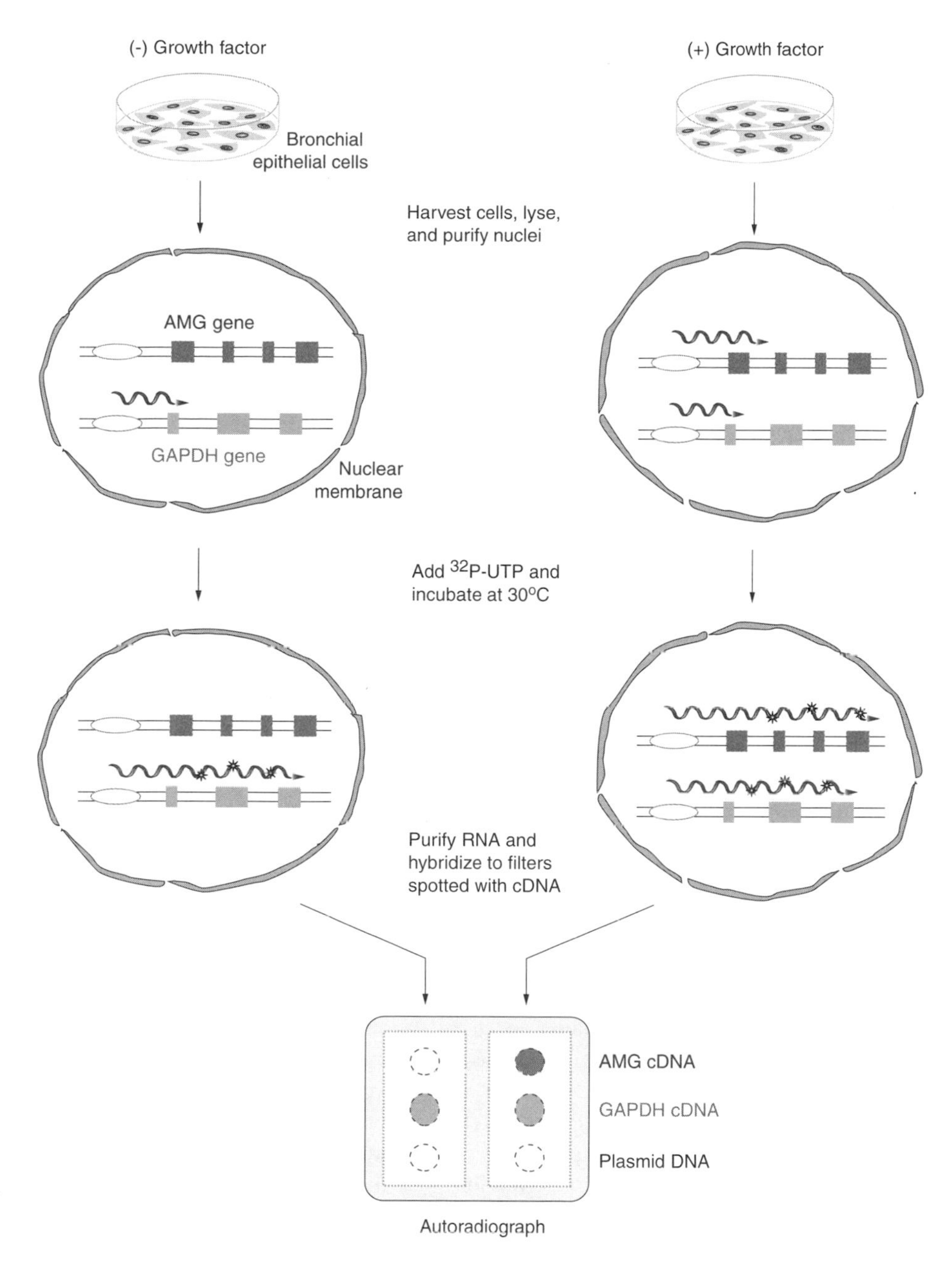

Figure 5.11 Nuclear run-on assays can be used to determine if observed changes in RNA steady-state levels reflect alterations in the number of newly initiated transcripts. In this example, nuclei were prepared from cultured bronchial epithelial cells that had been cultured in the absence (-) or presence (+) of a growth factor that had been shown to increase the steady-state RNA levels of the hypothetical *AMG* gene. By hybridizing the in vitro synthesized pool of ^{32}P-labeled RNA with filter membranes that have been spotted with denatured *AMG* cDNA and gene sequences from *GAPDH* as an internal control, it is possible to quantify the amount of hybridized probe RNA and determine that growth factor treatment does increase the number of transcriptional initiation events at the *AMG* gene. Nonspecific RNA-DNA hybridization can be monitored with a plasmid vector control.

Basic Strategy
Plasmid DNA is isolated from each pool using a 96 well mini plasmid prep format utilizing nucleic acid binding resins to purify plasmid DNA from *E. coli* lysates. The DNA is linearized with the rare-cutting NotI enzyme, which cleaves downstream of the 3′ cloning site to produce linear template for a coupled in vitro transcription–translation protein synthesis reaction. Aliquots of each in vitro reaction are added to a 96-well-formatted assay mixture containing extracts prepared from stimulated mutant cells, the cysteine protease-specific substrate, and protease inhibitors that do not block cysteine protease activity. Positive wells are identified by measuring fluorescence in each well following a 40 minute incubation at 37°C.

Sib selection is used to subdivide plasmid pools into individual *E. coli* strains by plating the corresponding bacterial stock from a positive well onto agar plates. The plates are incubated overnight and then each of 96 wells is inoculated with bacteria from single colonies that have grown up on the agar plates. Based on the estimated complexity of the plasmid pools (~50 recombinants/pool), each 96 well plate should contain two positive wells that represent duplicates of the same cDNA recombinant. Once individual cDNA clones are identified that encode protease activating function, the cloned inserts are sequenced and characterized. As a validation of the in vitro biochemical complementation assay, the candidate plasmids are transfected into the mutant neuronal cell line and in vivo protease activation is measured in stimulated cells. Figure 5.12 outlines the strategy for this IVEC protease screen.

Comments
A variation of in vitro expression cloning was used in the 1980s to identify a limited number of genes that could be assayed in *Xenopus* oocytes that had been microinjected with in vitro transcribed cDNA templates. By using sib selection strategies, researchers were able to identify genes encoding channel forming proteins and proteins required for early events in *Xenopus* development. The term IVEC and the idea of using this strategy to screen normalized cDNA libraries were proposed by Marc Kirschner and colleagues in a series of proof-of-principle papers. Their initial application of IVEC was to identify substrates for cyclin-dependent protein kinases by screening for the appearance of electrophoretic mobility shifts of in vitro labeled proteins. The strategy outlined in Figure 5.12 is slightly more demanding because it requires that the missing "signaling" protein be made in sufficient quantities to function as a soluble component in the complementation assay. Moreover, the protein activity would have to be encoded by a single polypeptide because it would be unlikely that two functional subunits would be encoded by the same initial plasmid pool. The cDNA sequence of the positive clone would presumably provide additional information regarding possible protein functions based on sequence homology searches using the GenBank database (Chapter 9).

Prospective
As shown in Figure 5.12, any cDNA clones identified in the in vitro assay will have to be tested by transient transfection into the mutant cell line. Assuming that expression levels are physiological, this in vivo assay should provide good evidence that cloned cDNA encodes the defective protein and that receptor stimulation is required for complementation. The next logical experiment would be to determine the nature of the genetic mutation that gave rise to the mutant cell phenotype. For example, a Northern blot could be performed to see if the gene is expressed at normal levels, and if it is, it would suggest that there is a mutation in the coding sequence that prevents the production of a functional protein. This would have to be tested by cloning and sequencing the corresponding cDNA. If the Northern blot shows that the gene is not expressed, then

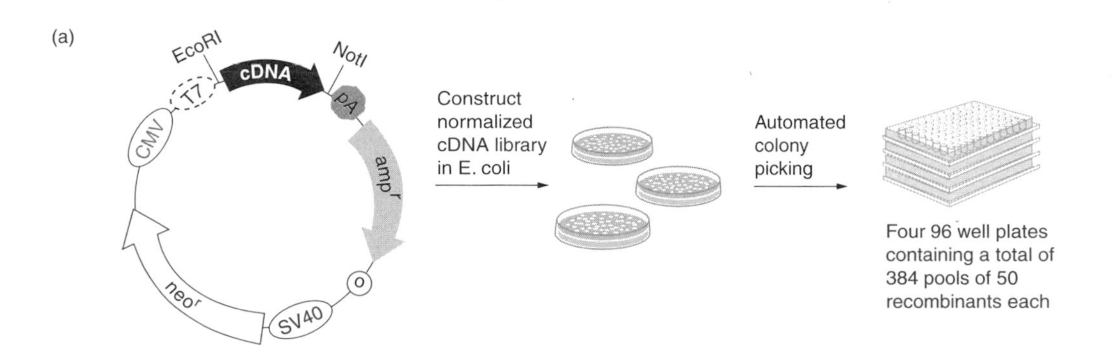

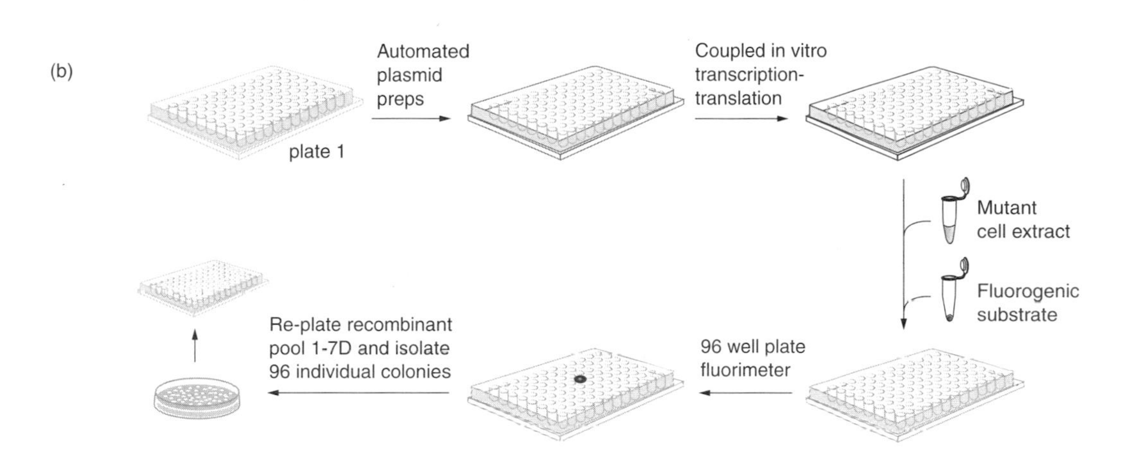

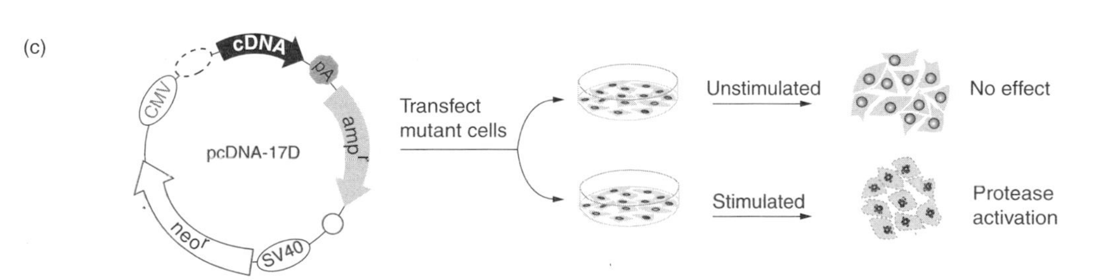

Figure 5.12 In vitro expression cloning (IVEC) could be used to isolate cDNA clones that complement a functional defect in protease activation in a mutant cell line. (*a*) Directional cDNA library is constructed in a plasmid vector that contains both a bacteriophage T7 promoter and a eukaryotic promoter (CMV) upstream of the EcoRI-NotI cloning site. This plasmid vector also contains a gene marker (*neo*^{r}) for stable selection in eukaryotic cells (see Chapter 7). The cDNA library is normalized and plated on LB+ampicillin. An automated robotic instrument is used to pick ~20,000 colonies in a way that inoculates 50 recombinants into a single well of a 96-well plate. Four plates are collected that together constitute 384 plasmid pools. (*b*) The pools are assayed using a 96-well-formatted set of reactions that permits plasmid purification and linearization, in vitro coupled transcription–translation, and protease activity measurements. A single positive clone from the plasmid pool can be identified by a second round of screening using *E. coli* transformants obtained from replating the plasmid pool from the original positive well. (*c*) A candidate plasmid (pcDNA-17D) can be characterized using an in vivo expression assay in transfected neuronal cells to determine if the encoded protein complements the mutant phenotype following stimulation.

genomic DNA would have to be analyzed for promoter mutations or chromosomal rearrangements. Another research direction would be to perform protein structure–function analyses using cDNA mutagenesis in conjunction with the in vivo complementation assay.

REFERENCES

Converting mRNA transcripts into cDNA

*Aviv, H., Leder, P. Purification of biologically active globin messenger RNA by chromatography on oligothymidylic acid-cellulose. Proc. Natl. Acad. Sci. USA 69:1408–1412, 1972.

*Baltimore, D. Viral RNA-dependent DNA polymerase. Nature 226:1209–1211, 1970.

Blackburn, P., Jailkhani, B.L. Ribonuclease inhibitor from human placenta: Interaction with derivatives of ribonuclease A. J. Biol. Chem. 254:12488–12492, 1979.

Blumberg, D.D. Creating a ribonuclease-free environment. Methods Enzymol. 152:20–24, 1987.

*Chirgwin, J.J., Przbyla, A.E., MacDonald, R.J., Rutter, W.J. Isolation of biologically active ribonucleic acid from sources enriched in ribonuclease. Biochemistry 18:5294–5299, 1979.

*Chomczynski, P., Sacchi, N. Single-step method of RNA isolation by acid guanidinium thiocyanate-phenol-chloroform extraction. Anal. Biochem. 162:156–159, 1987.

Gilham, P.T. Synthesis of polynucleotide celluloses and their use in fractionation of polynucleotides. J. Am. Chem. Soc. 86:4982–4989, 1964.

*Gubler, U., Hoffman, B.J. A simple and very effective method for generating cDNA libraries. Gene 25:263–269, 1983.

Kotewicz, M.L., Sampson, C.M., D'Alessio, J.M., Gerard, G.F. Isolation of cloned Moloney murine leukemia virus reverse transcriptase lacking ribonuclease H activity. Nucl. Acids Res. 16:265–277, 1988.

Land, H., Grez, M., Hauser, H., Lindenmaier, W., Schutz, G. Synthesis of ds-cDNA involving addition of dCMP tails to allow cloning of 5′-terminal mRNA sequences. Methods Enzymol. 100:285–292, 1983.

*Maniatis, T., Kee, S.G., Efstratiatis, S., Kafatos, F.C. Amplification and characterization of a β-globin gene synthesized in vitro. Cell 8:163–182, 1976.

*Okayama, H., Berg, P. High-efficiency cloning of full-length cDNA. Mol. Cell. Biol. 2:161–170, 1982.

*Rabbitts, T. H. Bacterial cloning of plasmids carrying copies of rabbit globin messenger RNA. Nature 260:221–225, 1976.

*Rougeon, F., Kourilsky, P., Mach, B. Insertion of the rabbit β-globin gene sequence into *E. coli* plasmid. Nucl. Acids Res. 2:2365–2378, 1975.

Seed, B. An LFA-3 cDNA encodes a phospholipid-linked membrane protein homologous to its receptor CD2. Nature 329:840–842, 1987.

Short, J., Fernandez, J. Sorge, J., Huse, W. λZap: A bacteriophage λ expression vector with in vivo excision properties. Nucl. Acids Res. 16:7583–7600, 1988.

*Soares, M.B., Bonaldo, M., Jelene, P., Su, L., Lawton, L., Efstratiadis, A. Construction and characterization of a normalized cDNA library. Proc. Natl. Acad. Sci. USA 91:9228–9232, 1994.

*Temin, H.M., Mizutani, S. Viral RNA-dependent DNA polymerase. Nature 226:1211–1213, 1970.

Screening representative cDNA libraries

*Benton, W.D., Davis, R.W. Screening λgt recombinant clones by hybridization to single plaques in situ. Science 196:180–182, 1977.

---

* Landmark papers in applied molecular genetics.

Briehl, M., Flomerfelt, F.A., Wu, X-P, Miesfeld, R.L. Transcriptional analyses of steroid-regulated gene networks. Mol. Endocrinol. 4:287–294, 1990.

*Davis, R.L., Weintraub, H., Lassar, A.B. Expression of a single transfected cDNA converts fibroblasts to myoblasts. Cell 51:987–1000, 1987.

Erlich, H.A., Cohen, S.N., McDevitt, H.O. A sensitive radioimmunoassay for detecting products translated from cloned DNA fragments. Cell 13:681–689, 1978.

*Goeddel, D. V., Yelverton, E., Ullrich, A. et al. Human leukocyte interferon produced by *E. coli* is biologically active. Nature 387:411–414, 1980.

*Hedrick, S.M., Cohen, D.I., Nielsen, E.A., Davis, M.M. Isolation of cDNA clones encoding T cell-specific membrane-associated proteins. Nature 308:149–153, 1984.

Helfman, D.M., Feramisco, J.R., Fiddes, J.C., Thomas, G.P., Hughes, S.H. Identification of clones that encode chicken tropomyosin by direct immunological screening of a cDNA expression library. Proc. Natl. Acad. Sci. USA 80:31–35, 1983.

Jacobs, K.A., Rudersdorf, R., Neill, S.D., Dougherty, J.P., Brown, E.L., Fritsch, E.F. 1988. The thermal stability of oligonucleotide duplexes is sequence independent in tetraalkylammonium salt solutions: Application to identifying recombinant DNA clones. Nucl. Acids Res. 16:4637–4650, 1988.

Kent, S., Hood, L., Aebersold, R., Teplow, D., Smith, L. et al. Approaches to sub-picomole protein sequencing. BioTechniques 5:314–321, 1987.

Kuang, W., Thompson, D., Hoch, R., Weigel, R. Differential screening and suppression subtractive hybridization identified genes differentially expressed in an estrogen receptor-positive breast carcinoma cell line. Nucleic Acids Res. 26:1116–1123, 1998.

Lathe, R. Synthetic oligonucleotide probes deduced from amino acid sequence data. Theoretical and practical considerations. J. Mol. Biol. 183:1–12, 1985.

Miesfeld, R., Okret, S., Wikstrom, A-C., Wrange, O., Gustafsson, J-A., Yamamoto, K.R. Characterization of a steroid receptor gene and mRNA in wild-type and mutant cells. Nature 312: 779–781, 1984.

Paterson, B.M., Roberts, B.E., Kuff, E.L. Structural gene identification and mapping by DNA:mRNA hybrid-arrested cell-free translation. Proc. Natl. Acad. Sci. USA 74:4370–4374, 1977.

Sagerstrom, C.G., Sun, B.I., Sive, H.L. Subtractive cloning: Past, present and future. Ann. Rev. Biochem. 66:751–783, 1997.

Skalka, A., Shapiro, L. In situ immunoassays for gene translation products in phage plaques and bacterial colonies. Gene 1:65–79, 1976.

*Suggs, S.V., Wallace, R.B., Hirose, T., Kawashima, E.H., Itakura, K.. Use of synthetic oligonucleotides as hybridization probes: Isolation of cloned cDNA sequences for human β2-microglobulin. Proc. Natl. Acad. Sci. USA 78:6613–6616, 1981.

Vernon, D., Bohnert, H. A novel methyl transferase induced by osmotic stress in the facultative halophyte *Mesembryanthemum crystallinum*. EMBO J. 11:2077–2085, 1992.

Wood, W.I., Gitschier, J., Lasky, L.A., Lawn, R.M. Base composition-independent hybridization in tetramethylammonium chloride: A method for oligonucleotide screening of highly complex gene libraries. Proc. Natl. Acad. Sci. USA 82:1585–1588, 1985.

Yang, J.H., Ye, J.H, Wallace, D.C. Computer selection of oligonucleotide probes from amino acid sequences for use in gene library screening. Nucl. Acids Res. 12:837–843, 1984.

*Young, R.A., Davis, R.W. Efficient isolation of genes by using antibody probes. Proc. Natl. Acad. Sci. USA 80:1194–1198, 1983.

Functional screening of cDNA expression libraries

*Brent, R., Ptashne, M. A eukaryotic transcriptional activator bearing the DNA specificity of a prokaryotic repressor. Cell 43:729–7361, 1985.

Caterina, M.J., Schumacher, M., Tominaga, M., Rosen, T., Levine, J., Julius, M. The capsaicin receptor: A heat-activated ion channel in the pain pathway. Nature 389:816–824, 1997.

*Chien, C.-T., Bartel, P.L., Sternglanz, R., Fields, S. The two-hybrid system: A method to identify and clone genes for proteins that interact with a protein of interest. Proc. Natl. Acad. Sci. USA 88:9578–9582, 1991.

Davis, S., Aldrich, T.H., Jones, P.F. et al. Isolation of angiopoietin-1, a ligand for the TIE2 receptor, by secretion-trap expression cloning. Cell 87:1161–1169, 1996.

Durfee, T., Becherer, K., Chen, P.L., Yeh, S.H., Yang, Y., Kilburn, A.E., Lee, W.H., Elledge, S.J. The retinoblastoma protein associates with the protein phosphatase type 1 catalytic subunit. Genes Dev. 7:555–569, 1993.

Ellman, J., Stoddard, B., Wells, J. Combinatorial thinking in chemistry and biology. Proc. Natl. Acad. Sci. USA 94:2779–2782, 1997.

Estojak, J., Brent, R., Golemis, E.A. Correlation of two-hybrid affinity data with in vitro measurements. Mol. Cell. Biol. 15:5820–5829, 1995.

Evangelista, C., Lockshorn, D., Fields, S. The yeast two-hybrid system: Prospects for protein linkage maps. Trends Cell Biol. 6:196–199, 1996.

*Fields, S., Song, O. A novel genetic system to detect protein-protein interaction. Nature 340: 245–246, 1989.

Fisch, I., Kontermann, R. E., Finners, R. et al. A strategy of exon shuffling for making large peptide repertoires displayed on filamentous bacteriophage. Proc. Natl. Acad. Sci. USA 93: 7761–7766, 1996.

Fukunaga, R., Hunter, T. MNK1, a new MAP kinase-activated protein kinase, isolated by a novel expression screening method for identifying protein kinases. EMBO J. 16:1921–1933, 1997.

*Gluzman, Y. SV40-Transformed simian cells support the replication of early SV40 mutants. Cell 23:175–182, 1981.

Hengen, P. N., False positives from the yeast two-hybrid system. Trends Biochem. Sci. 22:33–34, 1997.

Kaelin, W.G., Krek, W., Sellers, W.R. et al. Expression cloning of a cDNA encoding a retinoblastoma-binding protein with E2F-like properties. Cell 70:351–364, 1992.

Kaupmann, K., Huggel, K., Heid, J. et al. Expression cloning of GABA receptors uncovers similarity to metabotropic glutamate receptors. Nature 386:239–246, 1997.

Lee, F., Yokota, T., Otsuka, T., Gemmell, L., Larson, N., Luh, J., Arai, K.-I., Rennick, D. Isolation of cDNA for a human granulocyte-macrophage colony-stimulating factor by functional expression in mammalian cells. Proc. Natl. Acad. Sci. USA 82:4360–4364, 1985.

Legerski, R., Peterson, C. Expression cloning of a human DNA repair gene involved in xeroderma pigmentosum group C. Nature 359:70–73, 1992.

Littman, D.R., Thomas Y., Maddon, P.J., Chess, L., Axel, R. The isolation and sequence of the gene encoding T8: A molecule defining functional classes of T lymphocytes. Cell 40:237–246, 1985.

*Ma, J., Ptashne, M. A new class of yeast transcriptional activators. Cell 51:113–119, 1987.

*Matthews, D.J., Wells, J.A. Substrate phage: Selection of protease substrates by monovalent phage display. Science 260:113–117, 1993.

Munro, S., Maniatis, T. Expression cloning of the murine interferon γ receptor cDNA. Proc. Natl. Acad. Sci. USA 86:9248–9252, 1989.

Ruden, D.M., Ma, J., Li, Y., Wood, K., Ptashne., M. Generating yeast transcriptional activators containing no yeast protein sequences. Nature 350:426–430, 1991.

*Scott, J.K., Smith, G.P. Searching for peptide ligands with an epitope library. Science 249:386–390, 1990.

Seed, B., Aruffo, A. Molecular cloning of the CD2 antigen, the T-cell erythrocyte receptor, by a rapid immunoselection procedure. Proc. Natl. Acad. Sci. USA 84:3365–3369, 1987.

SenGupta, D.J., Zhang, B., Kraemer,B., Pochart, P., Fields, S., Wickens, M. A three-hybrid system to detect RNA-protein interactions in vivo. Proc. Natl. Acad. Sci. USA 93:8496–8501, 1996.

Singh, H., Clerc, R.G., LeBowitz, J.H. Molecular cloning of sequence-specific DNA binding proteins using recognition site probes. BioTechniques 7:252–261, 1989.

*Smith, G.P. Filamentous fusion phage: Novel expression vectors that display cloned antigens on the virion surface. Science 228:1315–1317, 1985.

Smith, W.C., Harland, R.M. Injected Xwnt-8 RNA acts early in Xenopus embryos to promote formation of a vegetal dorsalizing center. Cell 67:753–765, 1991.

Staudt, L.M., Clerc, R.G., Singh, H., LeBowitz, J.H., Sharp, P.A., Baltimore, D. Cloning of a lymphoid-specific cDNA encoding a protein binding the regulatory octamer DNA motif. Science 241:577–580, 1988.

Sternberg, N., Hoess, R.H. Display of peptides and proteins on the surface of bacteriophage λ. Proc. Natl. Acad. Sci. USA 92:1609–1613, 1995.

Tamura, S., Okumoto, K., Toyama, R. et al. Human PEX1 cloned by functional complementation on a CHO cell mutant is responsible for peroxisome-deficient Zellweger syndrome of complementation group I. Proc. Natl. Acad. Sci. USA 95:4350–4355, 1998.

Vidal, M., Brachmann, R.K., Fattley, A., Harlow, E., Boeke, J. Reverse two-hybrid and one-hybrid systems to detect dissociation of protein-protein and DNA-protein interactions. Proc. Natl. Acad. Sci. USA 93:10315–10320, 1996.

Vinson, C.R., LaMarco, K.L., Johnson, P.F., Landschulz, W.H., McKnight, S.L. In situ detection of sequence-specific DNA binding activity specified by a recombinant bacteriophage. Genes Devel. 2:801–806, 1988.

Wysocki, L.J., Sato, V.L. 1978. Panning for lymphocytes: A method for cell selection. Proc. Natl. Acad. Sci. USA 75:2844–2848.

Using cloned cDNA as a reagent

*Alwine, J.C., Kemp., D.J., Stark, G.R. Method for detection of specific RNAs in agarose gels by transfer to diazobenzyloxymethyl-paper and hybridization with DNA probes. Proc. Natl. Acad. Sci. USA 74:5350–5354, 1977.

Ashe, H.L., Monks, J., Wijgerde, M., Fraser, P., Proudfoot, N.J. Intergenic transcription and transinduction of the human β-globin locus. Genes Devel. 11:2494–2509, 1997.

Bailey, J.M., Davidson, N. Methylmercury as a reversible denaturing agent for agarose gel electrophoresis. Anal. Biochem. 70:75–85, 1976.

*Berk, A.J., Sharp, P.A. Sizing and mapping of early adenovirus mRNAs by gel electrophoresis of S1 endonuclease-digested hybrids. Cell 12:721–732, 1977.

Casey, J., Davidson, N. Rates of formation and thermal stabilities of RNA:DNA and DNA:DNA duplexes at high concentrations of formamide. Nucl. Acids Res. 4:1539–1552, 1977.

Elferink, C.J., Reiners, J.J. Quantitative RT-PCR on CYP1A1 heterogeneous nuclear RNA: A surrogate for the in vitro transcription run-on assay. BioTechniques 20:470–477, 1996.

*Greenberg, M.E., Ziff, E.B. Stimulation of 3T3 cells induces transcription of the c-fos protooncogene. Nature 311:433–438, 1984.

Groudine, M., Peretz, M., Weintraub, H. Transcriptional regulation of hemoglobin switching on chicken embryos. Mol. Cell. Biol. 1:281–288, 1981.

Kassevetis, G.A., Butler, E.T., Roulland, D., Chamberlin, M.J. Bacteriophage SP6-specific RNA polymerase. II. Mapping of SP6 DNA and selective in vitro transcription. J. Biol. Chem. 257: 5779–5787, 1982.

Lehrach, H., Diamond, D., Wozney, J.M., Boedtker, H. RNA molecular weight determinations by gel electrophoresis under denaturing conditions: A critical reexamination. Biochemistry 16: 4743–4751, 1977.

Marzluff, W.F. Transcription of RNA in isolated nuclei. Methods Cell Biol. 19:317–331, 1978.

*Melton, D.A., Krieg, P.A., Rebagliati, M.R., Maniatis, T., Zinn, K., Green, M.R. Efficient in vitro synthesis of biologically active RNA and RNA hybridization probes from plasmids containing a bacteriophage SP6 promoter. Nucl. Acids Res. 12:7035–7056, 1984.

*Southern, E.M. Detection of specific sequences among DNA fragments separated by gel electrophoresis. J. Mol. Biol. 98:503–517, 1975.

*Tabor, S., Richardson, C.C. A bacteriophage T7 RNA polymerase/promoter system for controlled exclusive expression of specific genes. Proc. Natl. Acad. Sci. USA 82:1074–1078, 1985.

*Thomas, P.S. Hybridization of denatured RNA and small DNA fragments transferred to nitrocellulose. Proc. Natl. Acad. Sci. USA 77:5201–5205, 1980.

*Zinn, K., DiMaio, D., Maniatis, T. Identification of two distinct regulatory regions adjacent to the human β-interferon gene. Cell 34:865–879, 1983.

In vitro expression cloning of genes encoding intracellular signaling proteins

King, R.W., Lustig, K.D., Stukenberg, P.T., McGarry, T.J., Kirschner, M.W. Expression cloning in the test tube. Science 277:973–974, 1997.

Li, H., Zhu, H., Xu, C., Yuan, J. Cleavagae of BID by caspase 8 mediates the mitochondrial damage in the Fas pathway of apoptosis. Cell 94:491–501, 1998.

Lustig, K.D., Stukenberg, P.T., McGarry, King, R.W., Cryns, V.L., Mead, P.E., Zon, L.I., Yuan, J., Kirschner, M. Small pool expression screening: Identification of genes involved in cell cycle control, apoptosis and early development. Methods Enzymol. 283:83–99, 1997.

Stukenber, P.T., Lustig, K.D., McGarry, T.J., King, R.W., Kuang, J., Kirschner, M.W. Systematic identification of mitotic phosphoproteins. Curr. Biol. 7:338–348, 1997.

CHAPTER

6

THE POLYMERASE CHAIN REACTION

One of the most important developments in applied molecular genetics in the last decade was the invention of an amazingly simple DNA amplification strategy called the polymerase chain reaction (PCR). The idea for PCR is credited to Kary Mullis, who was a research scientist in the 1980s at a California biotechnology company called Cetus. Mullis, along with five other researchers in the Human Genetics Department at Cetus, demonstrated that oligonucleotide primers could be used specifically to amplify defined segments of genomic DNA (or cDNA). Mullis was awarded the Nobel Prize in Chemistry in 1993 for his recognized contribution to the development of PCR. In many respects, the combined specificity and sensitivity of PCR make it the premier application of molecular genetic principles.

This chapter first describes the basic reaction parameters of PCR, with an emphasis on the factors that most affect the specificity and sensitivity of the assay. The next three sections contain general descriptions of various PCR strategies that were chosen to represent the most common applications. These PCR strategies are grouped into three categories: (1) PCR strategies using cDNA templates, (2) diagnostic applications of PCR, and (3) laboratory applications of PCR. Because it is not possible to cover all of the reported applications of PCR in each of these categories, the emphasis is placed on key conceptual strategies rather than specific details. Laboratory practicum 6 illustrates how PCR strategies often constitute an integral component of a larger research objective by describing how single-cell PCR can be used to isolate cell-specific cDNA transcripts.

BASIC PCR METHODOLOGY

The logic of the PCR protocol follows directly from well-understood principles of nucleic acid biochemistry. The basic components of a PCR reaction are one or more molecules of target DNA, oligonucleotide primers, thermostable DNA polymerase, and dNTPs. This reaction mix is repeatedly heated and cooled to 95, 55, and 72°C, in that order, a total of 25–35 times to produce a $>10^6$-fold amplification of the target DNA. Although the PCR reaction is relatively simple, the biochemical and kinetic parameters that affect PCR specificity and sensitivity are actually quite complex and it can sometimes be troublesome to define a reaction condition that generates reproducible results. In fact, during the first few years after PCR was described in the literature, numerous problems were encountered by other researchers, primarily with regard to reaction specificity and sample contamination. At the time, these drawbacks seem to indicate that PCR applications may not be well-suited for routine diagnostic procedures. This dire prediction did not hold up as continued refinement of the technique solved the most serious problems. Nevertheless, it is important to keep in mind that each time a new DNA target or PCR primer pair is designed for a specific research objective, a number of key parameters will have to be optimized.

The PCR Amplification Cycle

PCR requires an instrument that can quickly heat and cool the reaction tubes to the temperature optimums required for template denaturation (95°C), primer annealing (55–65°C), and DNA polymerization (72°C). The source of DNA template can be either purified DNA or relatively crude nucleic acid such as found in an in situ cell lysis, the pulp of a fossil tooth, or a hair follicle found at a crime scene. Moreover, reaction conditions can be optimized to amplify a target DNA sequence from as small as 100 bp to as much as 20 kb, depending on the application. Most PCR reactions use a DNA repair polymerase called *Taq* polymerase which is a thermostable enzyme that retains activity at 95°C. Figure 6.1 illustrates the relationship between cycle repetitions and overall DNA amplification. Note that the 5′ end of each PCR primer defines the DNA termini of the PCR product.

In theory, each amplification cycle should double the number of target molecules, resulting in an exponential increase in PCR product. However, even before substrate or enzyme becomes limiting, the efficiency of exponential amplification is less than 100% owing to suboptimal DNA polymerase activity, poor primer annealing, and incomplete denaturation of the templates. The predicted amount of PCR product can be calculated by a simple equation that incorporates terms for the number of target molecules at the onset, the efficiency of each cycle, and the number of cycles. This PCR efficiency formula can be expressed as

$$\text{PCR product yield} = (\text{input target amount}) \times (1 + \%\ \text{efficiency})^{\text{cycle number}}$$

This equation can be used to calculate that ~26 cycles are required to produce 1 μg of PCR product from 1 pg of a target sequence (10^6 amplification) using an amplification efficiency value of 70% [1 μg PCR product = (1 pg target) $\times$ $(1 + 0.7)^{26}$].

The polymerase chain reaction would not be as useful as it is without automated instrumentation. A standard piece of molecular genetic lab equipment is the "PCR machine," or more precisely, a thermal cycling instrument. The minimum requirement for a thermal cycler is that it be capable of rapidly changing reaction tube temperature to provide the optimal conditions for each step of the cycle. There are three basic designs for PCR-dedicated thermal cyclers, which differ in how the temperature "ramping" (°C change per second) is achieved. The most intuitive design is a mechanized robot which simply uses a programmable arm to move a set of PCR reaction samples between three separate heating blocks that are arrayed in a circle around the arm. A fourth block in this setup is used for sample storage at 4°C after the last cycle is completed. A more conventional thermal cycler, however, is one that uses refrigeration and

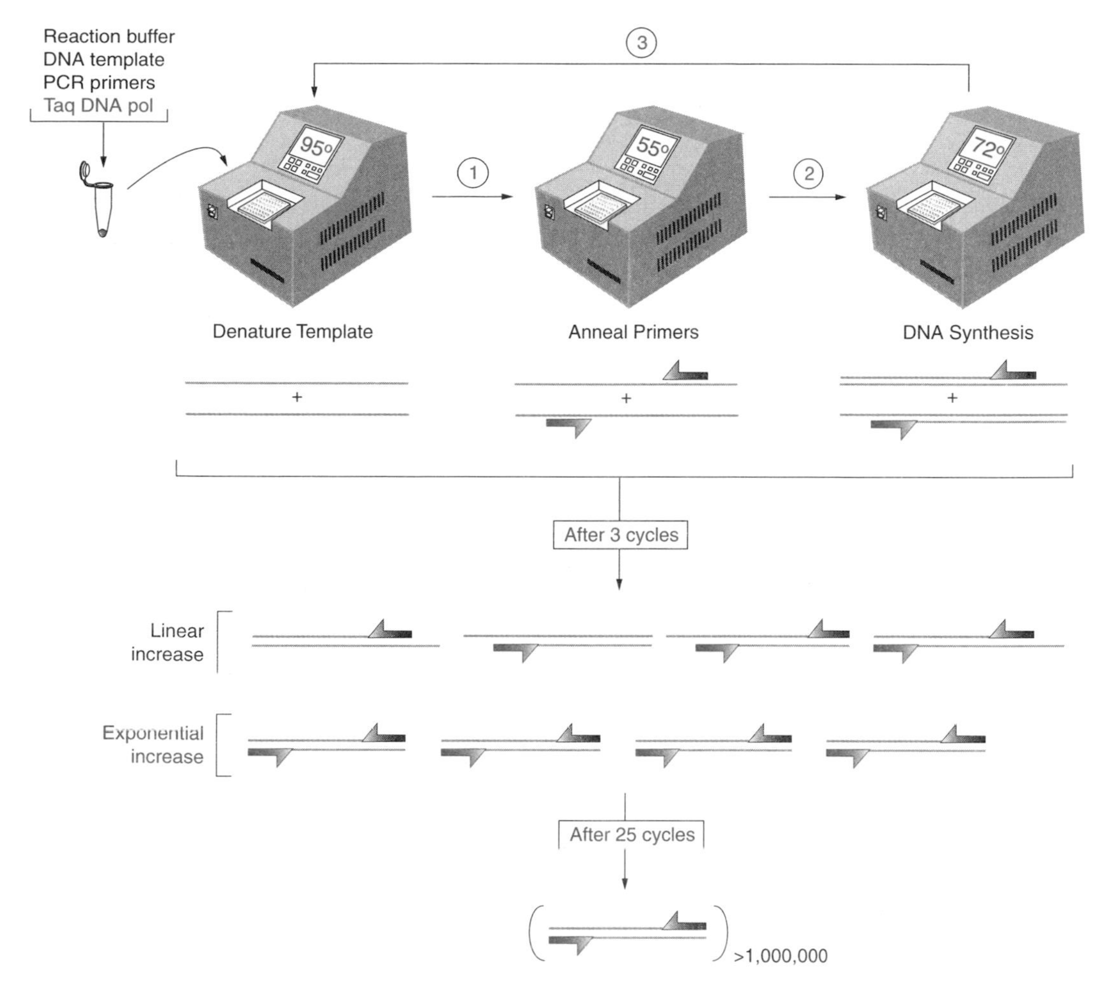

Figure 6.1 Each PCR cycle requires three temperature steps to complete one round of DNA synthesis. Before cycle 1 can be initiated, the double-strand DNA target must be heat-denatured to provide single-strand regions for primer annealing. Because the primer molecules are in a vast molar excess over the number of target DNA molecules, primer annealing is a rapid bimolecular reaction that is promoted by a temperature shift to ~55°C. The extension reaction is maximal at the thermal optimum for the DNA polymerase, which is 72°C. After three rounds of cycling, each template molecule in the reaction generates four complete PCR products that are flanked by primer sequences, as well as four partial PCR products. Complete PCR products accumulate exponentially until the reaction components become limiting (~30 cycles).

heating elements to change the temperature of a single block rapidly throughout the cycle. A third design, represented by the MJ Research DNA Engine™, utilizes the heating/cooling pump principle of the Peltier effect, which is based on heat exchange occurring between two dissimilar surfaces that are connected in series by electric current. All three of these thermal cycler designs function well for routine PCR applications.

Figure 6.2 shows the temperature profile for a typical PCR reaction cycle run on a thermal cycler. The ramp time and temperature accuracy are the most important parameters that must be controlled by the thermal cycler. A "hot start" is a method that can be used to increase primer specificity during the first round of DNA synthesis. In this procedure, the *Taq* polymerase is held inactive at temperatures below the annealing opti-

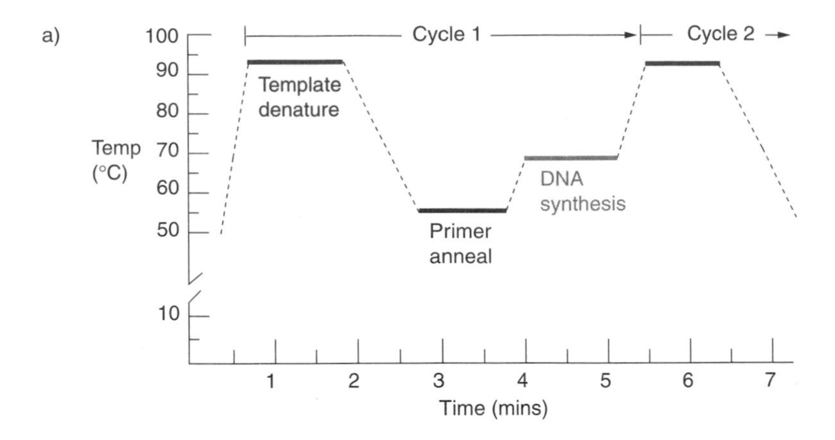

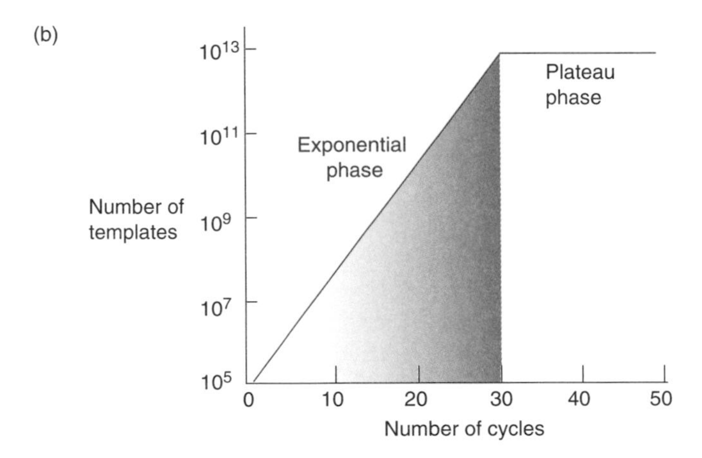

Figure 6.2 The temperature profile of a PCR cycle is controlled by the thermal cycler program, which results in a near exponential increase in PCR product accumulation for about the first 30 cycles. (*a*) Temperature profile of a typical PCR cycle. The dotted lines represent the ramping period which is usually Δ1°C per second with most thermal cyclers. (*b*) Accumulation of PCR target molecules as a function of cycle number. The exponential phase lasts for about 30 cycles under standard reaction conditions. The plateau phase results from limiting amounts of enzyme and reduced enzyme activity.

mum (<55°C) to avoid extending mismatched primers. One hot start technique uses a wax barrier to separate the *Taq* polymerase physically from the reaction mixture until the tube reaches 90°C, at which time the wax melts and the reaction components combine. The other strategy is to use a heat labile anti-*Taq* antibody that binds to the polymerase and inhibits activity at low temperatures. Figure 6.2 also shows a graph depicting the amount of PCR product produced as a function of cycle number. The PCR efficiency equation applies only to the exponential phase, which represents the linear range of the reaction when plotted on a log scale. The plateau effect results from enzyme limitations due to decreased enzyme activity from repeated exposure to 95°C and from stoichiometric limitations of active enzyme molecules, relative to DNA templates, within the time frame of the extension period.

PCR Requires a Thermostable DNA Polymerase

The enzymatic basis for PCR amplification (Fig. 6.1) is a DNA polymerization reaction that extends the annealed primers in the standard 5′ to 3′ direction. The original PCR protocol used the Klenow fragment of *E. coli* DNA polymerase I to perform the primer extension reaction; however, this meant that fresh enzyme had to be added after each round of denaturation because this DNA polymerase is easily heat-inactivated. A related problem with using an *E. coli* DNA polymerase is that the optimal activity level of the enzyme is 37°C, which greatly limits the specificity of the reaction owing to degenerate primer annealing at this low temperature. Both of these problems were solved by switching to a DNA polymerase that had been isolated from a thermophilic bacterium found near geothermal vents. Not only do these thermostable DNA polymerases retain activity for an extended period of time after repeated exposure to temperatures in excess of 90°C, but they are also fully active at ~75°C, which essentially eliminates degenerate heteroduplex formation.

The first commercially available thermostable DNA polymerase for PCR came from the thermophilic eubacterium *Thermus aquaticus*. More recently, several other thermostable DNA polymerase have been isolated and characterized that offer several advantages for specialized PCR assays. Table 6.1 lists some of the thermostable DNA polymerases that are commercially available for use in PCR assays. The biggest difference between these enzymes is whether or not they contain an inherent 3′ to 5′ exonuclease proofreading activity, which is important if the sequence of the PCR product must be error-free. DNA polymerases possessing 3′ to 5′ exonuclease activity, such as *Tli* from *Thermococcus litoralis,* and *Pfu* from *Pyrococcus furiosus*, have error rates that are 10–100 times lower than DNA polymerases lacking this activity. Thermostable DNA polymerases with DNA repair functions, such as *Taq,* which is similar to *E. coli* DNA polymerase I, encode a 5′ to 3′ exonuclease activity. Two other important parameters are processivity (number of nucleotides polymerized before template dissociation) and extension rate (nucleotides polymerized/second). Some thermostable DNA polymerases can use RNA templates as a substrate, which can be useful for PCR applications that require a separate cDNA synthesis reaction using viral reverse transcriptase. An example of this is the recombinant form of *Tth* polymerase (r*Tt*) from *Thermus thermophilus,* which can catalyze high-temperature reverse transcription of RNA in the presence of $MnCl_2$.

TABLE 6.1. Thermostable DNA polymerases differ in their enzymatic activities

Enzyme	Relative efficiency[a]	Error rate[b]	Processivity[c]	Extension rate[d]	3′ to 5′ exo	5′ to 3′ exo
Taq Pol	88	2×10^{-4}	55	75	no	yes
Tli Pol (Vent)	70	4×10^{-5}	7	67	yes	no
Pfu Pol	60	7×10^{-7}	n.d.	n.d.	yes	no
r*Tth*	n.d.	n.d.	30	60	no	yes

[a] Percent conversion of template to product per cycle.
[b] Frequency of errors per base pairs incorporated.
[c] Average number of nucleotides added before dissociation.
[d] Average number of nucleotides added per second.
n.d = not determined.

Design of PCR Primers

The amplification product of a PCR reaction is defined by the sequence of the PCR primers. The primers anneal to complementary sequences on the DNA template and thereby determine the boundaries of the amplified product. Typically, the DNA target can be of two types. The first type represents a general DNA region for which the precise location of the upstream and downstream PCR primers is flexible. An example would be when a genomic target sequence is being analyzed for potential mutations and the only requirement for primer site locations is that they flank the region to be sequenced. The second type of DNA target is highly specific in that one or both of the primers must be complementary to a specific target DNA sequence. This is usually the case when PCR is being used for cloning applications, such as in vitro mutagenesis methods or strategies involving coding sequence alterations. Optimal distance between primer pairs is also an important aspect of primer design. A good example of primer pair placement is using exon encoding primers that flank intronic sequence. The cDNA-derived PCR product of this reaction will reflect processed mRNA that would be smaller than the aberrant product derived from amplification of contaminating genomic DNA.

Once a DNA target has been chosen, there are several rules of thumb for primer design that are important to consider. These general rules follow.

PRIMER LENGTH

Choose primers that will anneal to complementary sequences that are 18–24 nucleotides long. This will maintain specificity (chance of finding the same sequence by chance is only 1 in 4^{20}) and provide sufficient base pairing for stable duplex formation. Note that many PCR primers have a stretch of nucleotides on the 5′ end of the primer that will not anneal to the template during the first cycle, for example, a restriction enzyme recognition sequence that gets incorporated into the PCR product. These modifying primers are usually 25–35 nucleotides long.

DUPLEX STABILITY

Both primers in a PCR reaction should have similar melting temperatures (T_m) to ensure that they will have the same hybridization kinetics during the template annealing phase. Primers with an overall G + C content of 45–55% are most common. The T_m of each oligonucleotide can be calculated using an empirical formula that takes into account base composition and the effect of nearest neighbor interactions (the T_m of AAG is different from AGA).

NONCOMPLEMENTARY PRIMER PAIRS

The two primers cannot share complementarity at the 3′ ends or else they will give rise to primer dimer products. For example, if one primer has the sequence 5′ – . . . GGCG – 3′ and the other primer terminal sequence is 5′ . . .CCGC – 3′, then they can form a short hybrid that will become a substrate for DNA synthesis. Once the primer dimer product is formed, it is a competing target for amplification.

NO HAIRPIN LOOPS

Each primer must be devoid of palindromic sequences that can give rise to stable intrastrand structures that limit primer annealing to the template DNA. Palindromic restriction enzyme sites located at the 5′ end of the primer are not a problem as long as they do not disrupt template annealing at the 3′ terminus.

OPTIMAL DISTANCE BETWEEN PRIMERS

This rule is very application specific, but for most diagnostic PCR assays, it is best when the opposing primers are spaced 150–500 bp apart. PCR assays with products in this size range are relatively easy to optimize. Longer PCR products can also be produced (up to 20 kb) using special "long distance" PCR conditions.

Fortunately, a variety of computer algorithms have been developed that take all of these parameters into account and the researcher only needs to provide a DNA target sequence. Figure 6.3 shows the results of running a PCR primer pair prediction program to find suitable primers for amplifying sequences within the hypothetical AMG gene.

(a) Analysis settings for AMG primer pair determination:

location: between 1200 and 1800
primer size: 24 - 25 nucleotides
Tm (°C): 55 - 75
% G+C: 45 - 55
product size: 400 - 600 bp

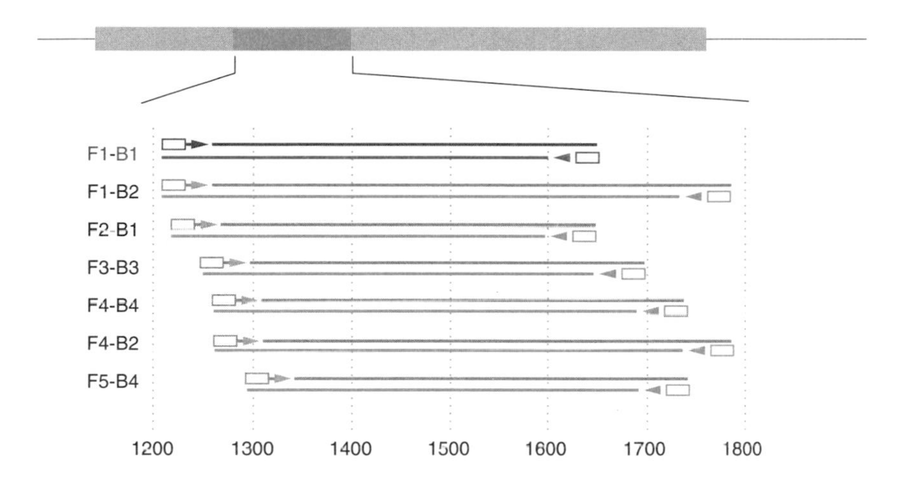

(b)

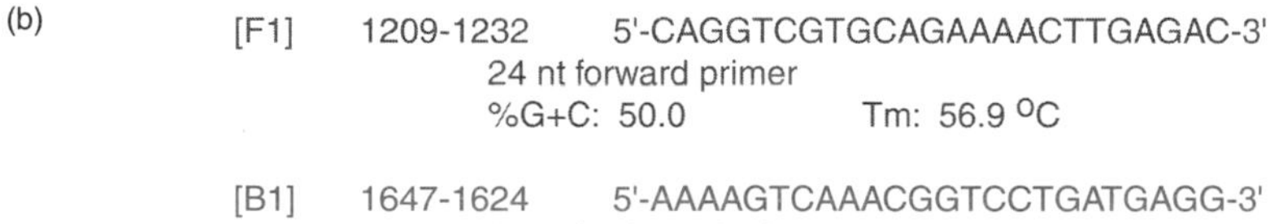

F1-B1 primer pair = 439 nt product from AMG region 1209-1647
optimal annealing temperature: 55.3 °C
%G+C: 42.8 Tm: 75.9 °C

Figure 6.3 Computer programs are often used to predict suitable PCR primer pairs for a given DNA target. The coding sequence of the hypothetical *AMG* gene was analyzed for optimal PCR primer pairs using the MacVector™ analysis program (Oxford Molecular Group). (*a*) A composite map of representative primer pairs for one region of the *AMG* sequence is shown. Note that multiple combinations of forward primers and backward primers are possible (e.g., F1-B1, F1-B2, F2-B2). (*b*) Primer sequences of individual oligonucleotides, predicted melting temperatures (T_m) of primers and PCR products, and predicted annealing temperature for optimal PCR amplification. The computed information for the F1-B1 primer pair is shown.

Optimizing a PCR Assay

A critical step in developing a new PCR assay is to optimize the reaction conditions to obtain maximum specificity and sensitivity. The primary reason for these optimization steps is to determine what deviations from the "standard" reaction conditions are necessary to promote functional primer annealing and extension. Through the trials and errors of early PCR researchers, a list of key optimization parameters were identified. The most common variables to test, and their effect on PCR specificity and sensitivity, are listed in Table 6.2. Without a doubt, the most critical parameter to vary when establishing a new PCR assay is to determine the optimal primer annealing temperature. Note that these optimization experiments must be performed under conditions where PCR products from a prior reaction do not contaminate subsequent assays. This is especially important in situations where the ultimate goal is to develop a diagnostic test. Two standard methods used to restrict template contamination problems are (1) performance of all PCR work in "clean" areas of the lab using dedicated pipettes and (2) employment of a target DNA decontamination step prior to each new round of PCR amplification.

The most common method used to analyze PCR products is gel electrophoresis. This is usually done with a high percentage mini-agarose gel that is stained with ethid-

TABLE 6.2. List of PCR reaction conditions that must be optimized

Parameter	Alteration	Effect on the reaction
Primer annealing temperature	Increase in temperature	Increases specificity of primer annealing by destabilizing base pair mismatches
	Decrease in temperature	Increases the sensitivity (and yield) of the reaction by stabilizing correct base pairing
DNA polymerase	Enzyme concentration	Enzyme concentrations affect the sensitivity and specificity; too little enzyme produces insufficient product and too much enzyme decreases specificity
	Type of DNA polymerase	*Taq* enzyme is the most efficient enzyme but it also has the highest error rate; in contrast, *Pfu* has a decreased error rate but synthesizes the least amount of product
Magnesium concentration	Varying the [$MgCl_2$]	Low [$MgCl_2$] increases specificity, high [$MgCl_2$] stabilizes primer annealing and can increase sensitivity, but can also decrease primer specificity
Cycle parameters	Denaturation temperature	Elevated denaturation temperature can increase sensitivity by allowing complete template denaturation, especially of G + C rich targets; however, *Taq* polymerase activity decreases rapidly above 93°C
	Duration time of primer extension	Longer primer extension times increase sensitivity in long distance PCR
	Cycle number	Assay sensitivity is determined by both the efficiency of the enzyme reaction and the initial number of DNA target molecules; it should only be necessary to increase the cycle number beyond 35 if the reaction contains $<10^3$ initial target molecules

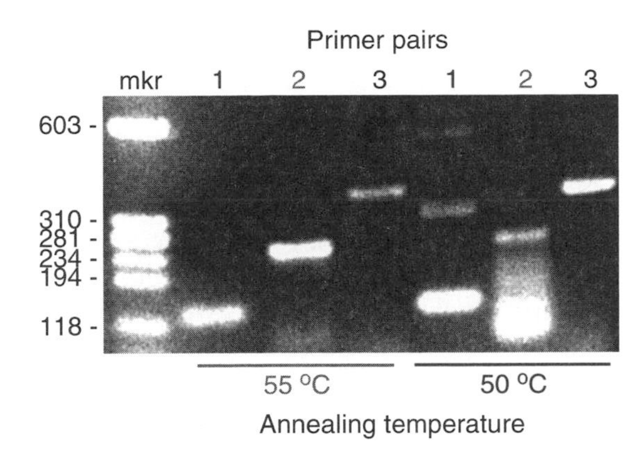

Figure 6.4 Agarose gel electrophoresis is used to characterize PCR products from a temperature optimization experiment. This photograph of an ethidium bromide stained gel illustrates the effect of temperature on product yield and reaction specificity. Primer pairs 1 and 2 generate higher levels of the correct PCR product at 55°C, whereas primer pair 3 functions best at 50°C.

ium bromide. Because the molecular weight of the PCR product is not always diagnostic, the reaction product can be characterized further by restriction enzyme cleavage, nucleic acid hybridization, or DNA sequencing. Figure 6.4 shows an agarose gel that was used to separate PCR products from an experiment performed to determine the optimal annealing temperature for three different primer pairs using the same DNA template.

REVERSE TRANSCRIPTASE-MEDIATED PCR

Amplification of double-strand DNA by PCR is not limited to genomic DNA targets; it can also be applied to double-strand cDNA that has been synthesized from RNA using a reverse transcription reaction. By combining the processes of cDNA synthesis and PCR amplification into a single application, called reverse transcriptase PCR (RT-PCR), it is possible to detect, and even isolate, very rare mRNA transcripts from small cell samples. As described in the preceding section, there are thermostable DNA polymerases such as r*Tth* that can utilize RNA templates for cDNA synthesis and thus permit single enzyme RT-PCR. Alternatively, one of the viral reverse transcriptase enzymes described in Chapter 5 (AMV RTase or MMLV RTase) can be used for the cDNA synthesis reaction, and then a second reaction is performed using one of the thermostable DNA polymerases. There are advantages and disadvantages to each of these methods and the deciding factor often depends on whether the RT-PCR reaction is being performed for a specialized research application or an automated diagnostic assay. At present, both classes of enzymes (viral RTases and r*Tth* DNA polymerase) are used for RT-PCR.

This section describes three basic molecular genetic strategies that utilize RT-PCR as a tool. The first approach, called rapid amplification of cDNA ends (RACE), is used to amplify segments of cDNA that extend in the 5′ or 3′ direction from a known sequence. The other two RT-PCR strategies, quantitative RT-PCR and differential cDNA cloning by RT-PCR, represent a large number of RT-PCR applications that loosely fit into these categories on the basis of the research objectives. Because there are too many variations of these approaches to describe in detail here, specific PCR protocols are identified by bold letters to emphasize that they are common "named" methods. Readers who would like more information about these strategies are directed to the reference section at the end of the chapter, which lists several excellent PCR protocol manuals that provide comprehensive explanations of each highlighted PCR application.

RACE: Rapid Amplification of cDNA Ends

As described in Chapter 5, cDNA synthesis and library screening can lead to the isolation of gene sequences that represent all or part of a processed mRNA transcript. It is not uncommon, however, to encounter significant problems when trying to isolate the 5′ end of low abundance transcripts because cDNA libraries often do not contain a full representation of all mRNA sequences. This is especially true of oligo dT-primed libraries. To circumvent this roadblock, RT-PCR can be used specifically to amplify sequences of a known gene using one gene-specific primer that is complementary to the previously cloned cDNA segment, and a second "anchored" primer that anneals to a sequence that has been covalently attached to the newly created cDNA terminus. This RT-PCR strategy was first described in 1988 by Michael Frohman and Gail Martin and was termed **RACE (rapid amplification of cDNA ends)**. The original protocol used a gene specific primer and polyadenylation of the extended cDNA at the 3′ end with terminal deoxytidyl transferase (TdT). Figure 6.5 outlines the basic Frohman and Martin RACE strategy. Note that 3′ RACE can also be performed by using anchored oligo-dT priming of mRNA to synthesize a cDNA target for PCR.

Several modifications of the original RACE protocol have been described. One of these, called **RNA ligase-mediated RACE** (RLM-RACE), involves the use of bacteriophage T4 RNA ligase to attach covalently a single-strand RNA anchor molecule to the decapped 5′ end of mRNA. First strand cDNA synthesis can then be performed using a gene-specific primer to produce a pool of cDNAs encoding the anchored primer sequence. A variation of this method is referred to as **ligation anchored PCR** (LA-PCR). In this strategy, T4 RNA ligase is used to attach a single-strand DNA molecule to the 3′ terminus of first strand cDNA (T4 RNA ligase can use either deoxyribonucleotides or ribonucleotides as substrates). A third type of adaptor-mediated RACE amplification is called **marathon RACE,** which uses standard first and second strand cDNA synthesis reactions (Chapter 5) to produce double-strand DNA molecules that are then ligated to oligonucleotide adaptors that serve as PCR priming sites. One advantage of marathon RACE is the increased efficiency of double-strand adaptor ligation by T4 DNA ligase, as compared to single-strand "anchor" primer ligation by T4 RNA ligase.

Quantitative RT-PCR

It is often important to know relative steady-state levels of specific gene transcripts as a means to investigate the role of a gene function in a particular cell phenotype. Chapter 5 described two common ways to measure steady-state mRNA levels, Northern blots and RNase protection assays. However, both of these assays are labor-intensive and may require up to 25 μg of total RNA for each assay. Moreover, many types of clinical studies have access to only small amounts of limiting tissue samples, and therefore most gene expression studies cannot be performed using conventional RNA-based assays. Researchers reasoned that with the proper internal controls, it should be possible to develop a quantitative RT-PCR assay that takes into account variation in the number of target molecules as well as amplification efficiency. We limit the discussion here to quantitative RT-PCR, but there are also diagnostic applications that use quantitative genomic DNA PCR to detect pathogens in patient samples.

Meaningful quantitative RT-PCR measurements require a standard RNA template against which the experimental RNA can be measured. Two types of RNA standards

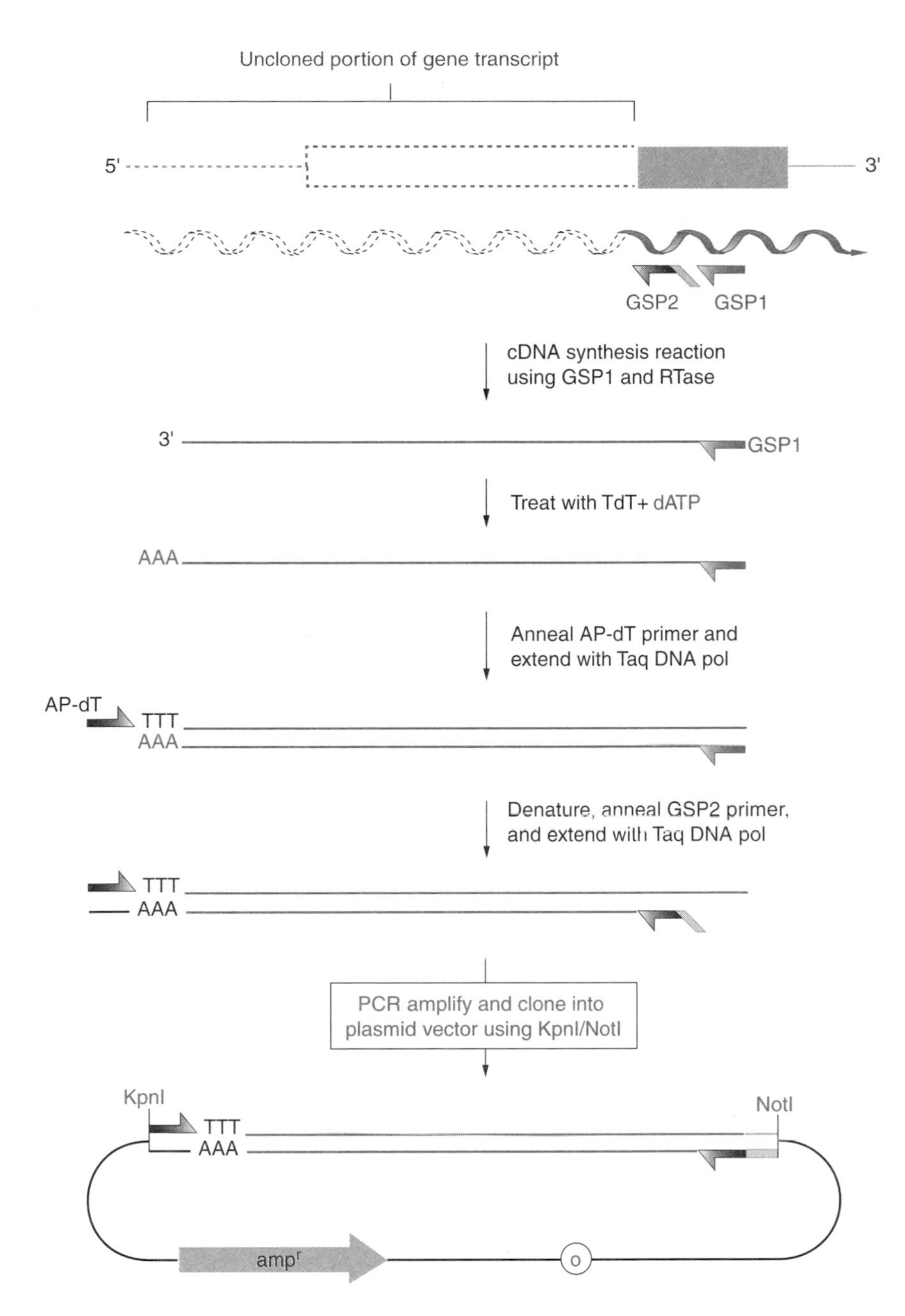

Figure 6.5 The Frohman and Martin RACE protocol utilizes a TdT-mediated polyadenylation reaction to create the upstream "anchored" cDNA priming site for PCR amplification. 5′ RACE is used to amplify uncloned upstream transcript sequences using a gene-specific primer (GSP1) during the reverse transcriptase reaction, and an oligo dT anchored primer (AP-dT) that anneals to the polyadenylated TdT extension of the cDNA. To increase specificity during PCR amplification, a "nested" gene-specific primer (GSP2) can be used in conjunction with an anchor primer lacking the oligo dT sequence (AP). Subcloning of the RACE product is facilitated by including restriction site sequences in the GSP and AP primers.

have been used for this purpose. One type is an endogenous gene product that is ubiquitously expressed in all cell types, for example, actin or glyceraldehyde-3-phosphate dehydrogenase (GAPDH), which are the same standards used in Northern blots and RNase protections assays (Chapter 5). The other type of standard is an exogenously added cRNA (complementary RNA) that is synthesized by in vitro transcription. These exogenous cRNAs are sometimes called RT-PCR "mimics" because they contain the same RT-PCR priming sites and overall sequence as the target RNA, but produce a PCR product that differs from the target RNA by a unique restriction site or shift in molecular weight. This makes it possible to compare directly the PCR product of the target RNA to the cRNA standard on the same agarose gel. Mimic cRNAs are transcribed from plasmids containing SP6 or T7 bacteriophage promoters and are constructed by in vitro mutagenesis of the cloned target cDNA.

There are two ways to use mimic cRNAs to measure the amount of target mRNA in a sample; one is called noncompetitive RT-PCR and the other is competitive RT-PCR. **Noncompetitive RT-PCR** is performed by adding mimic cRNA to the RNA sample at approximately the concentration of the target mRNA. By limiting PCR amplification to the linear range (midexponential phase), usually 25 cycles or less, it is possible to establish a direct correlation between the known amount of cRNA and the target mRNA. However, the absolute number of target mRNA molecules in the sample greatly affects the number of cycles needed to detect the product within the linear range. This is a significant problem if the noncompetitive quantitative RT-PCR reactions are going to be used for diagnostic applications in which the range of target mRNA varies between samples by more than about fivefold.

Competitive RT-PCR circumvents the problem of variable target copy number by using PCR product ratios that are independent of cycle number. This is done by performing a series of RT-PCR reactions that contain the same amount of sample RNA to which various known amounts of mimic cRNA have been added. The concentration of mimic cRNA that produces the same amount of mimic PCR product as the target RNA in the sample (a mimic:target product ratio of 1.0) represents the point at which both templates are competing equally for the primers. Figure 6.6 illustrates how competitive RT-PCR can be used to quantify the amount of *AMG* mRNA in a cellular RNA sample that has been doped with increasing amounts of an *AMG* mimic cRNA. Because competitive RT-PCR quantification is based on ratios, it is not necessary to restrict PCR amplification to the exponential phase, and therefore maximum sensitivity can be achieved by performing up to 35 cycles.

Amplification of Differentially Expressed Genes

Differential screening and subtraction hybridization are two techniques that can be used to isolate gene transcripts that are present at different levels in two RNA populations (Chapter 5). However, these approaches require large amounts of RNA to synthesize sufficient quantities of an enriched cDNA probe for library screening. In contrast, RT-PCR can be used to generate cDNAs from very small amounts of mRNA. This section describes the two basic RT-PCR cloning approaches that have been used to identify differentially expressed genes. The first type uses gel electrophoresis to survey randomly generated PCR products that display differential banding patterns owing to underlying differences in target copy number. The second strategy is based on C_0t hybridization

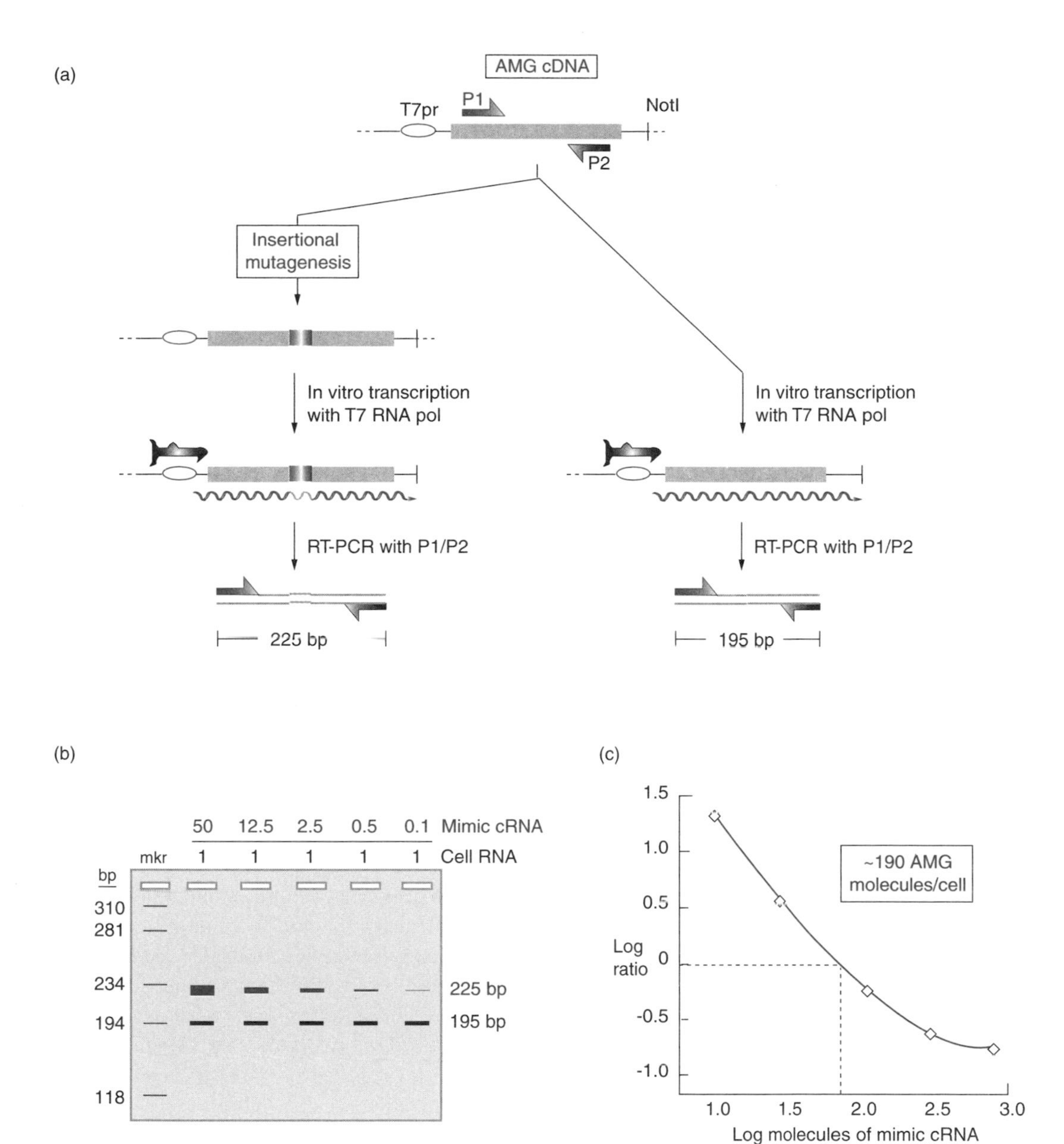

Figure 6.6 Competitive RT-PCR uses a mimic cRNA to quantify the amount of target mRNA in an experimental sample based on a ratio of PCR products. (*a*) Flow scheme showing how insertional mutagenesis can be used to construct an *AMG* mimic cRNA for use as an internal marker in RT-PCR reactions using *AMG* gene-specific primers (P1 and P2). The *AMG* mimic PCR product is 30 bp longer than the normal *AMG* PCR product. (*b*) Schematic representation of an agarose gel showing the relative amounts of RT-PCR products generated in reactions using a constant amount of cellular RNA and decreasing amounts of known in vitro synthesized AMG mimic cRNA. (*c*) Plot of $\log_{10}$ ratio of *AMG* mimic product/*AMG* target product versus $\log_{10}$ amount of *AMG* mimic cRNA added to the RT-PCR reaction expressed as molecules/cell. Assuming that both the *AMG* mimic and target templates compete equally for AMG primers, it can be deduced that there are ~190 AMG RNA molecules/cell based on the amount of *AMG* mimic RNA needed in the RT-PCR reaction to give a relative product ratio of 1.0 ($\log_{10}$ ratio = 0.0).

kinetics and adaptor primer differences to enrich underrepresented PCR products that are present at higher concentrations in one RT-PCR reaction, compared to another. As with quantitative RT-PCR, these PCR-based differential cloning techniques must be performed carefully in order to avoid aberrant amplification parameters that can significantly alter the abundance of some PCR products relative to the original RNA sample.

The technique of two-dimensional protein gel electrophoresis was developed by Pat O'Farrell as a rapid screening procedure for differential display of proteins that were metabolically labeled in vivo or in vitro. The appearance or disappearance of protein "spots" in the two-dimensional array could provide information about cellular changes in protein synthesis. Peng Liang and Arthur Pardee applied this same differential display approach to mRNA by developing a method to array RT-PCR products on standard DNA sequencing gels. Their technique, called **differential display reverse transcriptase PCR (DDRT-PCR),** provides the potential to clone "differential" RT-PCR cDNA products by physically excising radiolabeled bands from an acrylamide gel. The key to DDRT-PCR is the use of modified oligo dT primers for reverse transcription that anneal to a subset of poly A+ mRNAs owing to differences in dinucleotides at the 3′ end of the primer. Figure 6.7 illustrates how PCR amplification of cDNA subsets can be used in DDRT-PCR to identify putative differentially expressed mRNAs in RNA samples. Candidate RT-PCR products are excised from the gel, reamplified with the same primer pair, and used as probes on conventional Northern blots to verify that they correspond to differentially expressed gene transcripts. The false-positive rate with DDRT-PCR can be highly variable, and therefore, it is best described as a screening procedure rather than a cloning strategy. A similar RT-PCR screening technique called **RAP-PCR (RNA arbitrarily primed PCR)** has been described by John Welsh and Michael McClellend that is based on a genomic DNA fingerprinting strategy. The primary difference between these two methods is that RAP-PCR uses the same random primer for both reverse transcription and PCR amplification steps, which eliminates bias toward the noncoding 3′ poly A+ sequences.

Suppression subtraction hybridization (SSH) is a second PCR-based strategy that was developed by Paul Siebert and colleagues as a method to clone differentially expressed gene transcripts. Although it is technically not an RT-PCR approach, it does require the synthesis of double-strand cDNA as a starting point for exponential PCR amplification of gene transcripts present at higher concentrations in one of two mRNA pools. The basis for differential amplification in SSH is twofold. First, conditions are used that promote rapid reassociation kinetics with excess "driver" cDNA sequences to normalize two "tester" cDNA pools. Following this driver–tester hybridization reaction, the two tester pools are mixed and the unhybridized single-strand low abundance molecules are allowed to anneal under conditions that favor duplex formation. In the subsequent PCR amplification step, these unique "tester" duplexes are exponentially amplified by a pair of adaptor primers. Second, the PCR amplification of nondifferential cDNAs is suppressed because these undesirable sequences contain inappropriate priming sites, or are capable of forming intrastrand "panhandle" structures that are poor substrates for PCR. The combined PCR effect of exponential amplification of selective tester duplex cDNAs, with the resulting suppression of common sequence cDNAs, can potentially result in a 10^2–10^3-fold enrichment of differentially expressed tester cDNA sequences. The amplified cDNAs are used to make an enriched plasmid library which can be randomly sequenced or screened with a differential cDNA probe. Figure 6.8 outlines Siebert's SSH cDNA cloning strategy.

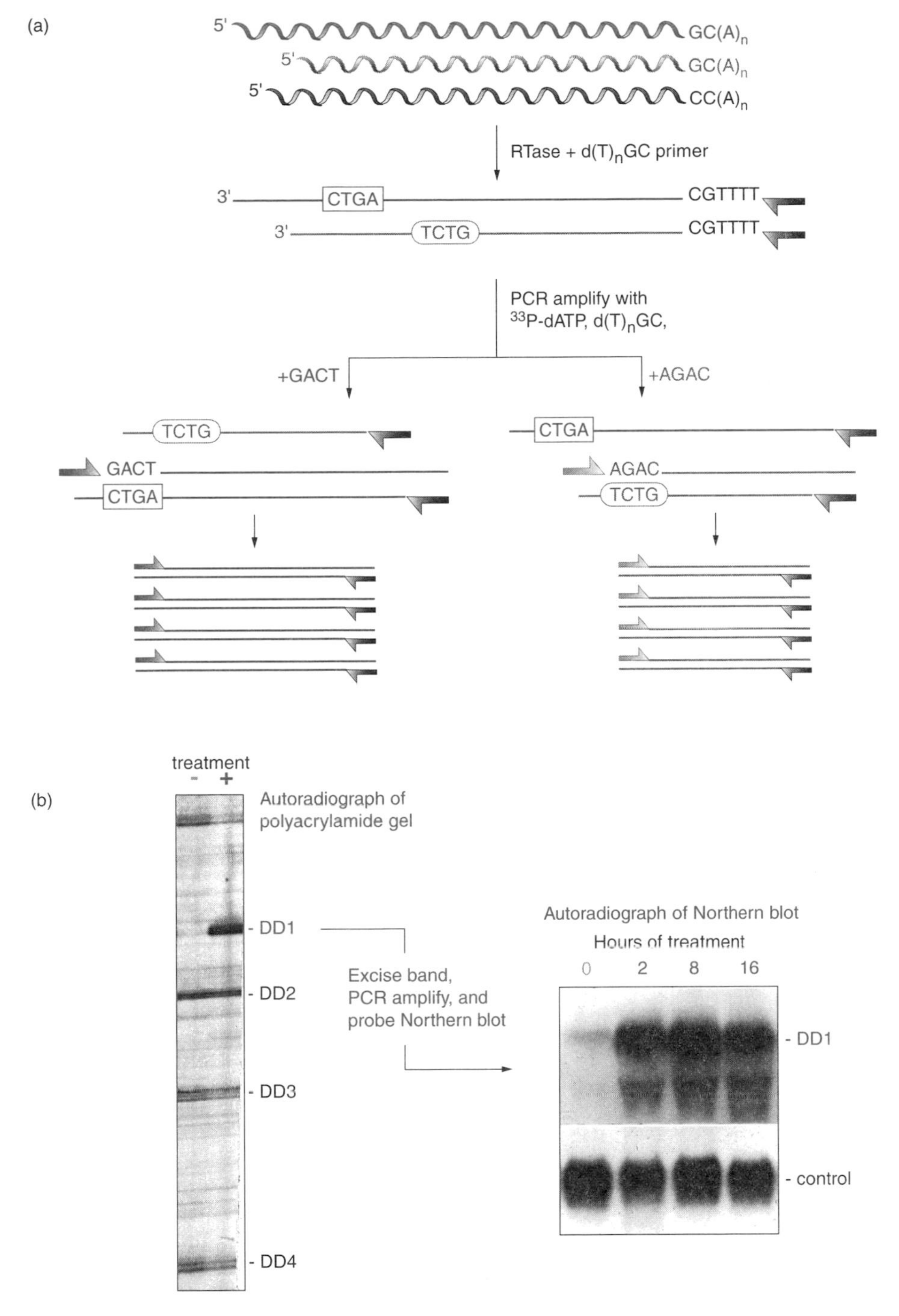

Figure 6.7 DDRT-PCR generates subsets of cDNA products based on differences in 3′ dinucleotides and the location of 5′ arbitrary priming sites for PCR. (*a*) Schematic diagram showing how a 5′-dT-GC-3′ primer is used to synthesize cDNA from a subset of RNAs containing the appropriate 3′ dinucleotide adjacent to the poly A tail. Of these first strand cDNAs, only a fraction will serve as appropriate templates for separate PCR reactions containing a known arbitrary primer. (*b*) A representative autoradiogram is shown on the left from a DDRT-PCR gel that contains reactions using RNA from cells that were untreated (–) or treated (+) with a steroid hormone. The representative Northern blot on the left demonstrates how a candidate DDRT-PCR product must be verified as a differentially expressed sequence using conventional RNA assays. In this example, the DD1 transcript displays a time-dependent increase in steady-state RNA levels following hormone treatment, whereas the control RNA levels are unaffected by steroid signaling.

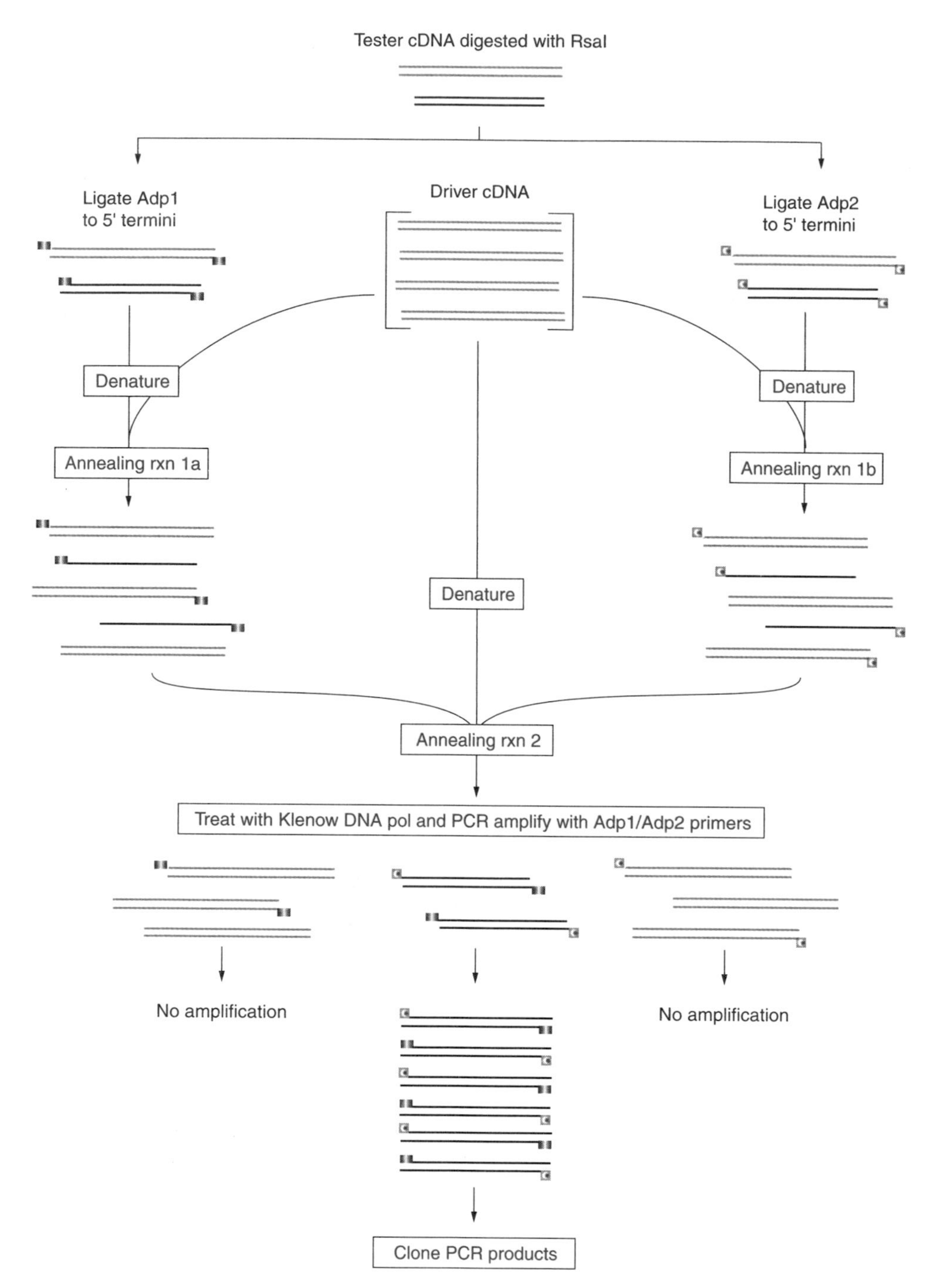

Figure 6.8 Suppression subtraction hybridization combines cDNA normalization with selective PCR amplification to achieve enrichment of differentially expressed gene transcripts. Double-strand cDNA is made from mRNA representing tester (containing gene transcripts of interest) and driver (lacks appreciable amounts of desired gene transcript) transcripts. The tester cDNA is first digested with the four-base cutting enzyme RsaI to create blunt-end termini and then divided into two equal portions and distinct primer adaptors (Adp1 and Adp2) are ligated to the two cDNA pools. In the next step, separate rapid reassociation hybridization reactions are performed (rxns 1a and 1b) with each tester cDNA pool and a mass excess of driver cDNA. Both pools of cDNAs are then mixed without prior denaturation and another aliquot of denatured driver cDNA is added to initiate a second hybridization reaction (rxn 2). In the last step, PCR is performed using the Adp1 and Adp2 primers selectively to amplify transcript sequences present in the tester cDNA pool.

DIAGNOSTIC APPLICATIONS OF PCR

Although PCR was developed by research scientists at a biotechnology company, its widespread use as a routine diagnostic tool has been limited. There are a number of reasons for this lag in PCR-based applications. First, there were problems adapting the conditions worked out for proof-of-principle PCR experiments as done in the Cetus laboratory to other applications that used novel primer pairs and crude template preparations. At the time, most labs used either manual temperature cycling methods or, at best, first-generation PCR machines that had insufficient ramping capabilities. In addition, cross-contamination of samples with PCR products was a continuing problem. Second, there were legal entanglements over PCR patent rights, and numerous litigations regarding reasonable terms for licensing agreements. Fortunately, by the early 1990s, extensive basic research revealed the critical reaction parameters needed to control the variability of PCR, and strategies were developed to restrict carryover contamination of PCR products severely. In this section, we describe three broadly defined applications of PCR to illustrate how this assay is now routinely used as a diagnostic tool.

Detecting Pathogens in Tissue Samples

Immunocytochemistry and histocytochemistry have long been used by pathologists to analyze cell phenotypes in histological specimens. These assays are amenable to standardization procedures and the reagents provide reproducible results for most general laboratory applications. However, a major limitation of these procedures is the inability to detect infectious agents such as viral pathogens and other microorganisms that persist at very low levels in infected cells. Compounding the sensitivity problem is the fact that most viruses undergo rapid evolution as a means to escape immune surveillance. When it became clear that PCR could offer increased specificity and sensitivity, researchers began to develop PCR strategies to detect as few as one viral genome in a tissue specimen. This PCR-based revolution in diagnostic methodology coincided with the worldwide outbreak of human immunodeficiency virus (HIV), which greatly accelerated the development process.

Two basic approaches have been taken to adopt PCR assays to pathogen detection procedures. The first has been to use quantitative solution PCR to detect HIV genomes in blood samples. The assays currently in use are based on the technique of **competitive quantitative PCR**, with an added emphasis on low level detection and strain identification. There are now PCR-based protocols in place that use strain-specific HIV primer pairs to detect latent viral infections that are well below the range previously achieved with direct and indirect immunological methods. The second approach has been to develop **in situ PCR (ISPCR)** procedures that allow the same level of sensitivity and specificity as solution PCR assays, with the advantage of cell-specific resolution. Although ISPCR is still considered more of a research tool than a diagnostic procedure because of the many problems associated with sample preparation, there are now several established protocols that demonstrate the utility of this method. Similar to solution PCR, the ISPCR amplification parameters must be optimized with regard to temperature, time, and reaction constituents. ISPCR products are detected by using labeled dNTPs (radioactive or biotinylated) in the reaction or by subsequently processing the ISPCR specimen using standard in situ hybridization conditions to detect the amplified hybridization target.

Identifying Genetic Mutations

Well over 150 human disease genes have been cloned and a large number of alleles have been identified. Obviously, it is not practical to clone and sequence 100 kb of genomic DNA each time an individual's genotype must be determined. Moreover, when a potential disease locus is identified by genome wide marker analysis, researchers need a way to scan the coding sequences of all genes in the region quickly to find disease-associated nucleotide alterations. The PCR technique of **single-strand conformational polymorphism (SSCP),** one of the most widely used methods for detecting single base pair changes in genomic DNA, is based on the principle that two single-strand DNA molecules of identical length but unlike sequence will migrate differentially in a nondenaturing acrylamide gel. This is due to sequence-dependent intramolecular folding reactions that contribute to the overall structure of the molecule. Single-strand conformational polymorphism is done by amplifying a specific DNA segment using a radiolabeled (or fluorescently labeled) primer, or by including trace amounts of ^{32}P-α-dNTP in the PCR reaction. The PCR products are then heat denatured and cooled on ice to promote rapid formation of intrastrand duplexes. Figure 6.9 shows how SSCP can be used to identify mutant alleles in known target DNA sequences.

Although SSCP is a rapid and easy procedure, studies have shown that it probably only detects about ~70% of possible mutations. Therefore other PCR-based methods have been developed to try and increase this efficiency. One such procedure is called **solid-phase chemical cleavage (SpCCM),** which detects cleavage products that result from chemical modification of mismatched bases that are formed when PCR products from control DNA and sample DNA are allowed to form heteroduplexes. Some studies have shown that SpCCM can be >90% efficient at detecting point mutations. Two other PCR-based methods for detecting single nucleotide differences are **base excision sequence scanning (BESS)**, which uses an enzymatic sequencing method to detect thymidine residues in PCR products, and **artificial mismatch hybridization (AMH)**, a method that exploits heteroduplex stability assays to detect single base pair mismatches.

PCR Genotyping Using Sequence Tagged Sites

As described in Chapter 4, STSs are defined as genomic DNA segments for which sequence information and physical mapping data are available. These markers can be defined by restriction fragment length polymorphisms (presence or absence of a restriction site), or more conveniently, by PCR primer pairs that can be used to generate distinct PCR products. Many of the STS markers that have been used for pedigree and forensic genotyping are VNTRs (variable nucleotide tandem repeats), STRPs, and SNPs (Chapter 4). PCR primers that specifically amplify across polymorphic repeat sequences can be used to determine genotype based on the pattern of maternal and paternal repeat length (see Fig. 4.8).

Another genotyping method is to analyze multiple unique sites in a single PCR reaction using a technique called **multiplex PCR**. In this method, a defined set of primer pairs is used simultaneously to amplify different genomic STSs that can subsequently be resolved by gel electrophoresis. Figure 6.10 illustrates how multiplex PCR can be used to screen for pathogenic strains of *E. coli* in food samples using strain-specific primer pairs. Multiplex PCR is also a convenient method to screen for DNA deletions across an entire gene by using exon-specific primer pairs.

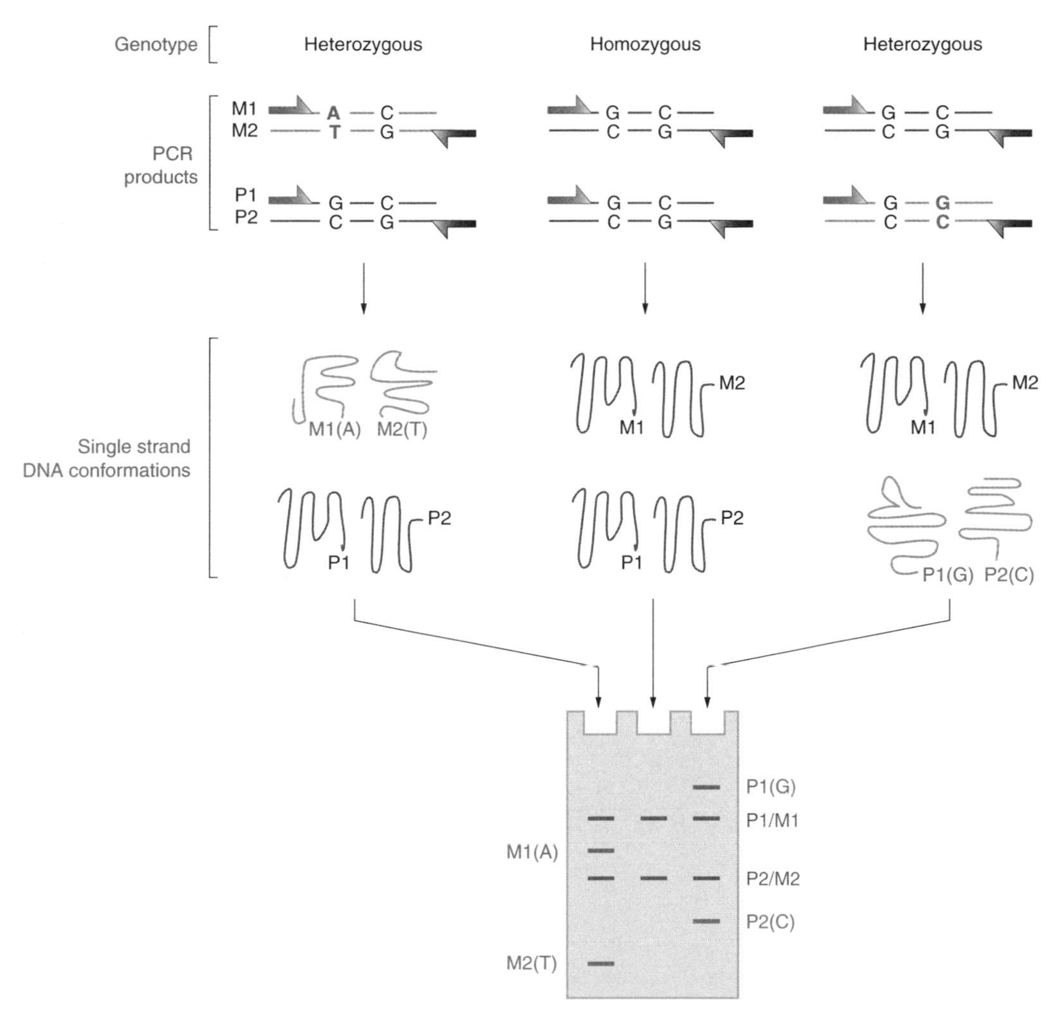

Figure 6.9 Point mutations in genomic DNA can be rapidly identified using the PCR technique of single-strand conformational polymorphism (SSCP). In this example, *AMG* primer pairs were used to detect nucleotide alterations in a specific region of the *AMG* gene known to co-segregate as dominant mutation. With DNA from three individuals (two disease heterozygotes and a homozygote normal), PCR amplification was performed with *AMG*-specific primers that flank a region where multiple dominant alleles have previously been identified. Because the primary nucleotide sequences of each of the DNA strands in a heterozygote individual are different, four separate bands are resolved in the corresponding lanes. Maternal (M1, M2) and paternal (P1, P2) DNA strands are shown.

LABORATORY APPLICATIONS OF PCR

PCR is a valuable laboratory tool that is used routinely for tasks that can be done other ways, but that are easier and quicker when using PCR. To illustrate further the versatility of PCR in applied molecular genetics, this last section gives a brief overview of PCR applications that are used in the lab. By the mid 1990s most all labs performing applied molecular genetic methods had ready access to PCR machines, and it was at this time that more and more researchers began finding ways to adapt their old cloning or screen-

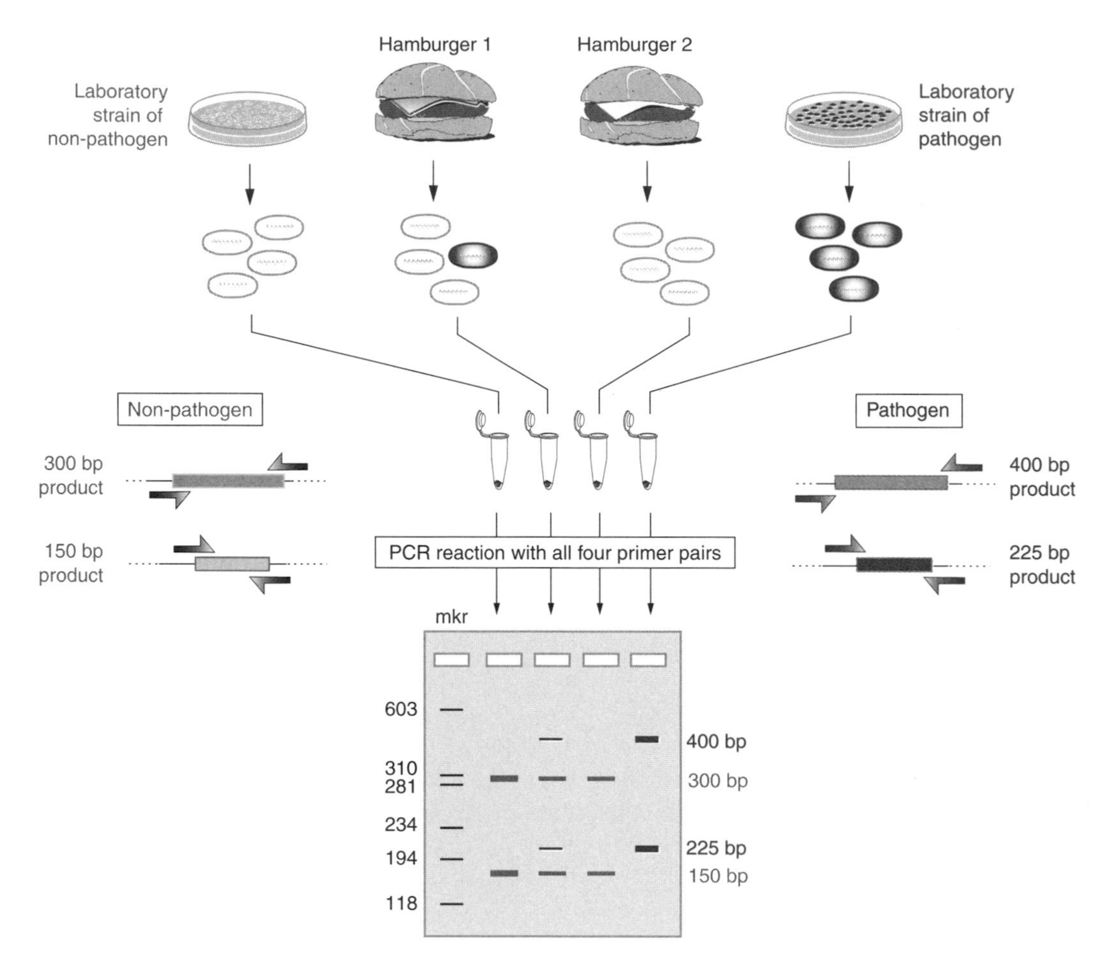

Figure 6.10 Multiplex PCR can be used simultaneously to screen for pathogenic and nonpathogenic strains of *E. coli* using strain-specific primer pairs. By using at least two strain-specific primer pairs for each organism, and replicate sample reactions, it is possible to detect reliably the presence of ≥10 genome equivalents. In this example, two food samples were analyzed for the presence of a specific pathogenic strain of *E. coli*. By using appropriate control samples for the two bacterial strains, it can be seen that hamburger 1 is contaminated.

ing strategies to PCR-based methods. It was not uncommon to hear, "Why not use PCR to do that, it would be much easier and quicker." This is even more true today because many of the screening procedures used in genomics are automated and incorporate one or more PCR amplification steps to minimize the amount of starting material required. Although only a few laboratory applications of PCR are described here, there are literally hundreds of variations of these published methods. Two general applications have been chosen as representative examples: (1) subcloning DNA targets using PCR and (2) PCR-mediated in vitro mutagenesis.

Subcloning DNA Targets Using PCR

There are numerous situations in which a specific DNA segment must be cloned into a plasmid vector, but there are no convenient restriction enzyme sites to facilitate the design of a subcloning strategy. Examples include moving the coding sequence of a cloned cDNA into an expression vector to accommodate transcriptional or translational start sites, or inserting a functional gene unit, such as a drug-selectable marker or a transcriptional promoter, into a newly constructed cloning vector. Alternatively, the isolation and characterization of a new gene may have been reported, but for one reason or another, the DNA reagent is not readily available. If the DNA sequence of this gene is available in the GenBank database, a researcher will simply design a primer pair to amplify directly the desired DNA target using a convenient source of target DNA or cDNA.

Three strategies are commonly used to subclone PCR products (Fig. 6.11). The first is **restriction site addition,** which is done by including the nucleotide sequence of an appropriate enzyme into the 5′ end of the primer. The restriction enzyme recognition site does not interfere with template annealing during the first PCR cycle. However, all subsequent cycles contain PCR products that have the restriction enzyme site incorporated into the DNA termini. By digesting the PCR products with the corresponding enzyme, the amplified DNA segment can be subcloned by a standard ligation reaction. The second cloning method, called **T/A cloning,** relies on the fact that thermostable DNA polymerases lacking 3′ to 5′ exonuclease activity (*Taq* and *Tth*; see Table 6.1) misincorporate a single unpaired deoxynucleotide into the 3′ end of PCR products. For reasons not totally understood, the additional deoxynucleotide is often dATP. This fortuitous one-base overhang facilitates ligation to plasmid vectors containing a thymidine nucleotide (TMP or TTP) overhang. Although not as efficient as restriction site addition PCR cloning, T/A cloning is easier because the PCR product can be ligated directly to the vector without additional purification steps or enzyme cleavage reactions. The third PCR cloning method is **blunt-end ligation,** which is required if a proofreading thermostable DNA polymerase is used and restriction sites have not been added to the primer. Blunt-ligated recombinants can be identified by loss of β-galactosidase activity using a blue-white colony screening protocol.

PCR-Mediated in vitro Mutagenesis

Mutagenic PCR is a random mutagenesis technique that exploits the elevated error rate of *Taq* polymerase in the presence of $MnCl_2$ and high $MgCl_2$. Mutagenic PCR is similar to random chemical mutagenesis with the added advantage that the chosen primer pair provides a convenient way to target the random base pair mutations to a defined segment of DNA. The development of optimal reaction conditions for **long-distance PCR** made it possible to design site-specific mutagenesis strategies that exploit unique primer designs to direct the amplification of entire plasmids. One example of this is to use **inverse PCR** to amplify a plasmid with a pair of tail to tail oligonucleotides, one of which contains a mismatched nucleotide. Following amplification, the mutated linear product is purified using agarose gel electrophoresis, recircularized, and used to transform *E. coli*. Strategies have also been developed that permit the construction of unique gene fusions using consecutive PCR reactions that utilized the PCR products of one reaction to prime DNA synthesis during the second amplification, as illustrated in Figure 6.12. This technique is sometimes called **PCR-mediated gene SOEing** (splicing by overlap extension). In addition to creating gene fusions, this method also provides a PCR-based approach for inserting nucleotide mutations in to the middle of a gene coding sequence. This is done by using a mutagenic primer in the first set of reactions that includes an overlap sequence for gene splicing in the subsequent PCR reaction.

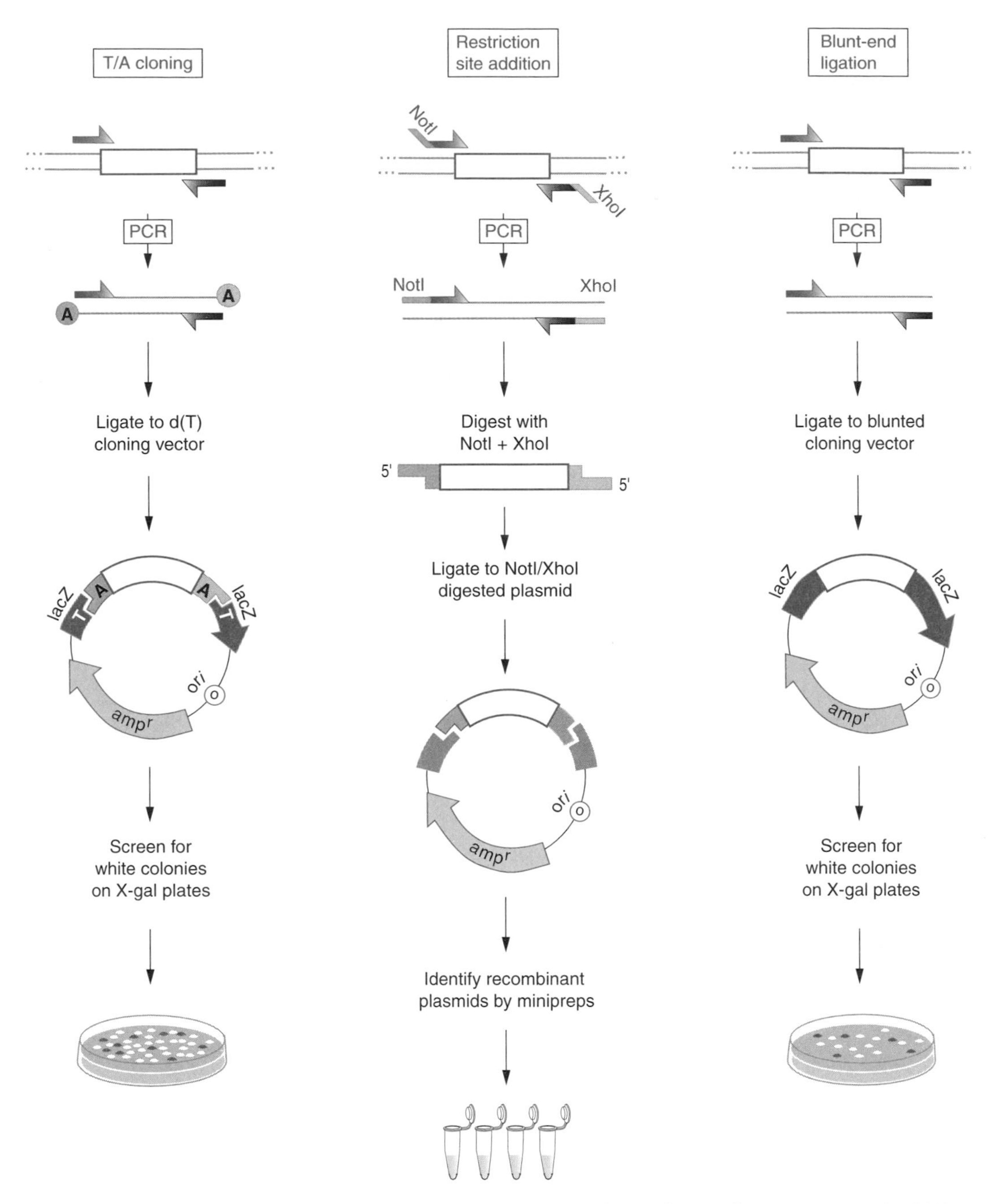

Figure 6.11 PCR products can be cloned in a variety of ways depending on the sequence at the DNA termini. The most common method is restriction site addition, which utilizes PCR primer pairs that incorporate restriction enzyme recognition sites into the 5′ end of the oligonucleotide. This method is often required for plasmid expression vectors, which usually lack the *lacZ* coding sequences needed for blue-white colony screening and require plasmid DNA analysis to identify recombinants. A second method is to take advantage of the propensity of *Taq* polymerase spuriously to add a dATP residue to the 3′ terminus of extended chains. A specially prepared dT cloning vector is often used that includes *lacZ* coding sequences for colony screening. Similarly, blunt-end cloning of PCR products can be done using a *lacZ* plasmid vector that has been digested with an appropriate enzyme such as SmaI.

Laboratory Practicum 6. *Cloning cell-specific transcripts using RT-PCR*

Research Objective

A large food processing conglomerate has recently acquired a small biotechnology company that specializes in molecular genetic neurobiology. The gustatory research team at the parent food company found that humans can taste many more flavors than previously known. The gustation researchers used high resolution human taste tests to map three new flavor-associated neuronal impulses to defined regions of the tongue. Based on data from olfactory research showing that neuronal cell functions often reflect developmental differences in the expression of outer membrane proteins, the molecular neurobiologists intend to clone putative flavor-detecting receptors using tongue needle biopsies from paid volunteers. The research team has decided to use RT-PCR and differential screening to identify genes expressed exclusively in the three distinct lingual regions.

Available Information and Reagents

1. Needle biopsies have been collected from the three mapped flavor zones (TX, TY, and TZ) using high resolution taste tests that were performed on five volunteers. Cell samples (~50 cells/sample) have been prepared from all 15 biopsies (TX1-5, TY1-5, and TZ1-5).
2. An EcoRI-ApaI-oligo dT primer, called EAT_{24}, with the nucleotide sequence 5′-ATTGGAATTCCTGAGGGCCC$(T)_{24}$ -3′, has been purified for use in the RT-PCR reaction and 10 μg of EcoRI-cleaved λZapII vector has been treated with calf intestinal alkaline phosphatase.
3. A cDNA probe corresponding to the 3′ end of *Tng1* (a tongue-specific developmental marker) has been prepared to use as hybridization control in Southern blots that will be used to verify putative flavor zone-specific transcripts identified by differential cDNA library screening.

Basic Strategy

The researchers hypothesize that flavor-specific receptors will be expressed at moderately high levels (~1% of the mRNA) in cells from the three lingual regions and that the 3′ sequence of the TX, TY, and TZ receptors will be different. Because the high resolution taste tests may not be precise due to subject variability, they will separately prepare RT-PCR products from each of the 15 samples to prevent cross-contamination. With methods worked out to clone cell-specific pheromone receptors, the EAT_{24} primer will be used to initiate first strand cDNA synthesis in cell extracts prepared with a reverse transcription lysis buffer containing RNase inhibitors. The 3′ end of single-strand cDNA products will be poly(A)-tailed with terminal deoxynucleotidyl transferase (TdT), and 30 cycles of PCR will then be performed with the EAT_{24} primer under appropriate buffer conditions. The RT-PCR cDNA products will be split into three portions: (1) cDNA that will be cleaved with EcoRI (the recognition site is GAATTC) and cloned into the λZapII vector to create 15 separate cDNA libraries; (2) cDNA that will be combined into separate TX, TY, or TZ cDNA pools and used for differential library screening; and (3) cDNA that will be saved for use as volunteer-specific cDNA probes on Southern blots.

A small amount of each λZap cDNA library will be used to infect *E. coli* cells, which will be plated at low plaque density. Triplicate filter lifts will be made from each master plate and screened with cDNA probes representing pooled TX, TZ, and TY RT-

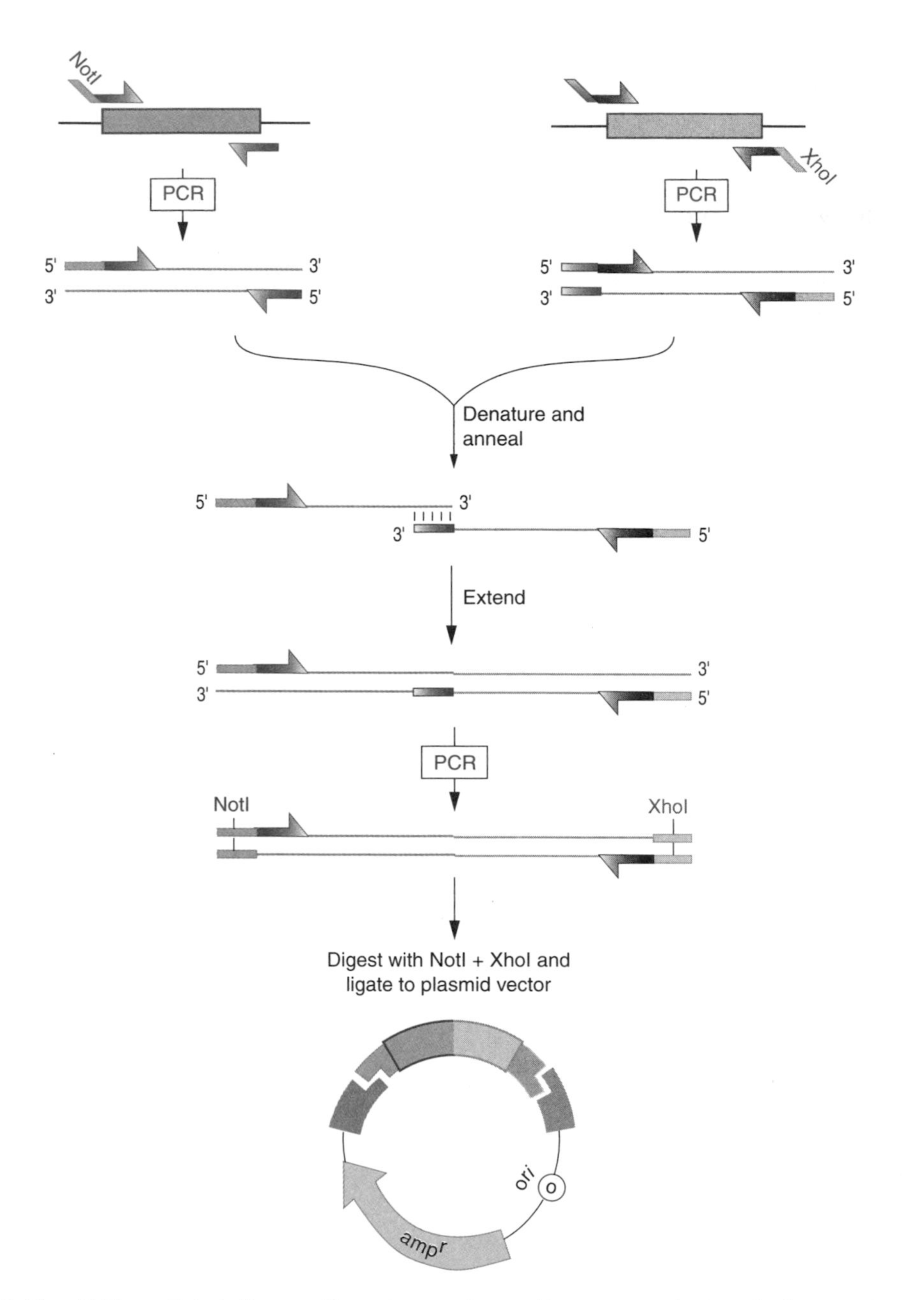

Figure 6.12 PCR-mediated site-specific mutagenesis provides a convenient method to create gene fusions and nucleotide substitutions rapidly. Novel gene fusions can be created by performing an initial pair of PCR reactions that introduce complementary DNA sequences at the ends of two sequences that will be spliced together. The overlapping segment functions as a bridge to generate a new template for a second amplification reaction that uses the two most outside primers from the initial reactions. The net result is a spliced PCR product that can be directly inserted into a cloning vector through the addition of restriction enzyme recognition sequences to the primers.

PCR products. Plaques that hybridize most strongly with the homologous, but not heterologous, T-mixed probes will be isolated (e.g., plaques from a TX library that hybridize with the TX mixed probe but not the mixed TY or TZ probes). Inserts from candidate recombinant phage will be recovered from phagemids following filamentous phage f1-mediated rescue, digested with ApaI restriction enzyme, and used for Southern blots that will be hybridized with separate cDNA probes. Figure 6.13 outlines the strategy that will be used to distinguish TX-specific phage inserts, and Figure 6.14 shows how Southern blot hybridization patterns can allow identification of putative flavor-determining gene sequences.

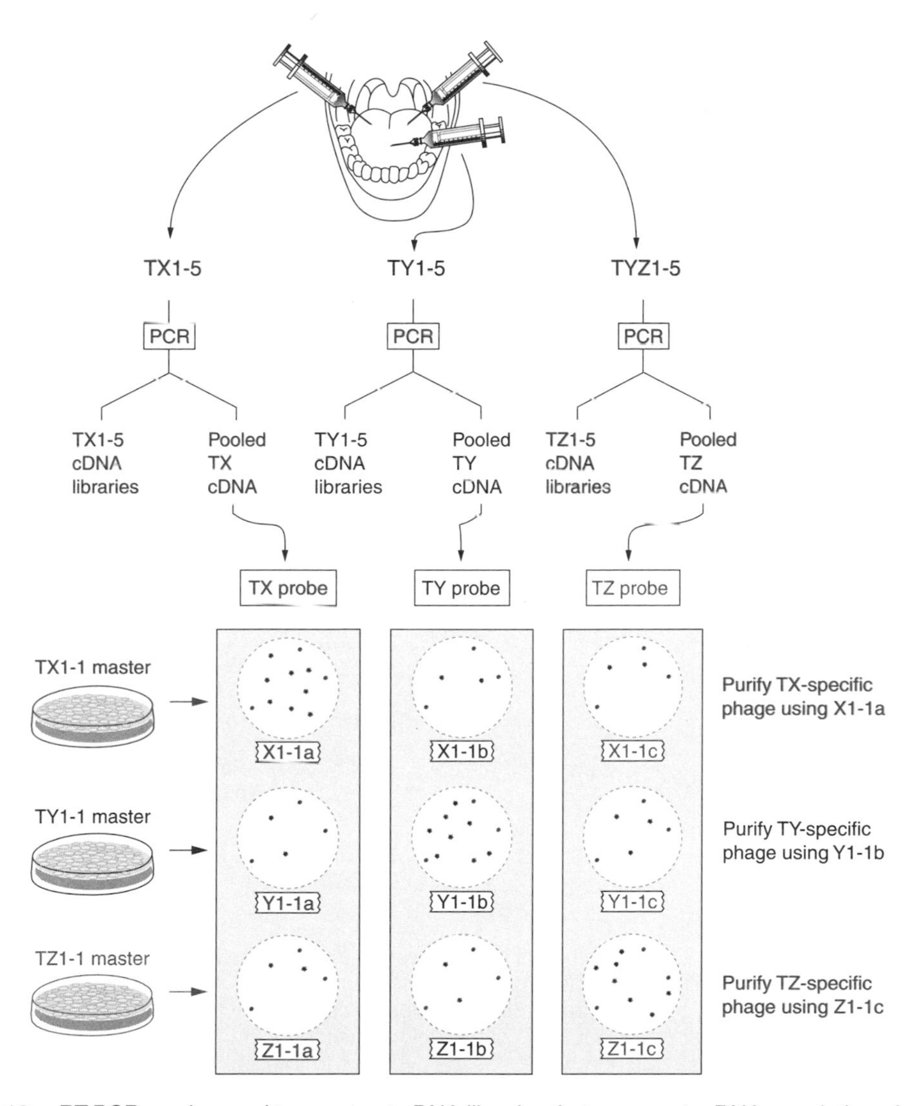

Figure 6.13 RT-PCR can be used to construct cDNA libraries that represent mRNA populations from single cells. Flow scheme showing how five samples of three representative cell types can be used differentially to screen single-cell RT-PCR cDNA libraries. Initial differential plaque screening is used to identify candidate clones that hybridize preferentially to pooled probes.

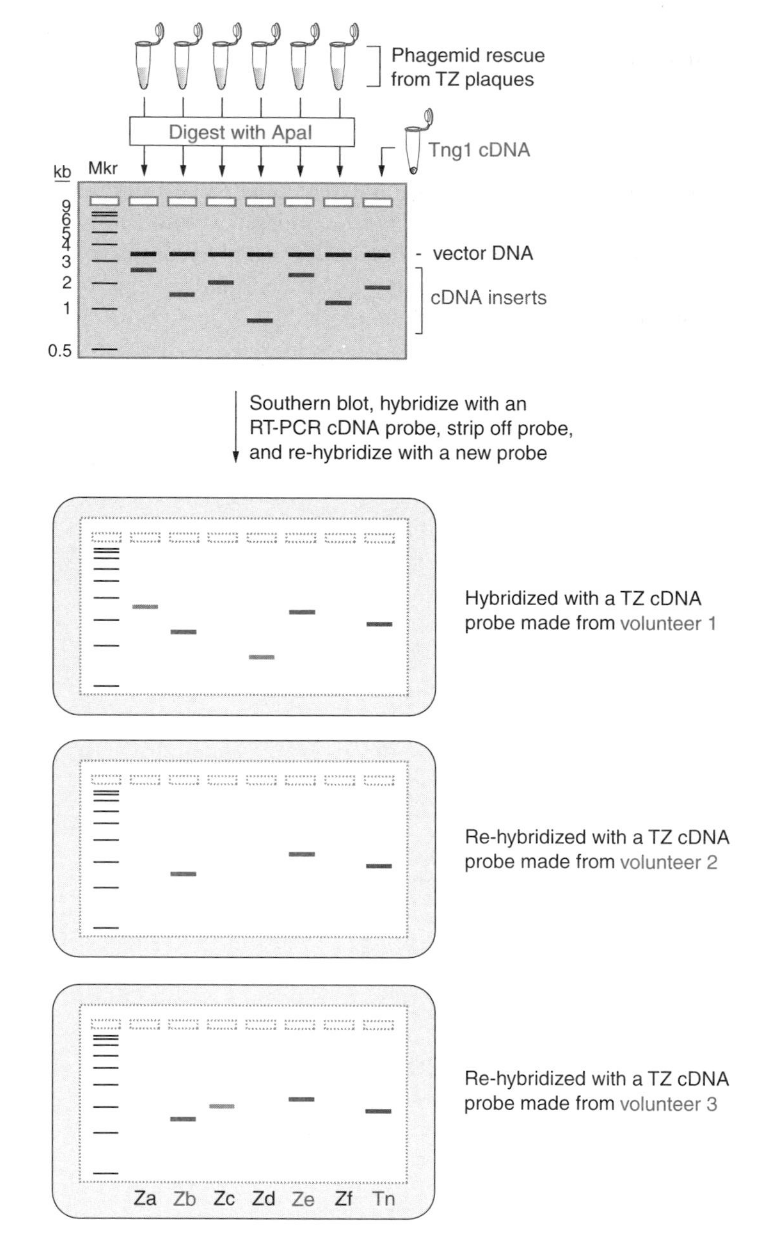

Figure 6.14 Candidate inserts are rescreened by Southern blot filters that are sequentially hybridized to RT-PCR probes from individual volunteers. The *Tng1* cDNA sequence is used as an internal control on these Southern blots to confirm that the individual RT-PCR cDNA samples contain sufficient levels of this ubiquitous tongue transcript. As shown in this example, the TZ1.Zb and TZ1.Ze inserts hybridize to transcript probes present in all three volunteer TZ samples, suggesting that these correspond to cell-specific gene products. The other four TZ1.Z cDNAs (Za, Zc, Zd, Zf) appear to represent less specific transcripts, for the complementary transcript probes are present in only one or none of the three volunteers.

Comments

The use of RT-PCR as a starting point for single-cell cDNA library construction has been used successfully to isolate cell-specific odorant and pheromone receptors. There are three major assumptions that must be met for the strategy outlined in Figure 6.13 to work. First, the TX, TY, and TZ cell types have to contain one or more gene transcripts that are totally unique for the differential screening to be informative. Second, these putative cell-specific transcripts have to be expressed at sufficiently high levels to permit screening of only a small portion of the library and to constitute a moderate amount of the probe (the example of 1% is what had been found for odorant receptors). Third, the most significant difference in gene expression profile must be directly related to the distinct cell phenotype, because there are no in vitro assays that can be performed to test for "taste transducing" genes.

Prospective

The most meaningful cDNA sequence information would be derived from the gene transcript ORF, and therefore additional cDNA cloning would be necessary because the RT-PCR products in this scheme are biased to the 3′ terminus (this strategy uses oligo dT primer to amplify poly A+ mRNA selectively). Two methods could be used to obtain 5′ coding sequence. One would be 5′ RACE as illustrated in Figure 6.5, and the other would be simply to screen a standard cDNA library. It would also be important to identify the putative TX, TY, and TZ cell types in the human tongue using in situ hybridization with the cell-specific cDNAs. This would have to be done with pathology samples, and depending on the sensitivity of the assay and type of sample, in situ RT-PCR may be required. The longer-term objective of a project like this would likely require the isolation of mouse TX, TY, and TZ genes for the development of genetic models to dissect the relative contribution of each receptor subtype to flavor "perception."

REFERENCES

Introduction

Audic, S., Beraud-Colomb, E. Ancient DNA is thirteen years old. Nature Biotech. 15:855–858, 1997.

*Jeffreys, A.J., Wilson, V., Neumann, R., Keyte, J. Amplification of human minisatellites by the polymerase chain reaction: Towards DNA fingerprinting of single cells. Nucl. Acids Res. 16:10953–10971, 1988.

*Kleppe, K., Ohtsuka, E., Kleppe, R., Molineux, I., Khorana, H.G. Studies on polynucleotides. XCVI. Repair replication of short synthetic DNA's as catalyzed by DNA polymerases. J. Mol. Biol. 56:341–361, 1971.

*Mullis, K.B., Faloona, F. Specific synthesis of DNA in vitro via a polymerase-catalyzed chain reaction. Methods Enzymol. 155:335–350, 1987.

Paabo, S. Ancient DNA: Extraction, characterization, molecular cloning and enzymatic amplification. Proc. Natl. Acad. Sci. USA 86:1939–1943, 1989.

Rabinow, P. Making PCR: A story of biotechnology. University of Chicago Press, Chicago, 190 pp., 1997.

*Saiki, R.K., Scharf, S., Faloona, F., Mullis, K., Horn, G., Erlich, H.A., Arnheim, N. Enzymatic amplification of β-globin genomic sequences and restriction site analysis for diagnosis of sickle cell anemia. Science 230:1350–1354, 1985.

* Landmark papers in applied molecular genetics.

Basic PCR methodology

Dieffendbach, C., Dveksler, G., eds. PCR primer: A laboratory manual. Cold Spring Harbor Press, Cold Spring Harbor, NY, 701 pp., 1995.

Eckert, K.A., Kunkel, T.A. High fidelity DNA synthesis by the *Thermus aquaticus* DNA polymerase. Nucl. Acids Res. 18:3739–3752, 1990.

Innis, M., Gelfand, D., Sninsky, J., eds. PCR strategies. Academic Press, New York, 373 pp., 1995.

Kellogg, D.E., Rybalkin, I., Chen, S., Mukhamedova, N., Vlasik, T., Siebert, P.D., Chencik, A. TaqStart antibody: "Hot start" PCR facilitated by a neutralizing monoclonal antibody directed against Taq DNA polymerase. BioTechniques 16:1134–1137, 1994.

Lawyer, F.C., Stoffel, S., Saiki, R.K., Myambo, K., Drummond, R., Gelfand, D.H. Isolation, characterization, and expression in *Escherichia coli* of the DNA polyermase gene from *Thermus aquaticus*. J. Biol. Chem. 264:6427–6437, 1989.

*Saiki, R.K., Gelfand, D.H., Stoffel, S., Scharf, S.J., Higuchi, R., Horn, G.T., Mullis, K.B., Erlich, H.A. Primer-directed enzymatic amplification of DNA with a thermostable DNA polymerase. Science 239:487–491, 1988.

Reverse transcriptase-mediated PCR

*Becker-Andre, M., Hahlbrock, K. Absolute mRNA quantitation using the polymerase chain reaction (PCR): A novel approach by a PCR aided transcript titration assay (PATTY). Nucl. Acids Res. 17:9437–9446, 1989.

*Diatchenko, L., Lau, Y., Campbell, A., et al. Suppression subtractive hybridization: A method for generating differentially regulated or tissue-specific cDNA probes and libraries. Proc. Natl. Acad. Sci. USA 93:6025–6030, 1996.

Dostal, D.E., Rothblum, K., Baker, K. An improved method for absolute quantitation of mRNA using multiplex polymerase chain reaction. Anal. Biochem. 223:239–250, 1994.

*Frohman, M.A., Dush, M.K., Martin, G.R. Rapid production of full length cDNAs from rare transcripts: Amplification using a single gene-specific oligonucleotide primer. Proc. Natl. Acad. Sci. USA 85: 8998–9002, 1988.

*Gilliland, G., Perrin, S., Blanchard, K., Bunn, H.F. Analysis of cytokine mRNA and DNA: Detection and quantitation by competitive polymerase chain reaction. Proc. Natl. Acad. Sci. USA 87:2725–2729, 1990.

Liang, P., Averboukh, L., Pardee, A.B. Distribution and cloning of eukaryotic mRNAs by means of differential display: Refinements and optimization. Nucl. Acids Res. 21:3269–3275, 1993.

*Liang, P. Pardee, A.B. Differential display of eukaryotic messenger RNA by means of the polymerase chain reaction. Science 257:967–971, 1992.

Matz, M. Usman, N., Shagin, D., Bogdanova, E., Lukyanov, S. Ordered differential display: A simple methods for systematic comparison of gene expression profiles. Nucl. Acids Res. 25: 2541–2542, 1997.

Ohara, O., Dorit, R.L., Gilbert, W. 1989. One-sided polymerase chain reaction: The amplification of cDNA. Proc. Natl. Acad. Sci. USA 86:5673–5677.

Schaefer, B.C. Revolutions in rapid amplification of cDNA ends: New strategies for polymerase chain reaction cloning of full-length cDNA ends. Anal. Biochem. 227:255–273, 1995.

Siebert, P., Larrick, J. Competitive PCR. Nature 359:557–558, 1992.

Souaze, F., Ntodue-Thome, A., Tran, C.Y., Rostene, W., Forgez, P. Quantitative RT-PCR: Limits and accuracy. Biotechniques 21:280–285, 1996.

Von Stein, O., Thies, W-G., Hofmann, M. A high-throughput screening for rarely transcribed differentially expressed genes. Nucl. Acids Res. 25:2598–2602, 1997.

Zhang, J., Byrne, C. A novel highly reproducible quantitative competitive RT PCR system. J. Mol. Biol. 274:338–352, 1997.

Diagnostic applications of PCR

Bagasra, O., Hansen, J. In situ PCR techniques. Wiley-Liss, New York, 142 pp., 1997.

Bej, A., McCarty, S., Atlas, R.M. Detection of coliform bacteria and Escherichi coli by multiplex polymerase chain reaction. Appl. Environ. Microbiol. 57:1473–1479, 1991.

Collier, M.C., Stock, F., DeGirolami, P., Samore, M., Cartwright, C. Comparison of PCR-based approaches to molecular epidemiologic analysis of Clostridium difficile. J. Clin. Microbiol. 34:1153–1157, 1996.

Cotton, R., Campbell, R. Chemical reactivity of matched cytosine and thymine bases near mismatched and unmatched bases in a heteroduplex between DNA strands with multiple differences. Nucl. Acids Res. 17:4223–4233, 1989.

*Edwards, A, Civitello, A., Hammond, H., Caskey, C.T. DNA typing and genetic mapping with trimeric and tetrameric tandem repeats. Am. J. Human Genet. 49:746–756, 1991.

Findlay, I., Taylor, A., Quirke, P., Frazier, R., Urquhart, A. DNA fingerprinting from single cells. Nature 389:555–556, 1997.

Guo, Z., Liu, Q., Smith, L.M. Enhanced discrimination of single nucleotide polymorphisms by artificial mismatch hybridization. Nature Biotech. 15:331–335, 1997.

Hawkins, G.A., Hoffman, L.M. Base excision sequence scanning. Nature Biotech. 15:803–804, 1997.

Kopp, M., de Mello, A., Manz, A. Chemical amplification: Continuous-flow PCR on a chip. Science 280:1046–1048, 1998.

*Kwok, S., Mack, D.H., Mullis, K.B., Poiesz, B., Ehrlich, G., Blair, D., Friedman-Kien, A., Sninsky, J.J. Identification of human immunodeficiency virus sequences by using in vitro enzymatic amplification and oligomer cleavage detection. J. Virol. 61:1690–1694, 1987.

*Li, H., Gyllensten, U.B., Cui, X., Saiki, R.K., Erlich, H.A., Arnheim, N. 1988. Amplification and analysis of DNA sequences in single human sperm and diploid cells. Nature 335:414–417.

Ochman, H., Gerber, A.S., Hartl, D.L. 1988. Genetic applications of an inverse polymerase chain reaction. Genetics 120:621–623.

Oehlenschlager, F., Schwille, P., Eigen, M. Detection of HIV-1 RNA by nucleic acid sequence-based amplification combined with fluorescence correlation spectroscopy. Proc. Natl. Acad. Sci. USA 93:12811–12816, 1996.

O'Leary, J., Chetty, R., Graham, A., McGee, J. In situ PCR: Pathologist's dream or nightmare? J. Pathol. 178:11–20, 1996.

*Orita, M., Suzuki, T., Sekiya, T., Hayashi, K. Rapid and sensitive detection of point mutations and DNA polymorphisms using the polymerase chain reaction. Genomics 5:874–879, 1989.

Pang, S., Koyanagi, Y., Miles, S., Wiley, C., Vinters, H.V., Chen, I.S.Y. High levels of unintegrated HIV-1 DNA in brain tissue of AIDS dementia patients. Nature 343:85–89, 1990.

Sheffield, V., Beck, J., Kwitek, A., Sandstrom, D., Stone, E. The sensitivity of single-strand conformation polymorphism analysis for the detection of single base substitutions. Genomics 16:325–332, 1993.

Laboratory applications of PCR

Aslandis, C., de Jong, P.J. Ligation-independent cloning of PCR products (LIC-PCR). Nucl. Acids Res. 18:6069–6074, 1990.

*Barnes, W. PCR amplification of up to 35-kb DNA with high fidelity and high yield from λ bacteriophage templates. Proc. Natl. Acad. Sci. USA 91:2216–2220, 1994.

Cadwell, C., Joyce, G. Randomization of genes by PCR mutagenesis. PCR Methods Appl. 2: 28–33, 1992.

*Cheng, S., Fockler, C., Barnes, W., Higuchi, R. Effective amplification of long targets from cloned inserts and human genomic DNA. Proc. Natl. Acad. Sci. USA 91:5695–5699, 1994.

Costa, G., Grafsky, A., Weiner, M.P. Cloning and analysis of PCR-generated DNA fragments. PCR Methods Appl. 3:338–345, 1994.

Ho, S., Hunt, H., Horton, J., Pullen, J., Pease, L. Site-directed mutagenesis by overlap extension using polymerase chain reaction. Gene 77:51–59, 1989.

Horton, R., Cai, Z., Ho, S., Pease, L. Gene splicing by overlap extension: Tailor made genes using the polymerase chain reaction. Biotechniques 8:528–535, 1990.

Leung, D., Chen, E., Goeddel, D. A method for random mutagenesis of a defined DNA segment using a modified polymerase chain reaction. Technique 1:11–15, 1989.

*Marchuck, D., Drumm, M., Saulino, A., Collins, F.S. Construction of T-vectors: A rapid and general system for direct cloning of unmodified PCR products. Nucl. Acids Res. 19:1154, 1991.

Rashtchian, A. Novel methods for cloning and engineering genes using the polymerase chain reaction. Curr. Opin. Biotechnol. 6:30–36, 1995.

Rashtchian, A., Buchman, G., Schuster, D., Berninger, M. Uracil DNA glycosylase-mediated cloning of PCR-ampified DNA: Application to genomic and cDNA cloning. Anal. Biochem. 200:91–97, 1992.

Rice, G., Goeddel, D., Cachianes, G. et al. Random PCR mutagenesis screening of secreted proteins by direct expression in mammalian cells. Proc. Natl. Acad. Sci. USA 89: 5467–5471, 1992.

Cloning cell-specific transcripts using RT-PCR

*Brady, G., Barbara, M., Iscove, N. Representative in vitro cDNA amplification from individual hemopoetic cells and colonies. Methods Mol. Cell. Biol. 2:17–25, 1990.

Brady, G., Billia, F., Knox, J., et al. Analysis of gene expression in a complex differentiation hierarchy by global amplification of cDNA from single cells. Current Biol. 5:909–922, 1995.

*Dulac, C., Axel, R. A novel family of genes encoding putative pheromone receptors in mammals. Cell 83:195–206, 1995.

Herrada, G., Dulac, C. A novel family of putative pheromone receptors in mammals with a topographically organized and sexually dimorphic distrubution. Cell 90:763–773, 1997.

Matsunami, H., Buck, L. A multigene family encoding a diverse array of putative pheromone receptors in mammals. Cell 90:775–784, 1997.

Ming, D., Ruiz-Avila, L., Margolskee, R. Characterization and solubilization of bitter-responsive receptors that couple to gustducin. Proc. Natl. Acad. Sci. USA 95:8933–8938, 1998.

SECTION

3

SPECIALIZED APPLICATIONS

CHAPTER

7

EXPRESSION OF CLONED GENES IN CULTURED CELLS

Molecular genetic methods used to characterize cloned genes fall into several basic categories: gene expression surveys, transcriptional regulation studies, ectopic expression experiments, and protein characterization assays. Methods required for gene expression surveys are described in Chapters 5 and 6. These include Northern blots, RNase protection assays, competitive RT-PCR, and in situ hybridization studies. Unlike earlier chapters, the material in this chapter is organized according to broadly defined research objectives, rather than the utilization of similar reagents and methodologies. In this way the relevant methods are introduced within the context of a research plan. The four research objectives in this chapter are (1) characterization of molecular determinants of gene regulation, (2) analysis of gene function in transfected cell lines, (3) exploitation of yeast as a single-cell eukaryotic model to investigate gene function, and (4) high level protein production in cultured cells. Laboratory practicum 7 ties many of these concepts together by describing how a researcher would go about characterizing a recently cloned promoter-selective transcription factor using transient transfection assays.

GENE REGULATION STUDIES

A major focus of molecular genetic studies during the 1980s was to understand the molecular basis of regulated gene expression. One reason for this was that many of the genes being cloned at the time were isolated on the basis of a unique developmental pattern of expression, and therefore much of the early work was to understand what determinants controlled this expression pattern. The emphasis on investigating gene regulatory mechanisms also arose because DNA binding assays were amenable to protein purification schemes, leading to the isolation and characterization of numerous transcription factors. During this "transcription-centric" time, a large number of experimental approaches were developed that provided powerful molecular genetic tools and sensitive gene expression assays. Many of these molecular genetic applications are now used to control the expression of heterologous genes that can be introduced into cultured cells and to regulate gene expression in whole organisms (Chapter 8). This first section gives an overview of the most commonly used gene expression vectors and promoter mapping strategies that have been used to investigate gene regulatory mechanisms. The term "DNA transfection" is used to refer to nonviral methods of introducing purified DNA into cells. This section also describes "antisense genetic" techniques that have been developed to inhibit gene function in vivo through the use of transcriptional interference approaches at the transcriptional and post-transcriptional levels.

Expression Vectors

Accurate transcriptional initiation requires DNA sequence elements that serve as binding sites for transcription factors. Broadly speaking, the regulatory regions of most genes contain two classes of transcription factor binding sites. The first type is those required for basal, or minimal, transcription of the linked gene. A well characterized DNA element of this type is the TATA box binding protein (TBP) recognition site that is located ~30 nucleotides upstream of the transcriptional start site. For genes that contain TATA box sequences, it is thought that TBP binding to the TATA box helps direct RNA polymerase II to the correct start site of transcription. The second class of transcription factor binding sites is those that function to recruit regulatory proteins to the promoter complex, which for lack of a better term we call enhancer binding proteins (EBPs). These special transcription factors serve to activate or repress the basal rate of transcriptional initiation. The molecular mechanism by which EBPs are able to modulate initiation rates at the transcription start site is not completely clear, but it is known that critical protein–protein interactions between the EBPs and the multi-subunit transcription initiation complex are important.

Expression vectors are a type of cloning vector that contains transcriptional promoter sequences on the 5′ side of the multiple cloning site. Similar to endogenous genes, expression vectors can contain DNA sequences that direct low level basal transcription or, more commonly, can have promoter elements linked to upstream regulatory sequences that serve as recognition sites for EBPs. Low level basal activity promoters, sometimes called minimal TATA-containing promoters, work in most all eukaryotic cells, whereas others are cell-specific owing to the requirement for cell-specific transcription factors. Cell-specific promoters are more commonly used in transgenic organismal studies (Chapter 8), because cultured cells often represent undifferentiated cell types. Expression vectors containing binding sites for modulatory EBPs come in two varieties. The first type are high activity constitutive expression vectors that usually contain potent viral regulatory sequences. The second type are represented by "on/off" regulated expression vectors that are controlled by ligand-regulated EBPs. Figure 7.1 summarizes the functional differences between these various types of expression vectors, and Table 7.1 lists representative gene promoters that are used to construct eukaryotic expression vectors. Note that expression vectors also contain transcriptional termination signals for polyadenylation to provide proper RNA processing signals.

Reporter Genes

It is not always convenient to monitor gene expression by assaying for the level of mRNA transcript, or even the protein product. This is especially true if the primary research objective is to identify gene regulatory sequences by testing a large number of deletion and point mutants in an in vivo cell transfection assay. One way to accelerate

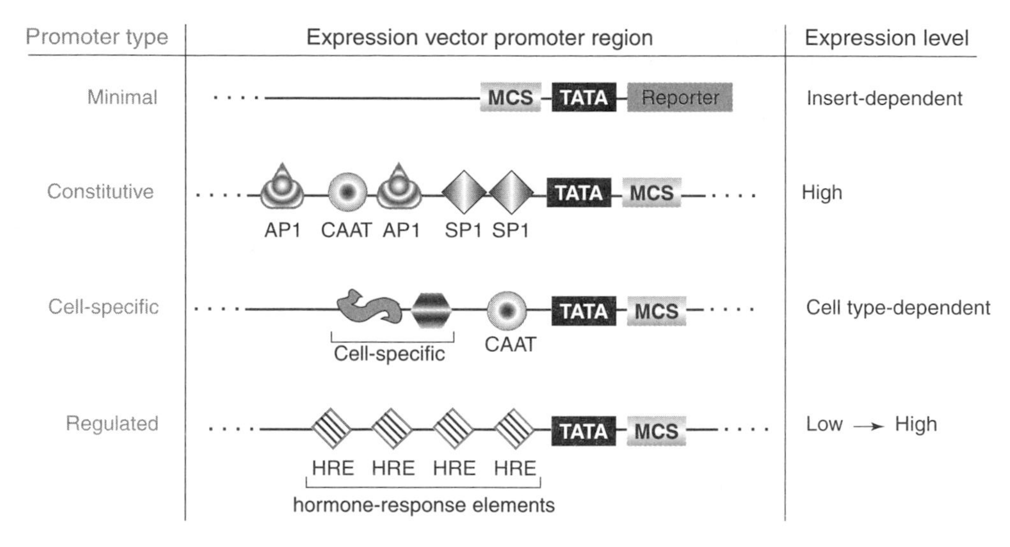

Figure 7.1 Expression vectors contain DNA sequences flanking the multiple cloning site (MCS) that function as sequence-specific binding sites for transcription factors. Four basic types of expression vectors used in gene expression studies in cultured cells and transgenic organisms are (1) minimal promoters used primarily as reporter genes that contain an MCS upstream of the TATA box for insertion of transcriptional regulator sequences, (2) strong constitutive promoters for high level gene expression in most all cell types, (3) cell-specific promoters used to target expression in selected cell types, and (4) regulated promoters that can be modulated by stimulus-activated transcription factors, such as intracellular hormone receptors. Common DNA sequence motifs found in many types of promoters are represented by SP1, CAAT, and AP1 binding sites.

the rate at which promoter mapping studies can be done is to create an artificial "reporter gene" by fusing the regulatory region of the test gene to a heterologous gene coding sequence that directs the synthesis of a readily detectable protein product. Assuming that the steady-state level of the reporter protein in the transfected cell is directly correlated to the steady-state level of reporter mRNA, then it is possible to use protein-based reporter gene assays for promoter mapping experiments. Figure 7.2 illustrates the principle of using reporter gene assays to demonstrate that a cloned transcriptional regulatory region functions as a cell-specific promoter. Chapter 4 describes how promoter mapping can be performed using in vitro mutagenesis and reporter gene assays (see Fig. 4.15).

Assuming that the reporter gene assays will be performed under conditions where the steady-state levels of mRNA and protein are directly correlated, reporter genes must also meet several other criteria if they are going to be reliable and useful for promoter mapping studies. First, the reporter gene must not encode a protein activity that is similar to one already present in the target cell. For example, if the reporter gene encodes an enzyme that is detected by a substrate modification assay, then the target cell must be devoid of that substrate modification activity. Second, the protein assay must be sensitive over several orders of magnitude, reproducible, and easy to perform. In this regard, it is also helpful if the assay is cost-effective. Third, the reporter protein function must not interfere with host cellular processes in a way that will alter intracellular signaling pathways or metabolic rates. Fortunately, a number of reporter genes have been developed that encode proteins fulfilling all of these criteria. A brief description of the most widely used reporter genes follows:

TABLE 7.1. Representative gene promoters commonly used for gene expression studies in cultured cells and transgenic animals[a]

Promoter type	Promoter name	Gene name	Molecular genetic application
Minimal promoters	Deleted *Drosophila* alcohol dehydrogenase promoter	*adh*	A ubiquitous low level promoter that is used to construct reporter genes
	Herpes simplex virus thymidine kinase promoter	*HSVtk*	A low level activity promoter used to direct the synthesis of marker genes in mammalian cells
High activity constitutive promoters	Cytomegalovirus immediate early promoter	*CMV*	High level gene expression in mammalian cells
	Simian virus 40 early enhancer/promoter	*SV40*	Moderately high level gene expression in mammalian cells
	Yeast alcohol dehydrogenase	*Adh1*	High level gene expression in yeast
	Cauliflower mosaic virus 35S promoter	*CaMV*	High level gene expression in plants
	Drosophila tubulin α1 promoter	*Tubα1*	Moderately high level gene expression in *Drosophila* cells
Cell-specific promoters	Whey acidic protein promoter	*WAP*	Targeted expression of genes to mammary tissue in animals
	Lymphocyte-specific tyrosine kinase promoter	*lck*	Targeted expression of genes to mouse thymocytes for immunological studies
	Tobacco root-specific gene promoter	*TobRB7*	Targeted expression of cloned genes to root cells in transgenic plants
Regulated promoters	Mouse mammary tumor virus long terminal repeat enhancer/promoter	*MMTV*	Steroid-regulated gene expression in mammalian cells
	Drosophila heat shock 70 promoter	*hsp*	Heat-regulated gene expression in *Drosophila* cells
	Yeast *GAL1-GAL10* promoter region	*GAL*	Galactose regulation of gene expression in yeast cells

[a]Artificial promoters that have been designed for regulated gene expression studies and special expression vectors used primarily for protein production in cultured cells are described below in this chapter.

CHLORAMPHENICOL ACETYL TRANSFERASE (CAT).
This is a bacterial enzyme that catalyzes the transfer of acetyl groups from acetyl-coenzyme A to the antibiotic chloramphenicol. This enzyme provides protection to the bacterium by inactivating chloramphenicol through acetylation. Most all eukaryotic cells lack a similar enzyme activity and cell extracts from transfected cells can be readily assayed for CAT enzyme activity using radioactive chloramphenicol and thin-layer chromatography (Fig. 7.2). A disadvantage of *CAT* reporter genes is that the CAT assay is laborious and expensive.

LUCIFERASE (luc).
Protein expressed from the *luc* gene of the firefly species *Photinus pyralis* can be easily detected in cell extracts using a specially designed luminometer that measures fluores-

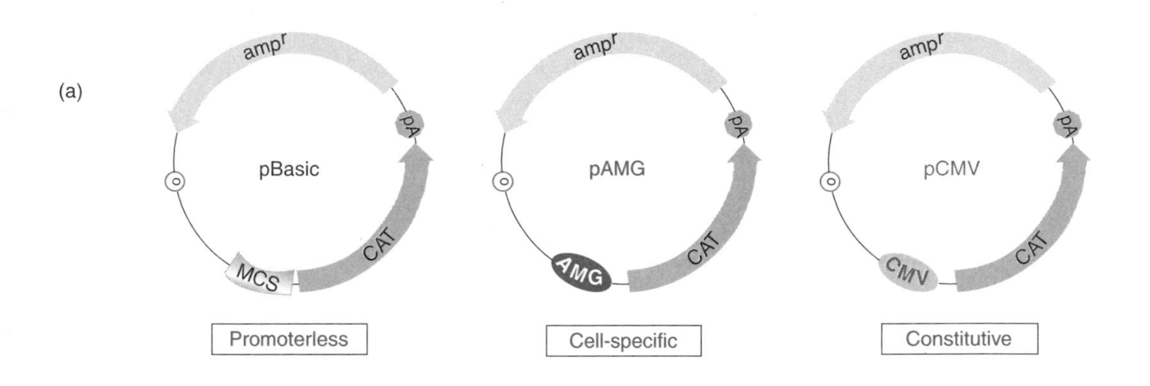

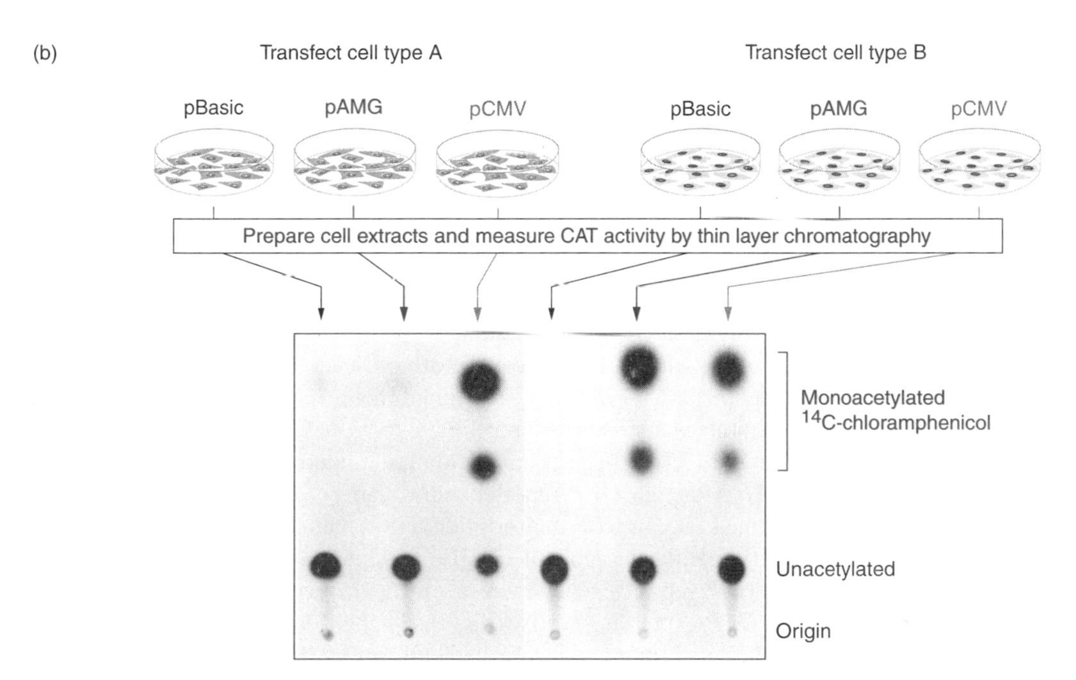

Figure 7.2 Reporter genes have been used to identify cell-specific transcriptional regulatory regions of cloned genes. (*a*) In this hypothetical example, the *AMG* gene 5′ regulatory region was cloned into a reporter gene vector containing the bacterial chloramphenicol acetyl transferase (*CAT*) coding sequence. (a) Transfection of two different cell types (A or B) with the promoterless *CAT* vector (pBasic), the AMG-*CAT* test plasmid (pAMG), or a constitutive *CMV-CAT* reporter plasmid (pCMV) would be used to determine if this cloned portion of the *AMG* regulatory region encodes a cell-specific promoter. (*b*) One way to measure the level of CAT enzyme activity in cell extracts is to quantify the production of acetylated ^{14}C-chloramphenicol using thin layer chromatography (TLC) followed by phosphorimaging. Two distinct monoacetylated chloramphenical products can be resolved by TLC. As depicted in this example, it would be concluded that the cloned fragment of the *AMG* promoter selectively directs transcription of the *CAT* reporter gene in cell type B. The constitutive cytomegalovirus (CMV) promoter functions in both cell types (Table 7.1).

cence emitted from the luc-catalyzed ATP-dependent oxidation of compounds called luciferans. Reporter genes containing the *luc* gene from the sea pansy *Renilla reniformis* can be used in combination with the firefly *luc* reporter gene as an internal control for DNA transfection efficiencies. The substrate specificity of these two luciferase enzymes is sufficiently different to permit parallel assays. A modified version of the firefly *luc* gene coding sequence has been developed (GL3) that produces increased levels of enzyme activity in cell extracts.

β-GALACTOSIDASE (β-gal OR lacZ).
This bacterial enzyme is one of the most versatile reporter genes available because β-gal enzyme activity can be measured both in cell extracts using a spectroscopic assay that detects cleavage of ONPG (*o*-nitrophenyl-β-D-galactopyranoside) and also by histocytochemical methods that measure the appearance of a blue precipitate produced from the cleavage of X-gal. β-gal histocytochemical assays have been invaluable for mapping cell-specific promoter functions in transgenic organisms (Chapter 8). β-*gal* reporter genes have been used as internal controls to monitor overall transfection efficiency in some mammalian cell transfection systems.

GREEN FLUORESCENT PROTEIN (GFP).
The gene for this autofluorescent protein has rapidly become one of the most useful reporter genes for marking transfected cells in vivo. Green flourescent protein, expressed in the outer dermal layer of the Pacific Northwest jellyfish *Aequorea victoria*, provides a green bioluminescent light to the jellyfish in response to a calcium-dependent energy transfer step involving a GFP-associated protein called aequorin. It was found that exposure to ultraviolet light causes GFP to autofluoresce a bright green color within living cells, even in the absence of aequorin, calcium, or any other co-factor or substrate. The molecular basis for this ultraviolet-induced autofluorescence is a cyclization process involving the tripeptide Ser65-Tyr66-Gly67, which functions as the chromophore in the intact protein. The molecular structure of purified *A. victoria* GFP reveals a striking barrel-like arrangement of eleven β-sheets surrounding the internal cyclic tripeptide chromophore (Fig. 7.3). Although *GFP* reporter genes can be used for gene expression analyses, the lag time between *GFP* transcription and protein autofluorescence and the inability to quantify readily the amount of GFP-mediated autofluorescence in transfected cells make *GFP* reporter genes more useful as in vivo markers of transfected cells. Individual *GFP*-expressing cells can easily be identified by fluorescence microscopy (Fig. 7.3), and fluorescent activated cell sorting (FACS).

Currently, the *CAT*, *luc*, β-*gal,* and *GFP* genes represent the most widely used reporter gene systems in molecular genetics, but there are also a number of other reporter genes that have been developed. One is the β-*glucuronidase* (*GUS*) gene, a bacterial enzyme that is primarily used as a reporter gene in plant cells that lack endogenous glucuronidase activity. In vitro cleavage assays and histocytochemical methods are used to detect GUS activity in transfected plant cells. Eukaryotic expression plasmids containing the *E. coli* β-*lactamase* (amp^r) gene have also been developed as reporter genes. Similar to GFP-based expression plasmids, amp^r reporter genes can be used to identify live transiently transfected cells using a real-time in situ assay. In this case, a fluorogenic substrate for β-lactamase, called CCF2, is incubated with transfected cells to distinguish between β-lactamase-expressing (blue) and nonexpressing (green) cells using fluorescence microscopy or FACS. Two other eukaryotic reporter genes have been used to monitor gene expression in live cell populations. One of these is the *secreted alkaline phosphatase* (*SEAP*) gene and the other is the *human growth hormone* (*hGH*) gene. *SEAP*

(a)

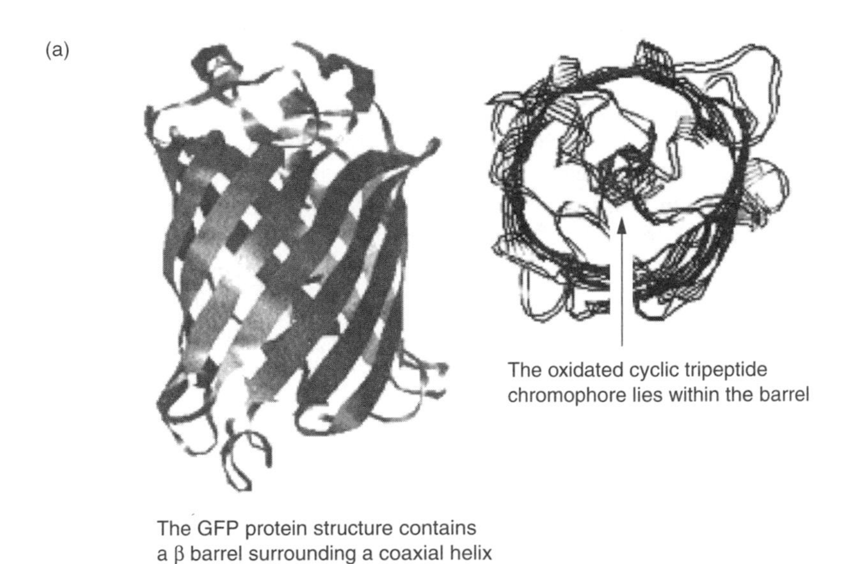

(b)

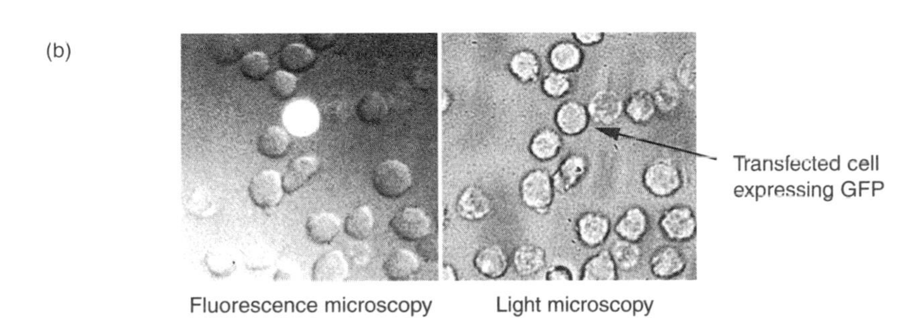

Figure 7.3 Molecular structure of the jellyfish green fluorescent protein (GFP) from *Aequorea victoria* and expression of *GFP* in a transfected mammalian cell. (*a*) The light-emitting chromophore of GFP variant consists of a tripeptide ring held in place by the surrounding structure of a barrel-like basket consisting of eleven β-sheets. (*b*) Fluorescence microscopy can be used to identify individual live mouse thymocyte cell that has been transiently transfected with a constitutive CMV-*GFP* expression vector. Note that this GFP-positive cell is indistinguishable from nearby untransfected thymocytes when viewed by phase contrast light microscopy.

and *hGH* genes direct the synthesis of secreted proteins that can be readily detected in cell culture media.

Antisense Genetics

Ectopic expression of cloned genes in transfected cells can provide gain-of-function activities, which are often informative and useful for understanding gene function. However, it is also important to test what happens to cell phenotype when the function of a specific gene is missing or inhibited. Classically this is done by selecting for genetic mutations in the endogenous gene, or by molecular genetic methods that are based on in vivo homologous recombination to produce gene knockouts (Chapter 8). As an alternative to these standard genetic approaches, which cannot be used with somatic cell models, several gene expression-based methods have been developed that use a form of antisense genetics, These methods are based on introducing sequence-specific nucleotide binding molecules into cells that cause a transient inhibition of gene expression at the transcriptional or post-transcriptional level. The broader implications of developing antisense reagents for gene-specific inhibition are their potential as therapeutic agents.

Two types of antisense genetic strategies have been developed. One involves targeting the inhibitory activity to mRNA transcript, whereas the other approach is aimed at blocking transcriptional initiation at the gene promoter. The first antisense experiments used oligodeoxynucleotides (ODNs) that were designed to form heteroduplexes with specific targeted mRNA transcripts through standard Watson-Crick base pairing. The basic idea was that formation of a double-strand DNA:RNA heteroduplex in a region near the translational start site would be sufficient to disrupt ribosome binding and thereby inhibit protein synthesis. It turned out that this rarely happened, and in fact most observed decreases in protein production were due to sequence-specific RNA degradation, presumably by cellular RNases that recognize RNA:DNA hybrids.

To overcome stability and permeability problems associated with standard antisense ODNs, researchers have tried to use modified ODNs, such as phosphorothioates or peptide-linked nucleic acids. Another strategy has been to transfect cells with expression vectors that contain portions of the gene coding sequence in the antisense orientation relative to the promoter. Antisense genetic strategies have also been reported using catalytic RNA enzymes called "ribozymes" that can be used to target specific cleavage of mRNA transcripts. Ribozymes are single-strand RNA molecules that contain a catalytic site, called a "hammerhead structure." Ribozymes can be targeted to an RNA substrate by complementary base pairs engineered into regions flanking the hammerhead structure. A number of proof-of-principle ribozyme experiments have been described using cultured cells and several clinical trials have been undertaken to determine if ribozyme-based gene therapy approaches are efficacious.

Few of these approaches have proved to be routinely effective for reasons related to insufficient levels of the antisense molecule or unexpected nonspecific effects. Phosphorothioates, for example, are highly charged molecules that do not readily cross the cell membrane and moreover have been found to interact nonspecifically with numerous cellular targets other than nucleic acids. A second, and perhaps more problematic, limitation to using ODN and ribozyme-based antisense approaches is that most RNA targets in the cell are folded into complex intramolecular structures that limit the accessibility of specific RNA sequences to sequence-specific heteroduplex formation. One way researchers are trying to overcome this barrier is to use combinatorial screening approaches to identify functionally regions of the RNA target that are the most likely to be sensitive to the inhibitory effects of these agents.

Researchers have recently begun to think of ways to block transcription directly at the gene promoter. One of these methods uses ODNs that are complementary to nontranscribed regions of the gene to target transcription factor binding sites. There is some evidence that the DNA double helix is able to form triplex structures at oligopurine–oligopyrimidine tracts when a mixed-tract ODN is present at high concentrations. If these triplex helical structures were able to form in vivo, then they could presumably block gene transcription by disrupting assembly of the transcription initiation complex. Peter Dervan and his colleagues are taking a slightly different approach by using a non-nucleic acid based molecule to block transcription at targeted gene promoters. This is being done by treating cells with polyamide molecules that possess sequence-specific binding properties. Because ODNs are negatively charged they are repelled by the overall net-negative charge of the cell membrane; in contrast, the hydrophobic properties of polyamides more readily permits cell penetration. Based on a chemical code that uses alternating *N*-methylimidazole (Imi) and *N*-methylpyrrole (Pye) moieties, Dervan's group has been able to design polyamides that inhibit gene expression by binding to specific DNA sequences. The best example so far of polyamide-mediated transcriptional interference has been the inhibition of 5S RNA gene expression in cultured *Xenopus* fibroblasts using an eight-ring polyamide [ImiPyePyePye-γ-ImiPyePyePye-β] that specifically binds to the DNA sequence 5′ - AGTACT - 3′ in the 5S RNA gene promoter.

METHODS TO EXPRESS GENES IN CELL LINES

Much of what we know about gene regulation in higher eukaryotes, especially mammalian systems, has come from studies using immortalized cell lines. As long as 100 years ago, biologists began to experiment with in vitro culture conditions that permitted dissected animal tissues to remain viable in culture for extended periods of time. Eventually, these studies gave rise to the development of various types of tissue culture media consisting of buffers, inorganic salts, trace elements, a source of carbon for energy production, amino acids, vitamins, and bovine serum. The serum provides necessary growth factors and other "enhancing properties" that permit most types of cells to proliferate under appropriate incubator conditions. More recently, serum-free media, also called defined media, have been developed that can greatly facilitate the ability to maintain the differentiated state of certain cell types in culture.

Early molecular geneticists, many of whom started as basic science oncologists or virologists, realized in the 1950s and 1960s that cells derived from disaggregated animal tumors, or isolated from the blood of patients with leukemia, could be adapted to growth in vitro using the same media developed for tissue culture. It was found that many types of cancer cells have an unlimited capacity to proliferate in serum-containing medium. With the discovery of viral oncogenes, and the isolation of mutated "transforming" genes from primary human cancers, it has become possible to immortalize most differentiated cell types by stably transfecting primary cells with one of these cloned immortalizing genes. We use the term *immortalizing* functions to refer to unlimited proliferative capacity in cell culture, whereas *transforming* functions refer to the ability of immortalized cells to form tumors in animals. Clonogenic immortalized cells grown in culture are called cell lines. Note that in the context of the gene expression analyses described here, there is no significant difference between studies performed in transformed versus untransformed immortalized cells.

Table 7.2 lists some of the mammalian cell lines that are most commonly used in molecular genetic applications. Other types of eukaryotic cell lines have been established from immortalized insect, amphibian, avian, and fish cells. Most cell lines in use today can be obtained for a nominal fee from the American Type Culture Collection (ATCC), a nonprofit organization in Rockville, Maryland that has collected an archive of more than 3500 immortalized cell lines representing at least 75 different species.

It is important to realize that immortalized cell lines are often dedifferentiated, and therefore have lost many of the cellular phenotypes associated with the normal differentiated state. Nevertheless, cell lines provide a convenient model system to study eukaryotic gene expression under defined conditions and have in many cases led to major breakthroughs in our understanding of human diseases. In this section, the molecular

TABLE 7.2 Common mammalian cell lines used in molecular genetic studies[a]

Cell line	Species	Cell type of origin	Year Established
MDCK	Dog	Kidney	1958
CHO	Hamster	Ovary epithelial	1957
CV-1/COS	Monkey	Kidney	1964
3T3	Mouse	Embryo fibroblast	1963
Rat-1	Rat	Embryo fibroblast	1968
HeLa	Human	Cervical carcinoma	1951
HEK-293	Human	Embryonic kidney	1977
LNCaP	Human	Prostate tumor	1977
MCF7	Human	Mammary tumor	1973

[a]These cell lines are normally maintained in serum-containing media at 37°C in a humidified incubator containing 5% CO_2 in air.

genetic methods commonly used to transfect cell lines with cloned DNA are described. The use of these techniques to study gene expression in transiently transfected cells, and to create isogenic variants of established cell lines by selecting for drug-resistant marker genes, is also discussed.

Gene Transfer Methods

The best way functionally to characterize a cloned gene is to put it into live cells and determine what effects the encoded gene product has on defined cellular processes. Similarly, to identify DNA elements required for the transcriptional control of a cloned gene, an appropriate reporter gene must be constructed and tested in vivo. Both these research objectives require an efficient means of transfecting experimental DNA into cells under conditions where the expression vector or reporter gene will be functional.

Three basic DNA transfection strategies have been developed to deliver cloned DNA to cells. We call them the (1) stealth, (2) attack, and (3) infection strategies of DNA transfection. The *stealth strategy* is based on the use of positively charged carrier molecules that are mixed with the experimental DNA in vitro and then applied directly to the cell culture media. These carrier–DNA complexes attach to cell membranes and stimulate the uptake of the "stealth" DNA molecules. The three most commonly used stealth transfection methods are calcium phosphate precipitation, DEAE dextran-mediated gene transfer, and liposome-mediated gene transfer. Each of these techniques is sensitive to cell type differences owing to the requirement for selective interactions between the carrier molecule and the cell membrane. The *attack strategy* uses physical methods literally to force experimental DNA into cells. Three attack DNA transfection methods have been developed: electroporation, biolistics (gene gun), and microinjection. The overall efficiency of these brute force transfection methods is relatively independent of cell type. The third method of DNA transfection is the *infection strategy,* which uses recombinant eukaryotic viruses to deliver DNA to host cells. The two most efficient mammalian cell viral vectors are derived from retroviruses and adenoviruses. Because viruses attach to cell surface receptors, viral-mediated transfection is very cell specific. Representative examples of stealth (liposome-mediated transfection), attack (electroporation), and infection (retroviral gene transfer) DNA transfection strategies are described below.

LIPOSOME-MEDIATED TRANSFECTION.
A mixture of a polycationic lipid, such as DOSPA (2,3-dioleoyloxy-*N*-[2-(spermine-carboxamido)ethyl]-*N*,*N*-dimethyl-1-propanaminium trifluoroacetate), and the neutral lipid DOPE (dioleoyl phosphatidylethanolamine), at a ratio between 4:1 and 1:1, results in the formation of unilamellar liposome vesicles that have a net positive charge due to the highly positive amine head groups on these molecules. When DNA is mixed with liposomes, it associates in such a way that the negative charge of the DNA phosphate backbone is neutralized by the lipids. Because cell membranes also have a net negative charge, these overall positively charged lipid–DNA complexes easily become associated with cells during a short treatment period (~2 hours) in culture medium. As a result of membrane fusion, which can be enhanced by exposing the liposome-treated cells to a brief incubation in 10% glycerol, the associated DNA molecules gain access to the inside of the cell. This straightforward technique was developed for use as a nonviral gene therapy approach by Philip Felgner and his colleagues. Liposomes have become the method of choice for carrier molecules in routine cell line DNA transfections because liposome-mediated gene delivery is technically easy, highly reproducible, and very efficient.

ELECTROPORATION.
This physical approach to DNA transfection is based on the finding that a short pulsed electric field can result in cellular uptake of DNA, presumably through a charge-induced pore-forming mechanism. Electroporation is the most versatile method of DNA transfection because it has been shown to work for such a wide variety of cell types, which include primary cells from tissue isolates, plant protoplasts, and bacterial cells. Electroporation of eukaryotic cells is performed by resuspending cells in a HEPES buffered saline solution containing DNA at a concentration of about 0.1 mg/ml. The buffered cell–DNA suspension is then placed in a special electroporation cuvette that contains a (+) and (–) electrode connected to a power supply, essentially the same type of electroporation device described in Chapter 2 (see Fig. 2.7) for bacterial transformation. By varying the electric field strength and the length of time the cells are exposed to the electric field, it is possible to optimize electroporation parameters for essentially any cell type. The main difference between the conditions used for eukaryotic and prokaryotic cell electroporation is the amount of voltage required to achieve optimal results. For example, high efficiency electroporation of *E. coli* cells requires 2.5 kV, as compared to 0.25 kV for most mammalian cells.

RETROVIRAL INFECTION.
One of the most efficient transfection procedures for mammalian cells is retroviral infection. Retroviral expression vectors are plasmid-based shuttle vectors that contain deleted versions of the viral genome. The cloned gene product is inserted into a multiple cloning site located between the retroviral transcriptional control elements, called the 5′ and 3′ long terminal repeats (LTRs). Because all the retroviral protein coding genes are replaced by the cloned DNA insert, the recombinant viral shuttle vector must be stably transfected into a special cell line that constitutively expresses the viral reverse transcriptase and capsid proteins. This "packaging" cell line is used to produce high titer viral stocks by harvesting recombinant virions that accumulate in the cell media. The two biggest drawbacks to retroviral infection is that not all cells contain the necessary

viral docking proteins on their cell membrane, and it is laborious to make separate packaging cell lines for each retroviral construct. As illustrated in Figure 7.4, the retrovirally encoded Ψ (psi) packaging sequence must be present in the transcribed RNA to initiate virion assembly. Because the integrated provirus in the packaging cell line does not contain a functional Ψ sequence, and the recombinant retroviral cloning vector lacks the necessary viral genes for virus replication and assembly (*gag*, *pol,* and *env*), no replicative virions are produced in this gene expression strategy. This is an important biosafety feature of retroviral expression systems designed for use in mammalian cells.

Transient Transfection Assays

For many types of gene expression assays, it is possible to transfect cell lines with an appropriate reporter gene, and then collect the data 24–48 hours later. Relative to the generation time of most cell lines (~16–24 hours), this represents a relatively short time period between DNA transfection and cell harvesting, hence the term "transient" transfection. Moreover, because the cells may divide only once during the transfection period, no genetic selection is required to enrich for cells that have taken up the DNA. It is important to optimize the transfection protocols by systematically testing key parameters of the chosen gene transfer method for each cell line. High transfection efficiencies (10% or more of the cells express the experimental DNA) are required for the development of a reliable transient assay. The usual way to determine optimal transfection conditions is to perform a series of transient transfections with a reporter gene that contains a high level constitutive promoter. Figure 7.5 shows the strategy for such an optimization experiment by illustrating how data from a GFP FACS analysis and luciferase enzyme assay can be used to define a standard transient transfection protocol.

A common research objective for transient transfection assays is to map gene regulatory regions using mutagenized genomic DNA segments linked to a reporter gene (Figs. 4.15 and 7.2). For example, transient transfection assays would be used to map regulatory elements in a cloned liver-specific gene promoter using a liver-derived cell line. Alternatively, a transient "co-transfection" assay could be used functionally to map protein domains of a cloned transcription factor by transfecting cells with both an expression vector and a reporter gene. This strategy is described in laboratory practicum 7 at the end of this chapter.

Stable DNA Transfection

One important feature of cell lines is that they provide a cell autonomous system to investigate the effect of cloned gene products on cell phenotype. For example, expression of wild-type and mutant variants of a cloned growth factor receptor gene in a cell line that does not express the receptor can provide a molecular genetic model system to study growth factor effects on cellular processes. Transient transfection strategies would not be appropriate in this case because they would require several weeks of culture time before the growth factor receptor gene effects on cell proliferation could be determined. By that time, the unintegrated plasmid DNA will have been lost as a result of dilution effects during cell division. One solution is to integrate stably the growth factor receptor cDNA expression plasmid into the host cell genome, and then isolate and characterize a new clonal cell line that expresses the receptor. This approach is called stable DNA transfection and it is done by including a drug-resistance marker gene on the expression

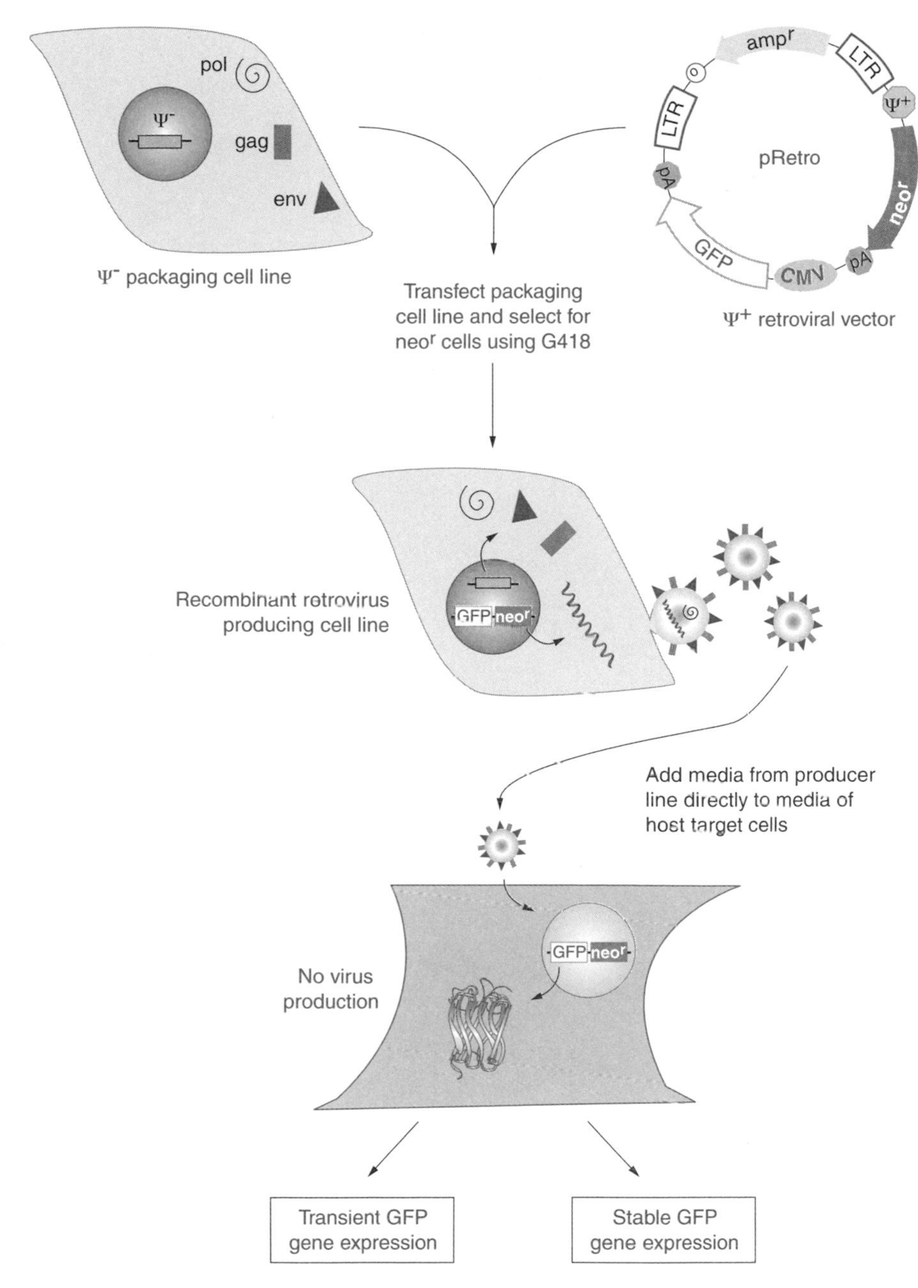

Figure 7.4 Retroviral expression vectors provide an efficient gene transfer method for expressing cloned genes in appropriate host cells. Nonreplicating recombinant retroviral stocks can be made by stably transfecting a Ψ– retrovirus packaging cell line, with a Ψ+ recombinant retroviral vector using liposome-mediated transfection. Transcription of the recombinant retrovirus genome results in the production of infectious virions that are released into the media by membrane blebbing. The recovered high titer recombinant viral stock is used to infect host target cells where the recombinant RNA genome is reverse transcribed into double-strand cDNA by the RTase enzyme contained in the viral capsid. Retroviral expression systems can be used for both transient and stable transfection strategies.

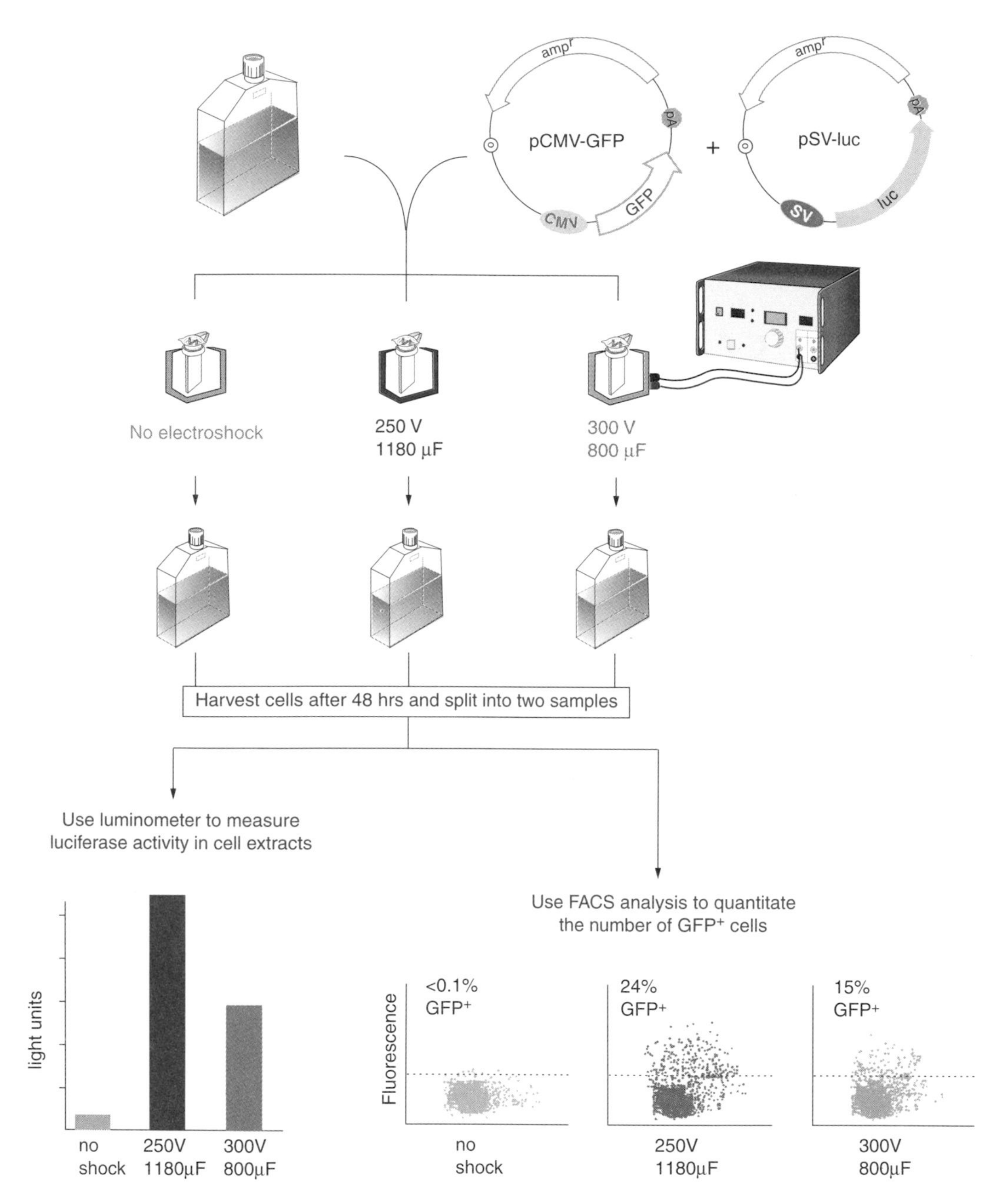

Figure 7.5 Transient transfection protocols must be optimized for each cell type being analyzed owing to inherent differences in DNA uptake efficiencies. In this example, two types of assays are being performed with the same pool of cells that have been electroporated with both a CMV-*GFP* and a SV40-*luc* expression plasmid under two conditions (including a negative control). The luciferase assay measures the amount of luminescence (light units) per microgram of protein using a luminometer, and the number of GFP^+ cells in the population is quantified by FACS analysis. In this example, the electroporation conditions of 250 V and 1180 μF would be chosen as optimal for efficient transient transfection of this cell type.

plasmid that can be used to select positively for transfected cells. In most cases, the resulting clonal cell line contains from 1 to 30 copies of the plasmid DNA randomly integrated into the host cell genome. Stable inheritance of a gene expression vector can also be accomplished using retroviral vectors that contain selectable marker genes.

Stable DNA transfection of cell lines is basically done the same way that transient transfections are, with the exception that stable integration of the transfected DNA must be selected for by including a marker gene on the expression vector. Most often the marker gene encodes an enzyme that inhibits the function of a toxic compound, similar to the way that the β-lactamase gene product is used positively to select transformed *E. coli* cells on the basis of ampicillin resistance. A second modification is that although some gene transfer methods work well for transient transfections, they are often less suitable for stable transfection because the cellular insult required for high transient transfection efficiencies can eventually cause cell death. For example, DEAE dextran-mediated transfection can be used to achieve up to 50% transfection efficiency in transient transfections, but within 5 days, all the transfected cells die from DEAE dextran toxicity. For this reason, liposome-mediated transfection and electroporation are the most common nonviral gene transfer methods used to transfect cells stably.

A description of the most commonly used marker genes for stable transfection of eukaryotic cell lines is given in Table 7.3. A standard kill curve experiment must be performed for each cell line prior to stable transfection to determine the minimum lethal dose of the compound. For most stable transfection experiments, the growth medium is replaced with selection medium 48 hours after transfection, and individual colonies are then isolated after an additional 10–14 days in culture. Assuming that expression of the experimental gene does not affect cell proliferation rates, the selection medium can usually be replaced with normal growth medium after the cell line has been expanded and characterized.

An important molecular genetic strategy for studying the cellular function of cloned genes is to use an expression vector that contains a transcriptionally regulated promoter. In this way, cell phenotypes can be characterized under conditions in which the expression level of the cloned gene is modulated by the presence or absence of extracellular signal. This strategy controls for unrelated variables that can result from com-

TABLE 7.3 Dominant marker genes commonly used for positive selection of stably transfected cell lines

Marker	Gene product	Selection method
neo	Aminoglycoside phosphotransferase; *neo* gene from the bacterial transposon Tn5	Select cells in G418 (0.1–1.0 mg/ml), an aminoglycoside that blocks protein synthesis and is similar to kanamycin
hyg	Hygromycin-B-transferase; *hyg* gene from *E. coli*	Select cells in hygromycin-B (10–300 μg/ml), an aminocyclitol that inhibits protein synthesis
pac	Puromycin-*N*-acetyl transferase; *pac* gene from *Streptomyces alboniger*	Select cells in puromycin (0.5–5 μg/ml), an antibiotic that inhibits protein synthesis
zeo	Bleomycin binding protein; a *zeo* gene (a.k.a. *bla*^{r})is located on the bacterial transposon Tn5	Select cells in bleomycin or commercially available Zeo (50–500 μg/ml), an antibiotic that binds DNA and blocks RNA synthesis
gpt	Xanthine-guanine phosphoribosyltransferase; *gpt* gene isolated from *E. coli*	Select cells in guanine-deficient media that contains inhibitors of de novo GMP synthesis and Xanthine; this selects for *gpt*+ cells that can synthesize guanine from xanthine

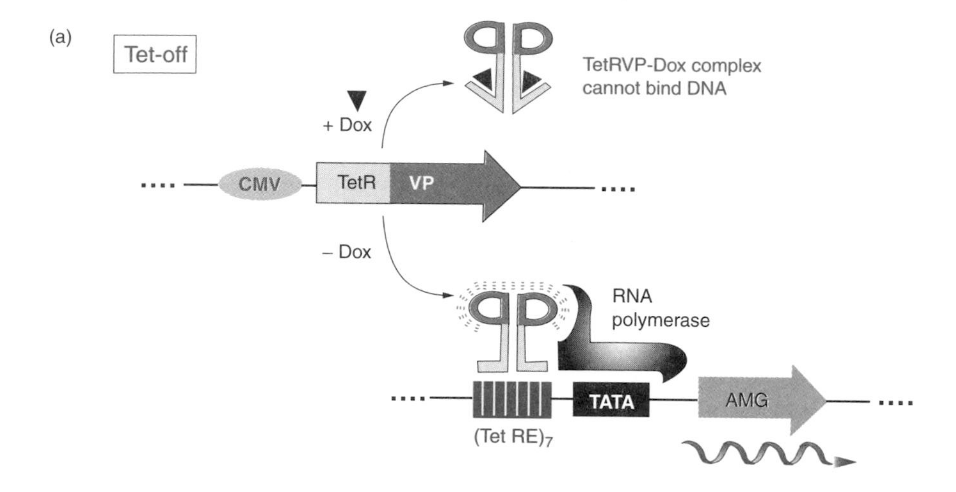

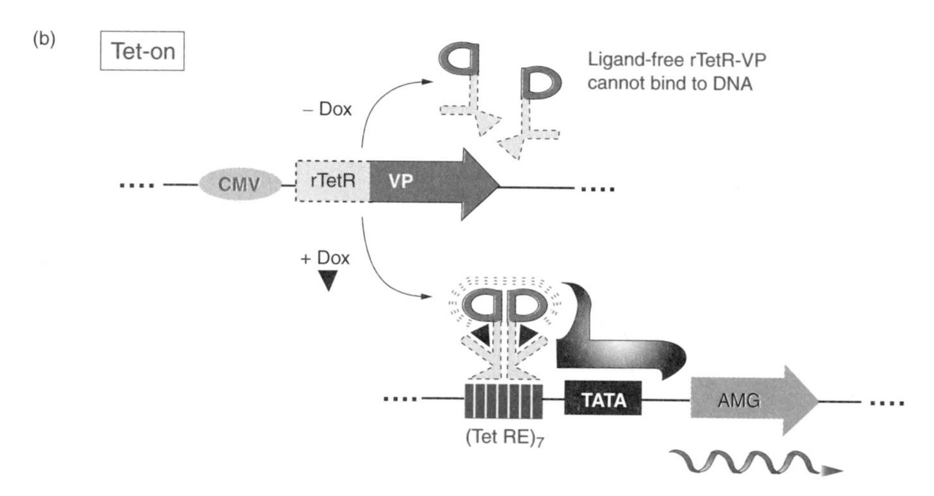

Figure 7.6 The tetracycline receptor can be used to control the expression of cloned genes in stably transfected mammalian cell lines. The Tet-off system is repressed in the presence of the tetracycline analogue doxycycline (*a*), whereas the Tet-on system is activated by doxycycline (*b*). The VP16 transcriptional activation domain (VP) is derived from the herpes simplex virus VP16 protein. The DNA binding domain of the Tet-on regulator (rTetR) contains mutations that convert it from a repressor that binds DNA only in the absence of ligand to a ligand-dependent DNA binding protein.

paring a stably transfected cell line expressing the cloned gene at high levels to another cell line that does not express the gene. In addition, acute transcriptional regulation of the expression vector promoter permits the use of careful time course experiments to study the effect of a gene product on cell cycle arrest or induction of apoptosis. Three regulated promoters that have been used for this purpose are the *S. cerevisiae Gal1* gene, the *Drosophila hsp70* gene, and the steroid-regulated *MMTV* promoter (Table 7.1). Although the *Gal1* and *hsp70* regulated promoters have been very useful for studies in yeast and flies, the *MMTV* promoter is less useful in mammalian cell lines because steroid treatment can cause cellular responses that are independent of the *MMTV* expression vector.

Fortunately, several new regulated expression vectors have been developed that can be controlled with extracellular stimuli that do not interfere with normal cellular processes in most mammalian cell lines. Both these approaches include the use of a cloned non-mammalian transcription factor that can be activated by ligands that are innocuous to mammalian cells. One of these regulated expression systems was developed by Manfred Gossen and Hermann Bujard. Their strategy uses the tetracycline-resistance operon of the *E. coli* Tn10 transposon to control the expression of genes that have been cloned downstream of a promoter containing tetracycline receptor (TetR) binding sites. As shown in Figure 7.6, Gossen and Bujard have actually developed two TetR-regulated expression systems, one that is repressed by tetracycline in the media (Tet-off) and one that is tetracycline-inducible (Tet-on).

More recently, Ron Evans and his colleagues used a similar strategy to exploit the transcriptional regulatory activities of the *Drosophila* ecdysone receptor (EcR). EcR is a member of the nuclear receptor superfamily of ligand-activated transcription factors, it has no mammalian homologue, and its sequence-specific DNA binding activity is induced by the synthetic ecdysteroid muristerone A. In this two-plasmid system, one expression vector encodes both a modified *EcR* gene and the EcR binding protein retinoid-X-receptor, whereas the other plasmid contains an EcR-regulated promoter upstream of the cloned experimental gene.

USING YEAST AS A MODEL EUKARYOTIC CELL

Over the past 20 years, *S. cerevisiae* has become the premier single-cell eukaryotic model for molecular genetic studies. Several reasons account for the immense popularity of yeast models. First, *S. cerevisiae* is a haploid organism that has been intensely studied by genetic analyses for over a century. This attention has provided a large array of genetic tools and mutant strains for powerful genetic screens. Second, there are a surprisingly large number of eukaryotic genes that have been evolutionarily conserved between yeasts and humans. This means that for cellular processes that do not depend on cell-specific phenotypes or intercellular signaling, there is likely to be a yeast gene homologue that can be studied by first principle methods. Third, the entire 12 megabase genome of *S. cerevisiae* has been sequenced, and the corresponding DNA for all ~6000 protein coding sequences has been mapped, cloned, and mutagenized. Moreover, this vast amount of *S. cerevisiae* biological information is available worldwide through the Internet.

Two basic molecular genetic approaches have been taken to exploit yeast as a model eukaryotic cell. The first has been to investigate the function of specific yeast genes that represent homologues of higher eukaryotic genes. Information gained from these yeast-based studies can often be applied to understanding gene functions in the multicellular organism. The second approach has been to use yeast as a "null genetic background" for protein structure–function studies of higher eukaryotic genes that do *not* have a known yeast counterpart. This experimental strategy can be a powerful biological screen for agents that modulate protein function in the broad context of a eukaryotic cell model. Representative examples of each of these two strategies are given below to illustrate the value of yeast-based molecular genetic studies.

Table 7.4 lists some of the human genes that have been characterized using genetic methods in *S. cerevisiae*. These include highly conserved eukaryotic genes involved in the control of cell cycle checkpoints, DNA repair, and signal transduction. Various genetic and biochemical analyses of yeast cell cycle regulatory proteins have revealed

TABLE 7.4 Yeast as a model eukaryotic cell to study the function of human gene homologues

Human genes	*S. cerevisiae* homologues	Gene function
RAS	*RAS1, RAS2*	Receptor-mediated signal transduction
NF1	*IRA1, IRA2*	Ras GTPase activating protein
ATM	*MEC1, TEL1*	Checkpoint control in response to DNA damage
XP-A, XP-B	*RAD14, RAD25*	Nucleotide excision DNA repair
MSH2, MLH1	*MSH2, MLH1*	DNA mismatch repair
MGMT	*MGT1*	O^6-methylguanine DNA repair
cyclin D, cyclin E	*CLN1, CLN2*	Cell cycle control
BLM	*SGS1*	DNA helicase

key mechanisms that operate in normal cells to limit cell division. Importantly, many of these same gene homologues have been found to be defective in human cancer cells, which characteristically display dysregulated cell cycle control. The study of salinity tolerance in plants is another experimental model in which yeast genes are being manipulated to develop single-cell models of multicellular processes. Numerous plant gene homologues have been identified in yeast by "mining" the *S. cerevisiae* DNA sequence database using sequence similarity algorithms (Chapter 9). This approach has allowed plant molecular biologists to use molecular genetic approaches in yeast to understand the biochemical basis for salinity tolerance in eukaryotes. With information gained from these yeast studies, the researchers intend to design transgenic crop plants that can thrive in agricultural areas where high salinity soil is an impedance to plant productivity (Chapter 8).

An example of how yeast null models have been exploited is the expression of mammalian steroid receptor genes in yeast as a means to screen genetically for novel steroid hormone antagonists/agonists. Genetic analysis of these novel yeast strains has also led to the identification of eukaryotic steroid transport proteins and permitted the characterization of yeast chromosomal proteins and transcription factors that modulate the transcriptional regulatory functions of steroid receptors.

It is important to appreciate that molecular genetic approaches utilizing "the power of yeast genetics" to study evolutionarily conserved *biological processes* are strategically quite different from yeast-based *technical applications,* such as characterizing megabase-sized genomic segments in YACs (Chapter 4) or screening cDNA libraries using yeast two-hybrid vectors (Chapter 5). The primary difference is that when yeast model systems are being used for biological studies, standard yeast genetic methods (multigenic crosses and tetrad analysis) are critical to the success of the project. In contrast, straightforward yeast-based laboratory methods require only minimal training in genetic analysis, just as plasmid cloning in *E. coli* does not depend on extensive knowledge of bacterial genetics.

PROTEIN EXPRESSION IN CULTURED CELLS

In this section, two cultured cell systems are described that are often used to produce large quantities of functional protein for biochemical studies. The first is an *E. coli* based expression system that utilizes affinity tagged fusion proteins to purify large amounts of protein rapidly. As described in laboratory practicum 2, high level heterologous protein expression in *E. coli* is a good source of antigen for antibody production. *E. coli* expression systems are also used as a ready source of milligram quantities of purified protein for the purposes of crystallographic studies of protein structure. A second protein expression system utilizes baculovirus expression vectors to produce eukaryotic proteins that require post-translational modifications to be fully functional. Although not described here, two other eukaryotic expression systems have also been developed. One is the use of yeast expression vectors to produce secreted fusion proteins, and the other is high density mammalian cell expression systems using viruses or stably transfected cell lines. However, because of problems associated with low protein yields and high scale-up costs, these alternative expression systems are used only if functional protein cannot be produced by the *E. coli* or baculovirus systems.

E. coli Expression Systems

Most of the *E. coli* expression vectors used for high level protein production contain regulated promoters that are based on the lac operon. Because many types of nonbacterial proteins are growth-inhibitory, or even toxic, when overexpressed in *E. coli*, protein expression is usually inhibited in these systems by lac repressor protein during culture scale-up. As described in Chapter 2, inactivation of lac repressor is usually done by adding an inducer compound to the media such as IPTG. The pET cloning vectors are a good example of this type of expression system (Fig. 2.13). Under optimal conditions, up to 1–10% of the total cellular protein in an IPTG-induced *E. coli* culture can be the protein product of the cloned gene.

Because high level expression alone does not guarantee that standard chromatographic methods will result in a homogeneous protein preparation, many types of *E. coli* expression vectors are designed to produce fusion proteins that contain affinity "tags" at the amino or carboxy terminus of the expressed protein. One example is the *polyhistidine tag (His)$_6$* that can be used to purify fusion proteins by binding to a Ni^{2+} affinity column as shown in Figure 2.14. Other examples of *E. coli* expression vectors utilizing fusion proteins with affinity tags are (1) pGEX vectors containing a 26 kDa *glutathione S-transferase* protein moiety that can be used to purify fusion proteins on glutathione affinity column, (2) pMAL vectors containing an *N*-terminal 42 kDa *maltose binding protein* region that can be purified on amylose resin columns, (3) pCAL vectors containing a 27 amino acid sequence from the *calmodulin binding protein* that can be used to purify fusion proteins using calmodulin affinity resin, and (4) pTrx vectors containing an *N*-terminal 12 kDa *thioredoxin* fusion protein that can be purified with thiobond resin. Each of these fusion proteins encodes sequence-specific protease cleavage sites located in the region between the affinity tag and the cloned gene to permit removal of the fusion protein tag following affinity purification. Figure 7.7 illustrates how a thioredoxin fusion protein can be purified using thiobond resin.

Baculovirus Expression Systems

E. coli expression systems are well-suited for the expression of small proteins (<50 kDa) that do not require special post-translational modifications, such as phosphorylation, glycosylation, or proteolytic processing. However, such limitations make it difficult to

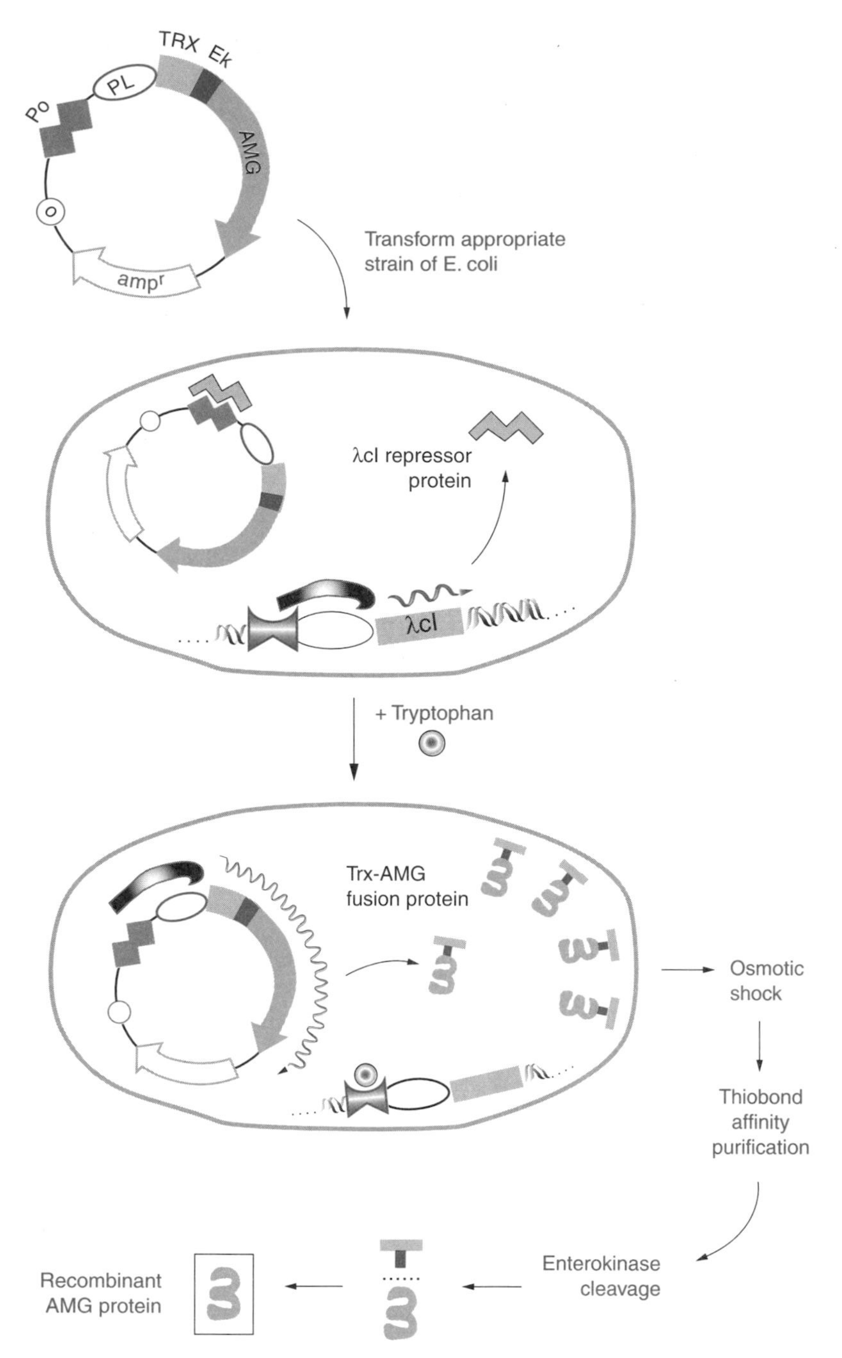

Figure 7.7 Fusion proteins made with the *E. coli* thioredoxin affinity tag at the *N*-terminus often remain soluble even when expressed at high levels. Flow scheme showing the strategy that would be used to express a Trx-AMG fusion protein in *E. coli* . In this system called ThioFusion™, commercially available from Invitrogen, expression of the Trx fusion protein is induced by tryptophan-mediated repression of the λcI repressor gene integrated into the bacterial chromosome. In tryptophan limiting media, λcI repressor levels are high and the P_L promoter on the pTrx plasmid is repressed by λcI binding to the operator site adjacent to P_L. Many types of thioredoxin fusion proteins expressed in *E. coli* localize to the cytoplasmic side of the inner membrane as a consequence of the thioredoxin moiety. This intracellular localization can be exploited as a means to purify recombinant proteins following osmotic shock. An enterokinase (Ek) protease cleavage site is included in the coding region of Trx, providing a means to release the AMG portion of the fusion protein following thiobond resin purification.

use bacterially expressed proteins for biochemical studies that require full protein function. The solution to this problem has been to use a cost-effective eukaryotic cell system that is based on recombinant viral vectors derived from a class of insect viruses called the *Baculoviridae*. The *Autographa californica* multiple nuclear polyhedron virus (AcMNPV) strain of baculovirus has a genome of 128 kb and normally infects the Lepidopteran alfalfa looper *Autographa californica*. The AcMNPV host cell line, Sf9, was derived from the ovaries of a *Spodoptera frugiperda* armyworm. The baculovirus-based expression systems were developed by Max Summers and his colleagues. As shown in Figure 7.8, recombinant baculovirus expression vectors are constructed in vivo by selecting for homologous recombination events that occur between a linear viral DNA fragment and baculovirus sequences contained on a plasmid shuttle vector. Once the correct viral recombinant is identified using a plaque assay, a high titer viral stock is made for the purpose of large-scale host cell infection and protein production. Similar to the *E. coli* expression vectors, protein affinity tags can be engineered into baculovirus vectors to facilitate protein purification from infected cell lysates (Figs. 2.13, 7.7).

There are several reasons why baculovirus vectors have become a useful alternative to *E. coli* based systems for the production of eukaryotic proteins: (1) baculovirus expressed proteins obtained from insect cell lysates are often fully active and soluble; (2) the viral genome is large and can accommodate cDNA inserts of up to 15 kb without compromising viral replication; (3) functional multi-subunit protein complexes can be assembled in vivo and expressed at high levels by co-infecting cells with two or more recombinant viral stocks; (4) the virus has a very restricted host range and is therefore safe to handle and poses minimal environmental risk; and (5) under optimal conditions, as much as 1–5 mg of protein can be produced per liter of infected cells. A potential drawback to using baculovirus expression systems is that the initial isolation and characterization of the primary viral recombinant can be challenging.

Laboratory Practicum 7. *Characterization of a promoter-selective transcription factor*

Research Objective

A developmental biologist has determined that a conserved DNA sequence (TTCAAT) located upstream of the insulin, glucagon, and somatostatin genes is required for coordinating gene expression in mouse pancreatic islet cells. Using a yeast one-hybrid screen (Chapter 5), she has isolated a mouse cDNA clone that encodes a DNA binding protein with high affinity for the TTCAAT sequence. She has named the mouse gene *Igs*, based on EMSA analysis that revealed high affinity binding activity of the encoded protein for the insulin-glucagon-somatostatin regulatory element. The sequence of the mouse *Igs* cDNA clone was used in an Internet-based GenBank BLAST search that led to the indentification of a human EST. This EST was later found to encode a portion of the human *Igs* gene homologue.

Recent evidence suggests that embryonic development of the mouse pancreas, and physiological insulin production in humans, requires the presence of stem-loop-helix (SLH) transcription factors. Because the DNA sequence of her mouse and human *Igs* cDNA clones revealed that Igs is *not* an SLH transcription factor, her immediate research objective is to understand how the Igs and SLH transcription factors may function together to regulate expression of the insulin gene. The long-term goal of this project is to elucidate the role of the human *Igs* gene in controlling pancreatic functions in the normal and diseased states.

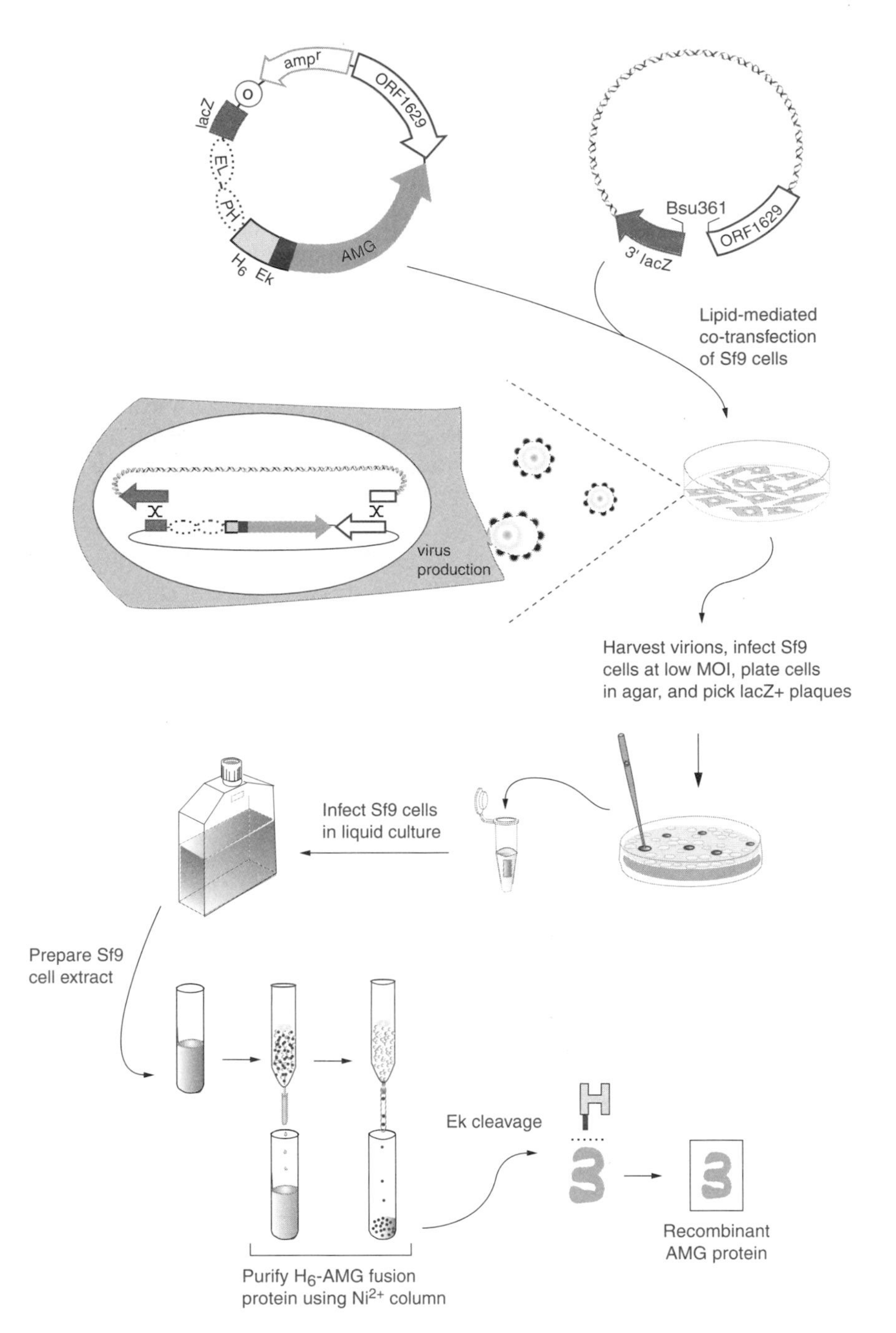

Figure 7.8 Baculovirus expression systems are used to produce large quantities of recombinant proteins that can be purified using affinity chromatography. In this example, the cloned *AMG* gene has been inserted into a plasmid shuttle vector that carries the coding sequence for both the baculovirus essential gene *ORF 1629* and the 3′ end of *lacZ*. Prior to liposome-mediated Sf9 transfection, the viral genome is linearized with the restriction enzyme Bsu361, which cuts once within the *ORF 1629* gene, and a second time in *lacZ*. In vivo recombination in Sf9 cells produces a viable virus that expresses both the *AMG* gene (from the baculovirus polyhedron gene promoter, PH) and the *lacZ* gene (from the baculovirus early-late gene promoter, EL). Infectious virions are harvested from the media and used at a low multiplicity of infection (MOI) in Sf9 cells to isolate individual blue plaques (*lacZ*$^{+}$) using agar plates. A large-scale liquid culture infection is then used to produce a polyhistidine-tagged (H_6) recombinant protein which would be purified by affinity chromatography (see Fig. 2.13). Enterokinase (Ek) cleavage releases the purified AMG protein from the affinity tag. This baculovirus expression system, called Bac-N-Blue™, is available commercially from Invitrogen.

Available Information and Reagents

1. A 2.5 kb human *Igs* cDNA clone has been sequenced and shown to contain a 2166 bp ORF encoding a 722 amino acid protein. This cDNA has been cloned in the sense orientation into two eukaryotic expression vectors: a conventional constitutive CMV plasmid vector and a retroviral vector containing a peptide affinity tag sequence (FLAG) in-frame with the full-length Igs ORF and three Igs deletants. Anti-FLAG antibodies to immunoprecipitate FLAG-Igs fusion proteins are commercially available.
2. A firefly luciferase reporter gene has been constructed that contains four copies of the Igs binding sequence (TTCAAT), referred to as IBS_4 (Igs binding sequence $\times$ 4), upstream of a minimal TATA box promoter. In addition, a *Gal4-Igs* fusion gene expression vector and *Gal4* reporter gene have been constructed to map Igs activation domain sequences.
3. A human pancreatic-derived cell line, called DiB-TC, is being cultured in the lab and will be used for all of the experiments. DiB-TC was isolated from an insulin-dependent diabetic patient. Northern blots were used to show that DiB-TC cells do not express the *Igs* gene, which makes them suitable for Igs functional mapping studies. Western blots with anti-SLH antibodies showed that DiB-TC cells do express normal levels of SLH proteins. Importantly, stable transfection of DiB-TC cells with full-length Igs coding sequences restores transcription of the endogenous insulin, glucagon, and somatostatin genes.

Basic Strategy

Because DNA sequence analysis of the *Igs* cDNA did not identify any previously characterized DNA binding domain protein motifs, the purpose of the first set of experiments is to define functional sequences within *Igs* that are required for transcriptional regulation of the IBS_4 and human insulin reporter genes. Co-transfection assays using DiB-TC cells transfected with CMV-*Igs* expression vectors, and either one of the two *F-luc* reporter genes, will be used to map Igs functional domains. In addition, yeast Gal4 DNA binding domain sequences have been fused to *Igs* coding sequences to facilitate the mapping of *Igs* transcriptional activation domains using a *Gal4-luc* reporter gene. The flow scheme for these *Igs* functional mapping experiments is shown in Figure 7.9.

Based on recently published data implicating SLH proteins in insulin gene expression, and results indicating that SLH proteins may function by forming heterodimers with other transcription factors (protein–protein interaction data from yeast two-hybrid studies), the developmental biologist predicts that Igs and SLH heterodimer formation is required for insulin gene expression. To test this model she immunoprecipitates FLAG-Igs fusion proteins from retroviral infected DiB-TC cells and uses Western blots with anti-SLH antibodies to map potential SLH interaction regions in the Igs protein. The experimental strategy for these Igs-SLH interaction studies are shown in Figure 7.10.

Comments

The *Igs* functional mapping experiments are relatively straightforward and, based on the analysis of other eukaryotic transcription factors, likely to reveal several distinct regions in the Igs protein that are required for full activity. The use of the DiB-TC cell line is essential for *Igs* mapping experiments because it lacks endogenous Igs but does contain the essential SLH transcription factors. The Igs:SLH interaction studies using anti-FLAG and anti-SLH antibodies would indicate physical interactions in DiB-TC cells, but functional tests would have to be performed to confirm the importance of this interaction. One type of functional study would be to attempt to disrupt SLH activity, that is,

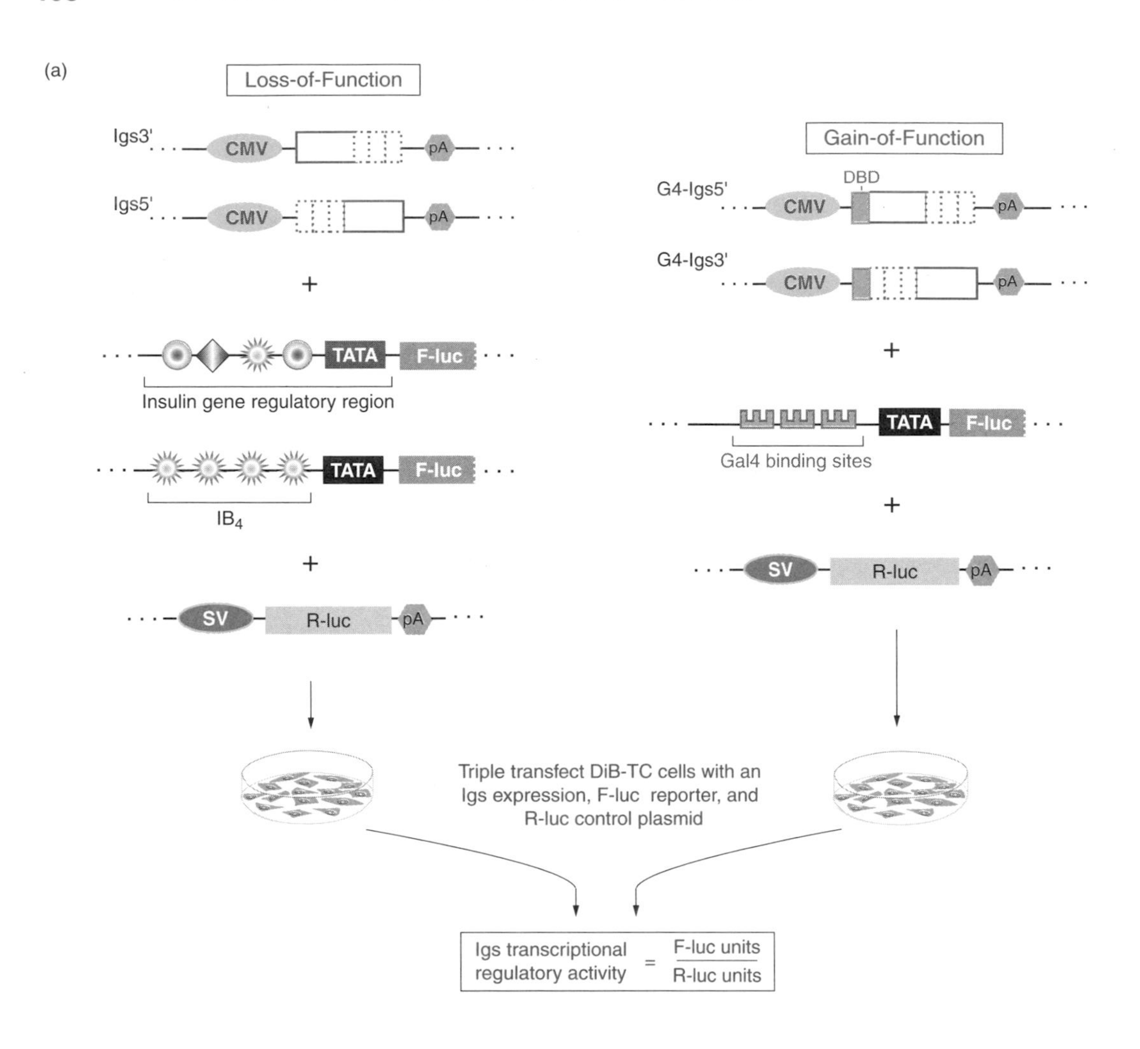

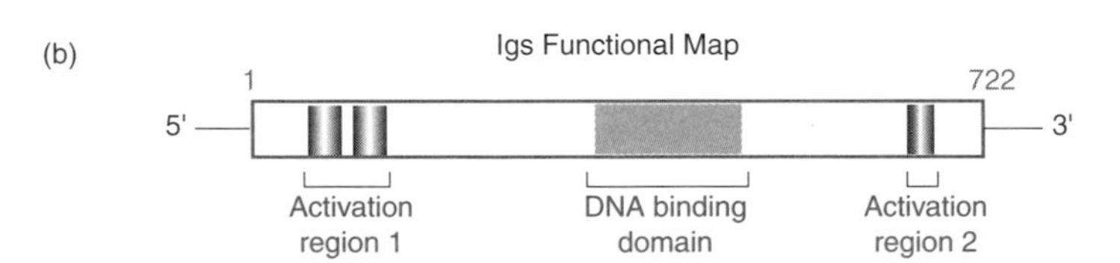

Figure 7.9 Transient co-transfection assays into DiB-TC cells, which lack *Igs* expression, can be used to characterize Igs transcriptional regulatory functions. (*a*) Flow scheme showing the experimental strategy for *Igs* functional mapping experiments using co-transfection assays. Two sets of experiments are shown: loss-of-function mapping using *Igs* deletions and the two *Igs-F-luc* (firefly luciferase) reporter genes, and *Gal4* DBD (DNA binding domain) gain-of-function mapping experiments to identify *Igs* transcriptional activation domain sequences. Both types of transient co-transfection studies also include a third plasmid encoding the *Renilla* luciferase (*R-luc*) gene to provide an internal control for variations in transfection efficiency between sets of data. In addition to functional promoter sequences, all the plasmids contain a polyadenylation signal sequence (pA) downstream of the coding regions. (*b*) Functional map of *Igs* transcriptional regulatory functions based on hypothetical luciferase data from transient co-transfection assays.

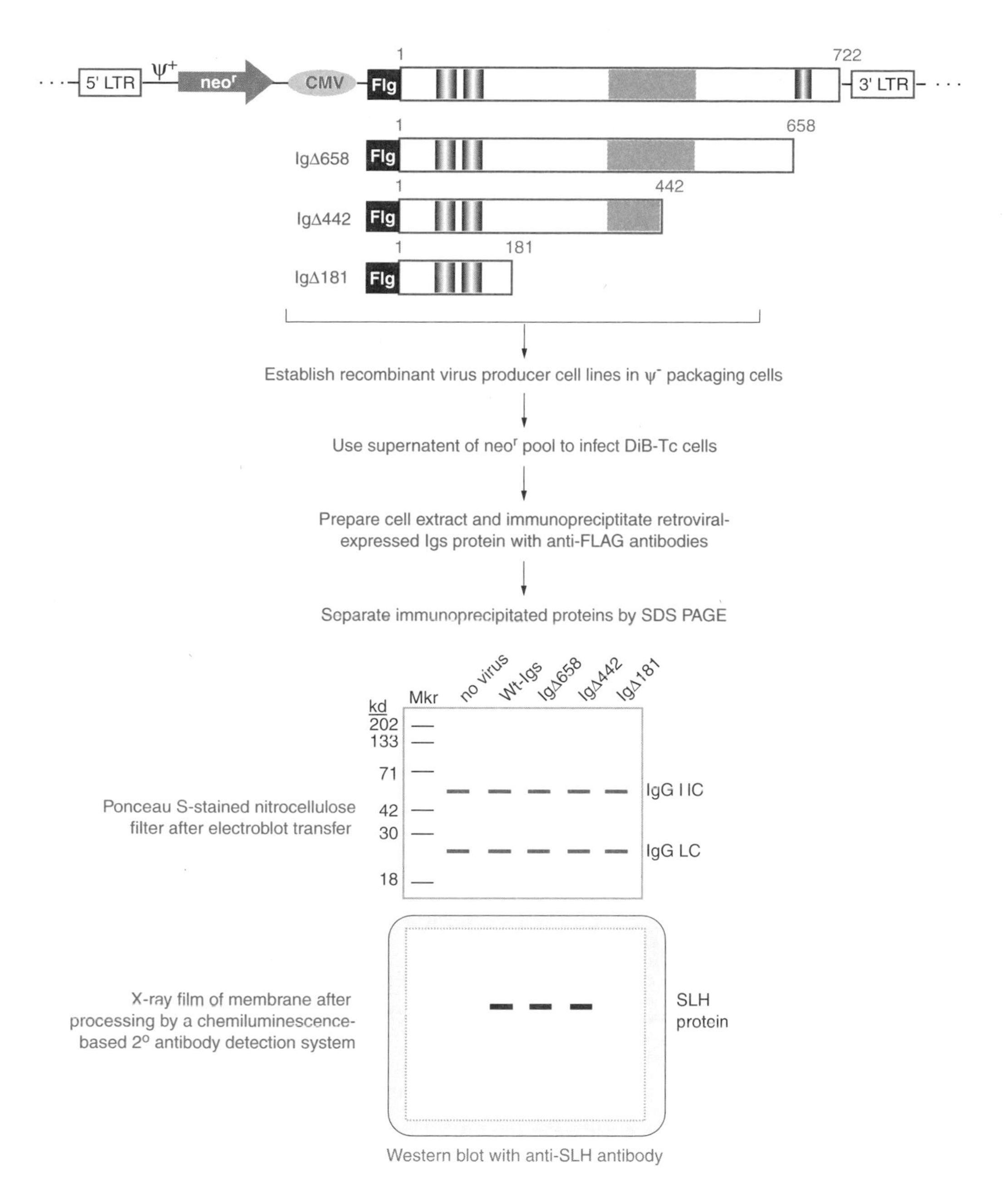

Figure 7.10 Co-immunoprecipitation assays can be used to map *Igs* residues that encode SLH interaction sites. Hypothetical results are shown from immunoprecipitation experiments using extracts from retrovirally infected cells (wild-type *Igs* and IgΔ658/IgΔ442/IgΔ181 carboxy terminal deletants) and anti-FLAG antibodies. The FLAG affinity tag is a short, highly immunoreactive, aspartate-rich amino acid sequence that can be incorporated into the amino or carboxy terminus of a heterologous protein. Proteins recovered from the immunoprecipitates are resolved by denaturing SDS PAGE and analyzed by Western blotting using the anti-SLH antibodies. A schematic representation of a reversible Ponceau S-stained nitrocellulose membrane is depicted showing the abundant immunoglobulin (IgG) heavy (HC) and light (LC) chains that would be observed. The anti-SLH antibodies would be detected with a secondary antibody chemiluminescence-based assay. In this example, the anti-SLH Western blot data would support the model that Igs and SLH proteins biochemically interact in DiB-TC cells, and that the Igs region between amino acids 442 and 181 is required for SLH binding.

SLH-dependent transcription of the insulin, glucagon, and somatostatin genes, using "dominant negative" forms of Igs protein. This could be done with some of the *Igs* deletion mutants that lack full function. A complementary approach for demonstrating Igs:SLH interactions in vivo would be to assay directly for protein–protein interactions using the yeast two-hybrid system.

Prospective

Based on the finding that DiB-TC cells are Igs-deficient, and that Igs and SLH function as heterodimers to regulate insulin gene expression, it would be important to investigate Igs function during different stages of mouse pancreas development using transgenic knockout experiments (Chapter 8) and to determine if pancreatic cells isolated from diabetic patients displayed any abnormalities in *Igs* and or *SLH* expression. Taken together, these studies may provide new insights into possible gene therapy strategies for the treatment of certain human metabolic diseases such as type 1 diabetes.

REFERENCES

Gene regulation studies

Agrawal, S., Jiang, Z.W., Zhao, Q.Y., Shaw, D., Cai, Q.Y., Roskey, A., Channavajjala, L., Saxinger, C., Zhang, R.W. Mixed-backbone oligonucleotides as second generation antisense oligonucleotides: *In vitro* and *in vivo* studies. Proc. Natl. Acad. Sci. USA 94:2620–2625, 1997.

Berger, J., Hauber, J., Hauber, R., Geiger, R., Cullen, B.R. Secreted placental alkaline phosphatase: A powerful new quantitative indicator of gene expression in eukaryotic cells. Gene 66:1–10, 1988.

Branch, A. D. A good antisense molecule is hard to find. Trends Biochem. Sci. 23:45–50, 1998.

Cambell, S., Rosen, J., Hennighuasen, L., Strech-Jurk, U., Sippel, A. Comparison of the whey acidic protein genes of the rat and mouse. Nucl. Acids Res. 12:8685–8697, 1984.

*Chalfie, M., Tu, Y., Euskirchen, G., Ward, W.W., Prasher, D.C. Green fluorescent protein as a marker for gene expression. Science 263:802–805, 1994.

*Chen, L., Glover, J., Hogan, P., Rao, A., Harrison, S. C. Structure of the DNA binding domains from NFAT, Fos and Jun bound specifically to DNA. Nature 392:42–48, 1998.

Cho, J.Y., Parks, M.E., Dervan, P.B. Cyclic polyamides for recognition in the minor groove of DNA. Proc. Natl. Acad. Sci. USA 92:10389–10392, 1995.

*De Wet, J.R., Wood, K.V., DeLuca, M., Helinski, D.R., Subramani, S. Firefly luciferase gene: Structure and expression in mammalian cells. Mol. Cell. Biol. 7:725–737, 1987.

Dickson, R.M., Cubitt, A.B., Tsien, R.Y., Moerner, W.E. On/off blinking and switching behaviour of single molecules of green fluorescent protein. Nature 388:355–358, 1997.

Escude, C., Nguyen, C., Kukreti, S. et al. Rational design of a triple helix-specific intercalating ligand. Proc. Natl. Acad. Sci. USA 95:3591–3596, 1998.

Flanagan, W.M., Wagner, R.W. Potent and selective gene inhibition using antisense oligodeoxynucleotides. Mol. Cell. Biochem. 172:213–225, 1997.

Garvin, A., Pawar, S., Marth, J., Perlmutter, R. Structure of the murine *lck* gene and its rearrangement in a murine lymphoma cell line. Mol. Cell. Biol. 8:3058–3064, 1988.

*Gorman, C.M., Moffat, L.F., Howard, B.H. Recombinant genomes which express chloramphenicol acetyltransferase in mammalian cells. Mol. Cell. Biol. 2:1044–1051, 1982.

Gottesfeld, J.M., Neely, L., Trauger, J.W., Baird, E.E., Dervan, P.B. Regulation of gene expression by small molecules. Nature 387:202–205, 1997.

Hein, R., Tsien, R.Y. Engineering green fluorescent protein for improved brightness, longer wavelengths and fluorescence resonance energy transfer. Curr. Biol. 6:178–182, 1996.

* Landmark papers in applied molecular genetics

Hennighausen, L., Fleckstein, B. Nuclear factor 1 interacts with five DNA elements in the promoter region of the human cytomegalovirus major immendiate early gene. EMBO J. 5: 1367–1371, 1986.

*Jefferson, R.A., Kavanagh, T.A., Bevan, M.W. GUS fusions: β-glucuronidase as a sensitive and versatile gene fusion marker in higher plants. EMBO J. 6:3901–3907, 1987.

Jones, J., Sullenger, B. Evaluating and enhancing ribozyme reaction efficiency in mammalian cells. Nature Biotech. 15:902–905, 1997.

Kriegler, M. Gene transfer and expression: A laboratory manual. Stockton Press, New York, 235 pp., 1990.

*Maher, L.J.,III, Wold, B., Dervan, P.B. Inhibition of DNA binding proteins by oligonucleotide-directed triple helix formation. Science 245:725–730, 1989.

Misteli, T., Spector, D. Applications of the green fluorescent protein in cell biology and biotechnology. Nature Biotech. 15:961–964, 1997.

*Ormö, M., Cubitt, A.B., Kallio, K., Gross, L.A., Tsien, R.Y., Remington, S.J. Crystal structure of the *Aequorea victoria* green fluorescent protein. Science 273:1392–1395, 1996.

*Prasher, D.C., Eckenrode, V.K., Ward, W.W., Prendergast, F.G., Cormier, M.J. 1992. Primary structure of the *Aequorea victoria* green fluorescent protein. Gene 111:229–233, 1992.

Reid, B.G., Flynn, G.C. Chromophore formation in green fluorescent protein. Biochemistry 36: 6786–6791, 1997.

Selden, R.F, Burke-Howie, K., Rowe, M.E., Goodman, H.M., Moore, D.D. Human growth hormone as a reporter gene in regulation studies employing transient gene expression. Mol. Cell. Biol. 6:3173–3179, 1986.

Vasquez, K., Wilson, J. Triplex-directed modification of genes and gene activity. Trends Biochem. Sci. 23:4–9, 1998.

Verrijzer,C.P., Tjian,R, TAFs mediate transcriptional activation and promoter selectivity, Trends Biochem. Sci. 21:338–342, 1996.

Wachter, R.M., King, B.A., Heim, R., Kallio, K., Tsien, R.Y., Boxer, S.G., Remington, S.J. Crystal structure and photodynamic behavior of the blue emission variant Y66H/Y145F of green fluorescent protein. Biochemistry 36:9759–9765, 1997.

Wenger, R.H., Moreau, H., Nielson, P.J. A comparison of different promoter, enhancer, and cell type combinations in transient transfections. Anal. Biochem. 221:416–418, 1994.

White, S., Szewczyk, J., Turner, J., Baird, E., Dervan, P. Recognition of the four Watson-Crick base pairs in the DNA minor groove by synthetic ligands. Nature 391:468–471, 1998.

Yamamoto, Y., Taylor, C., Acedo, G., Cheng, C., Conkling, M. Characterization of cis-acting sequences regulating root-specific gene expression in tobacco. Plant Cell 3:371–382, 1991.

Zlokarnik, G., Negulescu, P., Knapp, T., Mere, L., Burres, N., Feng, L., Whitney, M., Roemer, K., Tsien, R. Quantitation of transcription and clonal selection of single living cells with β-lactamase as reporter. Science 279:84–88, 1998.

Methods to express genes in cell lines

Cheng, L., Ziegelhoffer, P.R., Yang, N.-S. *In vivo* promoter activity and transgene expression in mammalian somatic tissues evaluated by using particle bombardment. Proc. Natl. Acad. Sci. USA 90:4455–4459, 1993.

*Chu, G., Hayakawa, H., Berg, P. Electroporation for the efficient transfection of mammalian cells with DNA. Nucl. Acids Res. 15:1311–1326, 1987.

Felgner, J., Kumar, R., Srider, C. et al. Enhanced gene delivery and mechanism studies with a novel series of cationic liposome formulations. J. Biol. Chem. 269:2550–2561, 1996.

*Felgner, P.L., Gadek, T.R., Holm, M., Roman, R., Chan, H.W., Wenz, M., Northrop, J.P., Ringold, G.M., Danielson, M. Lipofectin: A highly efficient, lipid-mediated DNA/transfection procedure. Proc. Natl. Acad. Sci. USA 84:7413–7417, 1987.

*Fromm, M., Taylor, L.P., Walbot, V. Expression of genes transferred into monocot and dicot plant cells by electroporation. Proc. Natl. Acad. Sci. USA 82:5824–5828, 1985.

*Gossen, M. and Bujard, H. Tight control of gene expression in mammalian cells by tetracycline-responsive promoters. Proc. Natl. Acad. Sci. USA 89:5547–5551, 1992.

Gossen, M., Freundlieb, S., Bender, G., Müller, G., Hillen, W., Bujard, H. Transcriptional activation by tetracyclines in mammalian cells. Science 268:1766–1769, 1995.

Gottesman, M.M. (ed.). Molecular genetics of mammalian cells, Vol. 151 of Methods in enzymology, 605 pp., 1987.

Gruber, D., Jayme, D. Cell and tissue culture media: History and terminology. In Cell Biology: A laboratory handbook. Academic Press, New York, pp 451–458, 1994.

Hinrichs, W., Kisker, C., Düvel, M., Müller, A., Tovar, K., Hillen, W., Saenger, W. Structure of the Tet repressor-tetracycline complex and regulation of antibiotic resistance. Science 264: 418–420, 1994.

Lopata, M.A., Cleveland, D.W., Sollner-Webb, B. High-level expression of a chloramphenicol acetyltransferase gene by DEAE-dextran-mediated DNA transfection coupled with a dimethysulfoxide or glycerol shock treatment. Nucl. Acids Res. 12:5707, 1984.

*Mann, R., Mulligan, R.C., Baltimore, D. Construction of a retrovirus package mutant and its use to produce helper-free defective retrovirus. Cell 33:153–159, 1983.

Miesfeld, R.L., Okret, S., Wikström, A.-C., Wrange, Ö., Gustafsson, J.K., Yamamoto, K.R. Characterization of a steroid hormone receptor gene and mRNA in wild-type and mutant cells. Nature 312:779–781, 1984.

Miller, A.D., Law, M-F., Verma, I.M. Generation of helper-free amphotropic retroviruses that transduce a dominant-acting, methotrexate-resistant dihydrofolate reductase gene. Mol. Cell. Biol. 5:431–437, 1985.

*Miller, D.A. Cell-surface receptors for retroviruses and implications for gene transfer. Proc. Natl. Acad. Sci. USA 93:11407–11413. 1996.

Neumann, E., Schaefer-Ridder, M., Wang, Y., Hofschneider, P.H. Gene transfer into mouse lymphoma cells by electroporation in high electric fields. EMBO J. 1:841–845, 1982.

No, D., Yao, T.P., Evans, R.M. Ecdysone-inducible gene expression in mammalian cells and transgenic mice. Proc. Natl. Acad. Sci. USA 93:3346–3351, 1996.

Parks, R., Chen, L., Anton, M., Sankar, U., Rudnick, M., Graham, F. A helper-dependent adenovirus vector system: Removal of helper viruse by Cre-mediated excision of the viral packaging signal. Proc. Natl. Acad. Sci. USA 93:13565–13570, 1996.

*Perucho, M., Hanahan, D., Wigler, M. Genetic and physical linkage of exogenous sequences in transformed cells. Cell 22:309–317, 1980.

*Potter, H., Weir, L., Leder, P. Enhancer-dependent expression of human κ immunoglobulin genes introduced into mouse pre-B lymphocytes by electroporation. Proc. Natl. Acad. Sci. USA 81:7161–7165, 1984.

*Robins, D.M., Ripley, S., Henderson, A.S., Axel, R. Transforming DNA integrates into the host chromosome. Cell 23:29–39, 1981.

*Wigler, M., Silverstein, S., Lee, L-S., Pellicer, A., Cheng, Y-C., Axel, R. Transfer of purified herpes virus thymidine kinase gene to cultured mouse cells. Cell 11:223–232, 1977.

Using yeast as a model eukaryotic cell

Adams, A., Gottschling, D., Kaiser, C., Stearns, T. Methods in yeast genetics: A laboratory course manual. Cold Spring Harbor Press, Cold Spring Harbor, NY, 200 pp., 1997.

Arnold, S.F., Klotz, D.M., Collins, B.M., Vonier, P.M., Guillette, L.J., Jr., McLachlan, J.A. Synergistic activation of estrogen receptor with combinations of environmental chemicals. Science 272:1489–1492, 1996.

Bassett, D.E., Jr., Boguski, M.S., Hieter, P. Yeast genes and human disease. Nature 379:589–590, 1996.

Baur, E., Harbers, M., Um, S.J., Benecke, A., Chambon, P., Losson, R. The yeast Ada complex mediates the ligand-dependent activation function AF-2 of retinoid X and estrogen receptors. Genes Dev. 12:1278–1289, 1998.

Bohnert, H., Jensen, R. Strategies for engineering water-stress tolerance in plants. Trends Biotechnol. 14:89–97, 1996.

Dana, S.L., Hoener, P.A., Wheeler, D.A., Lawrence, C.B., McDonnell, D.P. Novel estrogen response elements identified by genetic selection in yeast are differentially responsive to estrogens and antiestrogens in mammalian cells. Mol. Endocrinol. 8:1193–1207, 1994.

*Goffeau, A., Barrell, B.G., Bussey, H., et al. Life with 6000 genes. Science 274:546–567, 1996.

Hartwell, L.H., Szankasi, P., Roberts, C.J., Murray, A.W., Friend, S.H. Integrating genetic approaches into the discovery of anticancer drugs. Science 278:1064–1068, 1997.

Johnston, M. Genome sequencing: The complete code for a eukaryotic cell. Curr. Biol. 6: 500–503, 1996.

Lyttle, C.R., Damian-Matsumura, P., Juul, H., Butt, T.R. Human estrogen receptor regulation in a yeast model system and studies on receptor agonists and antagonists. J. Steroid Biochem. Mol. Biol. 42:677–685, 1992.

Nelson, D., Shen, B., Bohnert, H. Salinity tolerance: Mechanisms, models, and engineering of complex traits. In Genetic engineering, Vol. 20 (J. Setlow, ed.). Plenum Press, New York, pp. 153–176, 1998.

Schena, M. and Yamamoto, K.R. Mammalian glucocorticoid receptor derivatives enhance transcription in yeast. Science 241:965–967, 1988.

Thompson, E.B. Steroid hormones: Membrane transporters of steroid hormones. Curr. Biol. 5: 730–732, 1995.

Winzeler, E. Functional genomics of Saccharmyces cerevisiae. Am. Soc. Microbiol. News 63: 312–317, 1997.

Yoshinaga, S.K., Peterson, C.L., Herskowitz, I., Yamamoto, K.R. Roles of SWI1, SWI2, and SWI3 proteins for transcriptional enhancement by steroid receptors. Science 258:1598–1604, 1992.

Protein expression in cultured cells

de Boer, H.A., Comstock, L.J., Vasser, M. The tac promoter: A functional hybrid derived from the trp and lac promoters. Proc. Natl. Acad. Sci. USA 80:21–25, 1983.

*Edman, J.C., Hallewell, R.A., Valenzuela, P., Goodman, H.M., Rutter, W.J. Synthesis of Hepatitis B surface and core antigens in *E. coli*. Nature 291:503–506, 1981.

Guan, C., Li, P., Riggs, P.D., Inouye, H. Vectors that facilitate the expression and purification of foreign peptides in Escherichia coli by fusion to maltose-binding protein. Gene 67:21–30, 1987.

*Itakura, K., Hirose, T., Crea, R.M, Riggs, A.D., Heyneker, H.L., Bolivar, F., Boyer, H.W. Expression in *E. coli* of a chemically synthesized gene for the hormone somatostatin. Science 198:1053–1056, 1977.

*Kitts, P.A., Ayres, M.D., Possee, R.D. Linearization of baculovirus DNA enhances the recovery of recombinant virus expression vectors. Nucl. Acids Res. 18:5667–5672, 1990.

LaVallie, E.R., DiBlasio, E.A., Kovacic, S., Grant, K.L., Schendel, P.F., McCoy, J.M. A thioredoxin gene fusion system that circumvents inclusion body formation in the *E. coli* cytoplasm. Bio/Technology 11:187–193, 1993.

LaVallie, E.R., Rehemtulla, A., Racie, L.A., DiBlasio, E.A., Ferenz, C., Grant, K.L., Light, A., McCoy, J.M. Cloning and functional expression of a cDNA encoding the catalytic subunit of bovine enterokinase. J. Biol. Chem. 268:23311–23317, 1993.

Lunn, C.A., Pigiet, V.P. Localization of thioredoxin from Escherichia coli in an osmotically sensitive compartment. J. Biol. Chem. 257:11424–11430, 1982.

*Mackett, M., Smith, G.L., Moss, B. Vaccinia virus: A selectable eukaryotic cloning and expression vector. Proc. Natl. Acad. Sci. USA 79:7415–7419, 1982.

Miller, L.K. Baculoviruses as gene expression vectors. Ann. Rev. Microbiol. 42:177–199, 1988.

Robinson, M., Lilley, R., Little, S., Emtage, J.S., Yarranton, G., Stephens, P., Millican, A., Eaton, M., Humphreys, G. Codon usage can affect efficiency of translation of genes in Escherichia coli. Nucl. Acids Res. 12:6663–6671, 1984.

Schein, C.H. Production of soluble recombinant proteins in bacteria. Bio/Technology 7:1141–1148, 1989.

Smith, D.B. Purification of glutathione-S-transferase fusion proteins. Methods Mol. Cell Biol. 4: 220–229, 1993.

*Smith, D.B., Johnson, K.S. Single-step purification of polypeptides expressed in Escherichia coli as fusions with glutathione S-transferase. Gene 67:31–40, 1988.

*Stormo, G.D., Schneider, T.D., Gold, L. Characterization of translation initiation sites in *E. coli*. Nucl. Acids Res. 10:2971–2996, 1982.

*Strauch, K.L., Beckwith, J. An Escherichia coli mutation preventing degradation of abnormal periplasmic proteins. Proc. Natl. Acad. Sci. USA 85:1576–1580, 1988.

*Studier, F.W., Moffatt, B.A. Use of bacteriophage T7 RNA polymerase to direct selective high-level expression of cloned genes. J. Mol. Biol. 189:113–130, 1986.

*Summers, M.D., Smith, G.E. A manual of methods for baculovirus vectors and insect cell culture procedures. Texas Agricultural Experiment Station Bulletin No. 1555, College Station, Texas, 1987.

Characterization of a promoter-selective transcription factor

Brizzard, B.L., R.G. Chubet, D.L. Vizard. Immunoaffinity purification of FLAG epitope-tagged bacterial alkaline phosphate using a novel monoclonal antibody and peptide elution. BioTechniques 16: 730–734, 1994.

Epstein, J.A., T. Glaser, J. Cai, L. Jepeal, D.S. Walton, R.L. Maas. Two independent and interactive DNA-binding subdomains of the Pax6 paired domain are regulated by alternative splicing. Genes Devel. 8: 2022–2034, 1994.

Miesfeld, R.L., Godowski, P.J., Maler, B.A., Yamamoto, K.R. Glucocorticoid receptor mutants that define a small region sufficient for enhancer activation. Science 236:423–427, 1987.

Sander, M., A. Neubüser, J. Kalamaras, H.C. Ee, G.R. Martin, M.S. German. Genetic analysis reveals that PAX6 is required for normal transcription of pancreatic hormone genes and islet development. Genes Devel. 11:1662–1673, 1997.

St-Onge, L., B. Sosa-Pineda, K. Chowdhury, A. Mansouri, P. Gruss. Pax6 Is required for differentiation of glucagon-producing a-cells in mouse pancreas. Nature 387:406–409, 1997.

Zhang, X., Tran, P.B.-V., Pfahl, M. DNA binding and dimerization determinants for thyroid hormone receptor α and its interaction with a nuclear protein. Mol. Endocrinol. 5:1909–1920, 1991.

CHAPTER

8

CONSTRUCTION OF TRANSGENIC MULTICELLULAR ORGANISMS

The term genetic engineering is a label adopted by the popular media to describe the method of transferring human genes into bacteria using applied molecular genetic techniques. The public's view of molecular genetic research over the past 20 years has been a mixture of awe and fear. This perception is reinforced by the many books and movies that use a modern day version of "Frankenstein" to project a sinister side of molecular genetic research. In this context, nothing epitomizes the concept of genetic engineering more to the lay audience than the generation of transgenic multicellular organisms. In this chapter, we examine the four major types of transgenic organisms that have been engineered using standard molecular genetic techniques. The selected examples are (1) P element transformation of the laboratory fruit fly *D. melanogaster*, (2) DNA transfection of plants by bacterial infection and biolistics, (3) generation of transgenic mice as models of human disease, and (4) production of transgenic livestock.

Although not described here, two other often studied transgenic model organisms are the roundworm nematode, *Caenorhabditis elegans*, and the small tropical zebra fish, *Danio rerio*. The methods used to make transgenic worms and fish are similar to those for other organisms and involve microinjection of plasmid expression vectors into fertilized eggs. *C. elegans* and *D. rerio* transgenesis has primarily been used to investigate early developmental processes in vertebrates. The last section of this chapter describes a method called nuclear transfer. This developmental biology technique has recently been used to generate mice, sheep, and cattle from diploid somatic cells. Nuclear transfer has been used for animal "cloning" and can be coupled with standard molecular genetic methods using stably transfected somatic cells.

MOLECULAR GENETICS OF DROSOPHILA DEVELOPMENT

No single multicellular organism exemplifies the power of molecular genetics more than the fruit fly *D. melanogaster*. This innocuous insect has been studied for more than a century as a laboratory model of embryonic development. As is so elegantly described in Peter Lawrence's book *The Making of a Fly*, the process of embryonic development in *Drosophila* begins with the translation of maternally derived mRNAs that are present in the egg. These maternal gene products encode signaling proteins and transcription factors that induce a highly orchestrated set of molecular events at the onset of embryogenesis. Ultimately, this transcription factor cascade leads to the regulated expression of structural proteins and intercellular signaling molecules, which together specify the framework of the *Drosophila* embryo.

Up until the late 1970s, geneticists were able only to infer what nuclear processes were responsible for the developmental pathways that directed individual cells to become neurons, myoblasts, or structural components of fly appendages. With the advent of molecular cloning techniques, however, many of these same geneticists became adept at using their mutant fly stocks to pinpoint the physical location of specific gene coding sequences within the *Drosophila* genome. Although gene expression studies using Northern blots and in situ hybridization studies were informative, it wasn't until it became possible to integrate cloned DNA into *Drosophila* germ cells that we really began to understand how developmental processes are controlled. In this section, we describe the *Drosophila* gene transfer technique that was worked out by Allan Spradling and Gerry Rubin to create fertile transgenic flies. This method, called P element transformation, has allowed *Drosophila* geneticists unequivocally to demonstrate which *Drosophila* genes are responsible for the host of developmental mutants that have intrigued biologists for decades.

P Element-Mediated Transformation

Transposable elements are movable genetic "info-bytes" that exist in the genomes of most organisms as self-sustaining entities. The process of integration and excision of transposable elements is called transposition. Transposons contain two functional regions: (1) DNA repeat sequences at the termini of the transposon that are required for genomic integration and excision and (2) an internal segment encoding a transposase enzyme that is required for transposition. Examples of other transposons are the Ac/Ds elements of maize, the IS1 and Tn10 elements found in *E. coli*, and the *S. cerevisiae* transposon Ty1. Through a genetic analysis of a phenomenon known as hybrid dysgenesis, Margaret Kidwell and others discovered that most *Drosophila* species contain transposons, the best characterized of which is the P element transposon. The P element is a 2.9 kb DNA segment encoding an 87 kDa transposase enzyme flanked by an inverted repeat sequence of 31 bp. *Drosophila* in the wild contain about 30–50 genomic copies of randomly integrated P element sequences.

Based on transposition studies in other systems, Spradling and Rubin reasoned that cloned P element sequences could be used to construct *Drosophila* expression vectors that would induce "terminal" transposition of cloned DNA into *Drosophila* germ cells. The trick they used was to replace the transposase coding sequence of a cloned P element with both a marker gene to identify transgenes and the experimental gene to be tested. By co-injecting *Drosophila* embryos with this modified P element transposition plasmid and a second expression plasmid encoding the P element transposase, Spradling and Rubin pioneered a reliable method to promote integration of cloned genes into the *Drosophila* germ line. Because the transposition plasmid lacks transposase, and the co-injected transposase expression plasmid is not capable of integration, a stable transgenic *Drosophila* strain can be isolated following one round of mating. Figure 8.1 illustrates how P element transformation can be used to produce a transgenic fly that expresses the *lacZ* gene in specific cells known to be important for establishing pattern formation during *Drosophila* embryogenesis. These transgenic reporter gene experi-

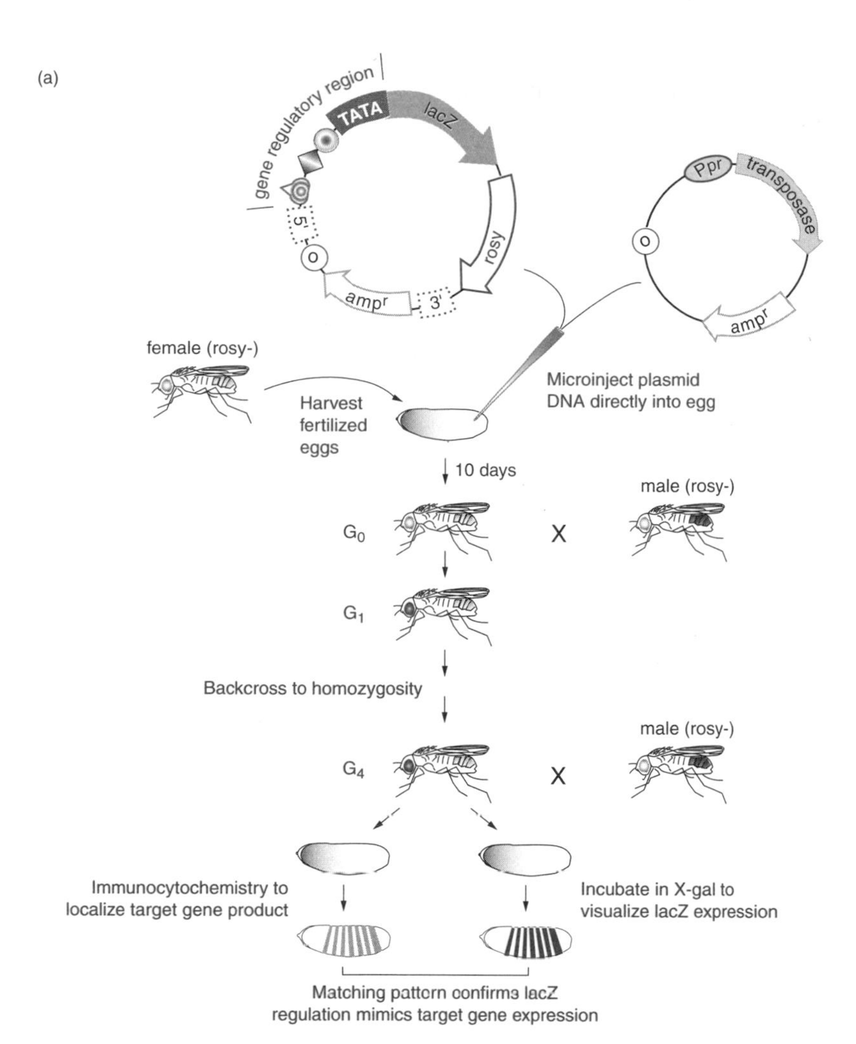

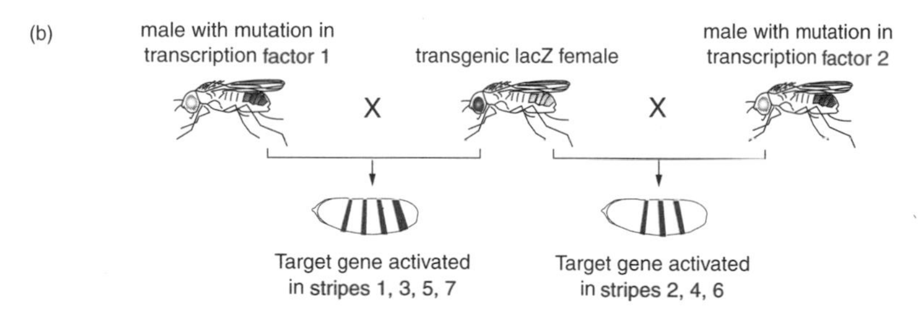

Figure 8.1 P element transformation is a method to obtain germ-line DNA integration following co-injection of preblastoderm embryos with purified plasmid DNA. Typically, the marker gene on the P element transformation plasmid encodes an eye color marker (e.g., the rosy gene, *ry*) that can be used to identify transgenes by virtue of their red eyes (*ry*+), against a background of nontransformed flies with white eyes (*ry*–). (*a*) In this example, the 5′ regulatory region of a developmentally expressed gene is inserted upstream of the *lacZ* reporter gene. Second generation females (G_1) with *ry*+ eyes contain the P element insertion in every cell and can be used to monitor expression of the *lacZ* reporter gene in early embryos. Identification of strains with single P element insertions and genetic backcrossing is performed with G_1 flies to obtain a homozygous isogenic strain. As shown in this example, the protein expression patterns of the endogenous target gene and *lacZ* reporter gene overlap in early embryos, indicating that the cloned target gene regulatory region is fully functional. (*b*) Matings between transgenic *lacZ* females and males with defects in genes encoding transcription factors required for embryo development can be used to determine which transcription factors control cell-specific expression of the target gene.

ments were invaluable in deciphering the transcriptional circuitry that underlies early events in *Drosophila* development. The frequency of obtaining fertile transgenes by P element transformation is ~5% based on the number of embryos injected.

P Element Enhancer Trap Vectors

Walter Gehring and his colleagues designed a special P element cloning vector that contained the *lacZ* gene fused in-frame with the second exon of the transposase. This "enhancer-less" *lacZ* P element was used to identify cell-specific genes by identifying transgenic flies that express β-galactosidase in subsets of cells based on histocytochemical staining with X-gal. Because enhancer regulatory elements modulate transcriptional promoters in a manner independent of distance and orientation, this P element enhancer trap vector need only integrate in the vicinity of a gene regulatory region. Moreover, the weak promoter of the transposase-*lacZ* fusion gene is insufficient to allow significant transcription in the absence of a linked enhancer element. Enhancer trap strategies have been used to mark specific cells for fate analysis during different stages of development, as well as to clone novel genes based on enhancer activation of the *lacZ* marker gene.

A variation of the original enhancer trap strategy was developed by Andrea Brand and Norbert Perrimon using P element insertions to identify cell-specific enhancers that selectively direct expression of the yeast *Gal4* gene contained on the enhancer trap vector. By crossing one of these *Gal4* transgenes with a fly containing an integrated copy of a *Gal4*-responsive *lacZ* expression vector, they showed it was possible to obtain cell-specific expression of β-galactosidase activity. This elegant *Gal4* enhancer trap strategy provides fly strains that can be used in a variety of experimental crosses to direct cell-specific expression of experimental target genes. Figure 8.2 shows how the *Gal4* enhancer trap strategy is initially used with the *lacZ* target gene to identify fly strains that contain insertions of the *Gal4 P* element vector in genomic regions controlling transcription of tissue-specific genes.

CONSTRUCTION OF TRANSGENIC CROP PLANTS

Two approaches have traditionally been taken to increase the quality and quantity of crop yields. First, plant breeders have incorporated desirable traits into seed stocks using standard genetic manipulations. However, because not all desirable traits are found within a single plant type, it has not been possible to breed a "super" plant crop with this approach. A second way to increase crop yields is to inhibit the deleterious effects of insects and weeds by spraying crops with selective pesticides and herbicides. The overuse of these agents on a long-term basis may cause a negative impact on the environment. One way to circumvent the limitations of plant breeding and chemical treatment is to use molecular genetic strategies to engineer transgenic crop plants that are of higher quality owing to increased resistance to drought, salinity, herbicides, and insects. In this section, two strategies are described that have been used to create transgenic crop plants: (1) transformation of dicotyledon (dicots) plants such as soybean, squash, and tomato by *Agrobacterium*-mediated bacterial conjugation, and (2) particle bombardment of monocotyledon (monocots) plants, represented by rice, corn, and wheat, using biolistic devices.

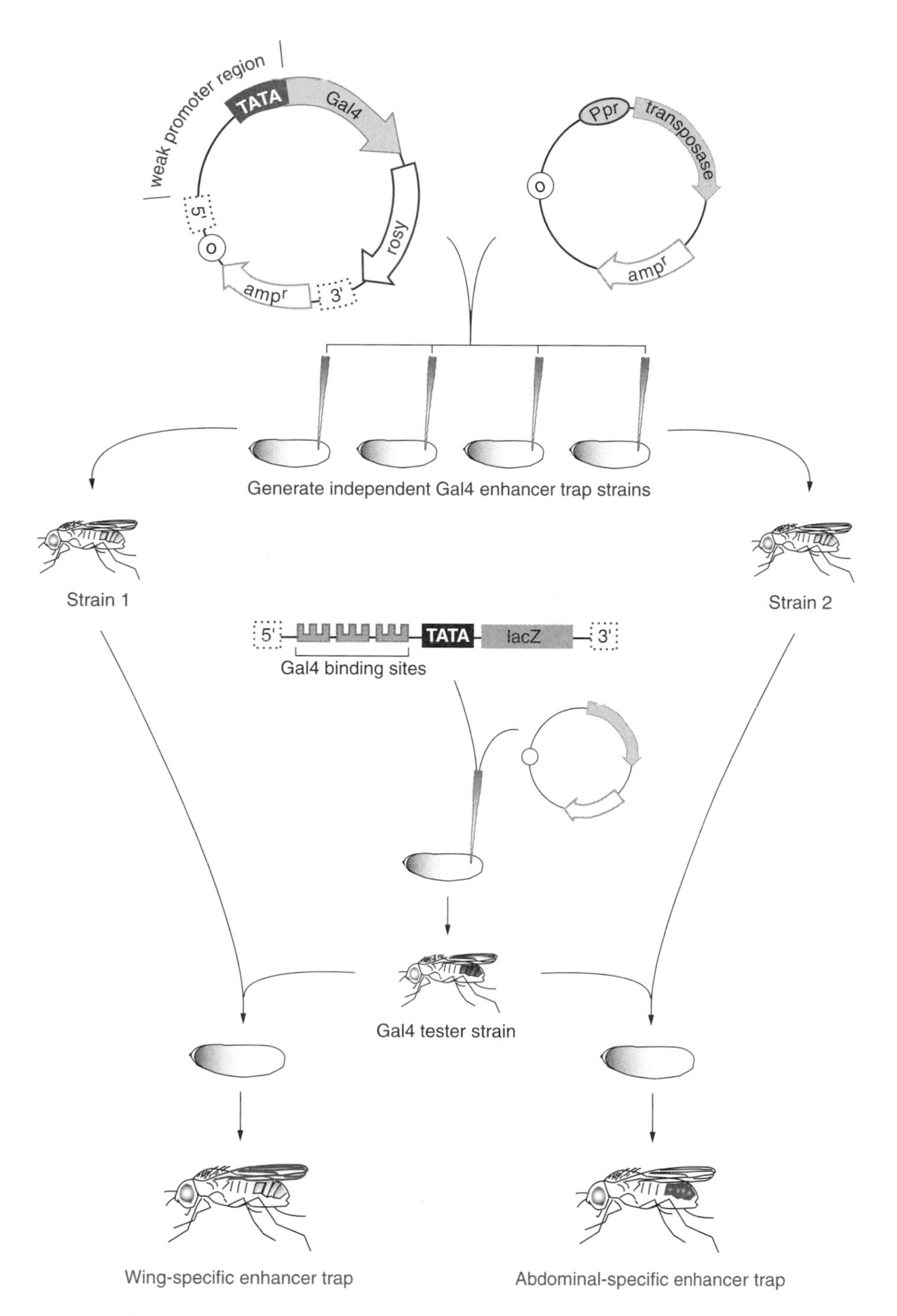

Figure 8.2 The *Gal4* enhancer trap strategy is used to identify cell-specific enhancers by screening for transgenic flies that display restricted expression of a Gal4-responsive *lacZ* gene. In this example, strain 1 contains a *Gal4* enhancer trap insertion in the vicinity of a wing-specific gene, and strain 2 identifies a transcriptional regulatory region for an abdominal segment gene. Cell-specific enhancer trap fly strains identified in this way have proved to be useful tools for studying dysregulated gene expression of a variety of experimental target genes. The *Gal4* enhancer trap insertion can result in a mutant phenotype by gene disruption, which provides a way to identify developmental genes by recovering genomic sequences adjacent to the *Gal4* coding sequence.

Agrobacterium-mediated Gene Transfer

Plants have several unique characteristics that influence their suitability for transgenesis. Most importantly, many types of dicots can be manipulated in tissue culture by hormones and nutrients to permit whole plant regeneration from a dedifferentiated tissue mass called a callus. Moreover, plants cannot move, they undergo self-fertilization, and they produce a large number of progeny that are easy to harvest. Genetically, however, plants can present some problems with regard to gene copy number and clonality. First, some plant genomes are polyploid, meaning that they contain many copies of the same genome, which can make it difficult to interpret genetic crosses. Second, although plant tissue culture is a powerful means to regenerate whole plants, the resulting material can display somaclonal variation, which indicates that genomic alterations occur in somatic cells. Nevertheless, researchers have been able to define conditions that provide a "window of opportunity" for gene transfer through molecular genetic intervention. This can be followed by tissue regeneration and genetic backcrossing to produce a true-breeding transgenic plant.

The most common gene transfer method in dicot plants is to expose callus tissue to special laboratory strains of the soil bacterium *Agrobacterium* as a means to transfer DNA material directly from a recombinant bacterial plasmid into the host plant cell. *Agrobacterium* was first identified as the pathogen responsible for the appearance of crown gall tumors in areas of plant wounding. The best characterized strain of *Agrobacterium* is *A. tumefaciens*, which transfers a ~25 kb segment of DNA, called T-DNA, into plant cells by a transfer mechanism that resembles conjugation and appears to involve a bacterial pilus structure. The conjugation process is stimulated by plant hormones that are released from areas of tissue wounding. The T-DNA "transforming" genes are encoded on a ~200 kb single copy *A. tumefaciens* plasmid known as the Ti-plasmid. Three plant biologists, Jef Schell, Marc Van Montagu, and Gene Nester, have been credited with the development of T-DNA-mediated plant transformation methods.

To mediate the gene transfer process, *Agrobacterium* proteins recognize and cleave Ti plasmid DNA at short repeat sequence elements called the right border (RB) and left border (LB), which define the termini of T-DNA. Because T-DNA is stably integrated into the plant cell genome following gene transfer, cloned DNA sequences contained within the region between the LB and RB repeats are transferred to the host plant cell. Recombinant T-DNA plasmid vectors contain a number of functional units in addition to the RB and LB elements; these include (1) a broad host range replication origin such as the RK *ori*, (2) an antibiotic-resistance gene for plasmid selection in bacteria, (3) a multiple cloning site (MCS) located between LB and RB for insertion of the cloned gene, and (4) a dominant selectable marker gene with a plant transcriptional promoter for selection in regenerating plants. Figure 8.3 outlines the key steps in a binary *Agrobacterium*-mediated plant transformation strategy based on electroporation of the recombinant T-DNA plasmid into *Agrobacterium*.

Transgenesis of Rice and Corn with Biolistics

The majority of important food crops in the world are monocots, such as rice, corn, and wheat. With few exceptions, monocot plants are resistant to *A. tumefaciens* transformation owing to inappropriate wound responses and hormonal differences between dicots and monocots. Therefore, DNA transfection strategies for monocots initially depended on the use of DNA microinjection and electroporation methods targeted to isolated pro-

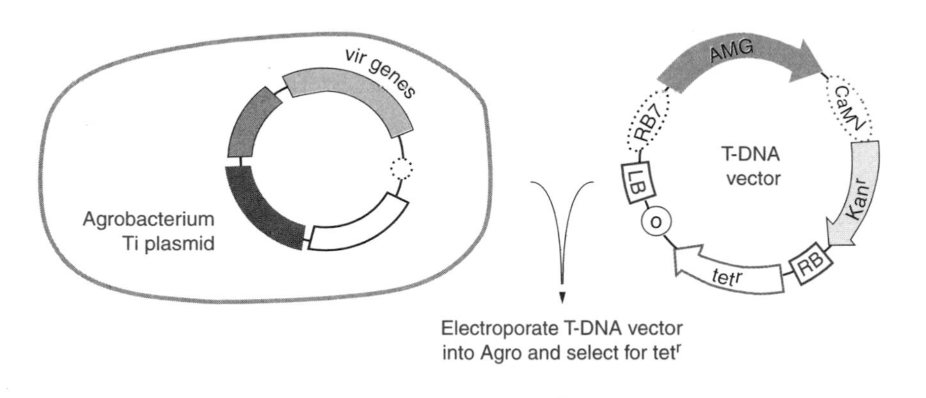

Electroporate T-DNA vector
into Agro and select for tet^r

Expose wounded plant cells
to transformed Agro strain

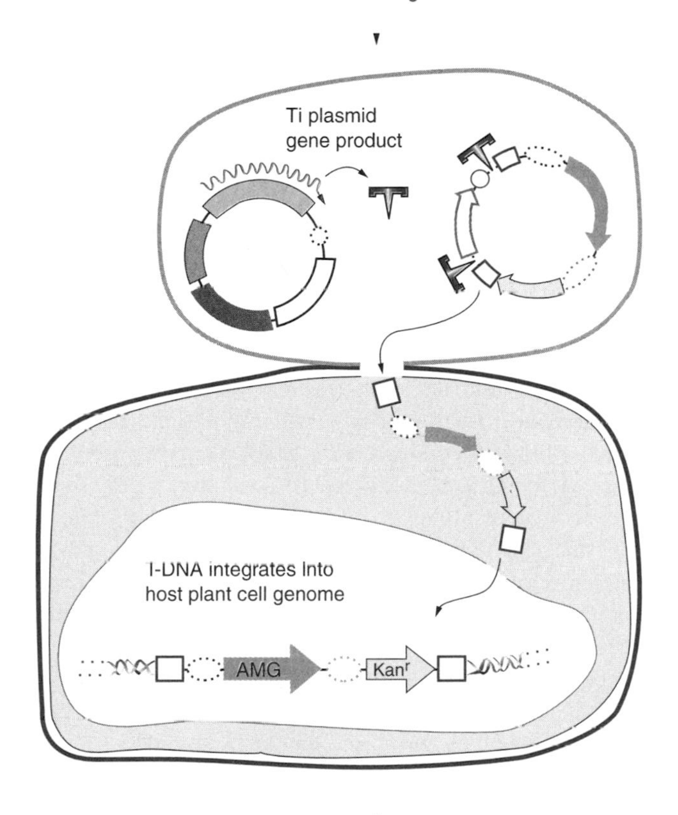

Induce plant regeneration and
select for Kan^r cell growth

Figure 8.3 *Agrobacterium*-mediated plant transformation has been exploited as an efficient gene transfer method in a variety of dicot plants. The T-DNA vector plasmid shown here contains sequences required for integration into the plant genome (LB and RB), a root-specific plant promoter (RB7) driving expression of the hypothetical *AMG* gene, the constitutive CaMV plant gene promoter upstream of the kanamycin resistance gene (*kan^r*), tetracycline resistance (*tet^r*) gene downstream of bacterial Tn10 promoter, and a broad host replication origin (RK). Electroporation of the recombinant T-DNA plasmid into an *Agrobacterium* strain containing a Ti-plasmid encoding *vir* genes, but lacking LB and RB sequences, establishes a binary plasmid system in the presence of tetracycline. Exposure of wounded plant tissue to this *Agrobacterium* strain promotes T-DNA transfer and random integration of the cloned genes into the plant genome. Plant regeneration in the presence of kanamycin would produce a transgenic plant with root-specific *AMG* expression.

toplasts (plant cells lacking cell walls). However, although these methods do result in DNA uptake and limited gene expression by protoplasts, insufficient regeneration protocols for monocot protoplasts severely restrict the usefulness of these transfection approaches. Molecular biologists are legendary for their brute force solutions to experimental problems, and the strategy behind developing a successful monocot transgenesis protocol is no exception. The goal was to use meristematic plant tissue as a target for DNA transfection, which would avoid problems associated with protoplast-derived plant regeneration. It was reasoned that by stably integrating expression vector DNA into plant cells that normally give rise to the floral meristem, and eventually gametes, marker gene selection could be used to identify even rare integration events using first-generation seedlings. John Sanford developed what can only be described as a "shotgun approach" to monocot transgenesis. This *attack* DNA transfection strategy (see Chapter 7) was first done by shooting DNA-coated metal projectiles out of a 22 caliber gun that was aimed directly at meristematic regions of whole plants. Amazingly, the gene gun approach works relatively well, and it is now known by the more technical term of biolistic-mediated DNA transfection.

Biolistic devices have become much more sophisticated and are routinely used to create transgenic rice, corn, and sorghum plants. Figure 8.4 diagrams two commercially available gene guns that are based on the principle of particle propulsion.

Agricultural Impact of Transgenic Crop Plants

It is inevitable that within the next decade, almost all economically important plant crops will be derived from transgenic strains. Within the past 5 years alone, transgenic varieties of more than 25 commercially grown plants have been approved for agricultural purposes by the U.S. Department of Agriculture (USDA). It is estimated that by 1999, well over 50% of all soybean plants grown in the United States will contain a herbicide-resistance gene that protects the crop from treatment with the broad-spectrum herbicide Round-Up™, a shikimic acid pathway inhibitor biochemically known as glyphosate. This rapid introduction of foreign genes into food producing plants has not been without controversy. Consumer advocate groups as well as concerned scientists have raised two major issues regarding the wholesale use of transgenic crops on a worldwide basis. The first is assessing the rate of horizontal gene transfer from crop plants to weedlike relatives and to indigenous wild-type varieties. One could imagine that without proper crop management practices, the transfer of herbicide- and pest-resistance genes to weeds could be problematic if it were to occur at a high rate. USDA sponsored research in the fields is now underway to determine the actual rates of horizontal gene transfer to address these concerns. A second issue is the possible long-term effects of using antibiotic-resistance genes as transgenic markers. It has been suggested that antibiotic-resistant strains of bacteria may develop as a result of gene transfer from plants to insect-borne bacteria and other soil microbes, or by antibiotic selection in livestock and humans through residual antibiotics retained in processed food. To confront this potential consequence, plant molecular geneticists have designed transgenesis strategies that include marker gene inactivation following the initial genetic screen.

Transgenic crop varieties are quickly becoming common choices for farmers; however, some aspects of the agricultural industry are still driven by consumer demand. An example of this economic principle is the FLAVR SAVR™ tomato. This transgenic variety of an essential dinner salad ingredient can withstand the adverse effects of mechanical processing because these tomatoes express an antisense RNA that is directed

(a)

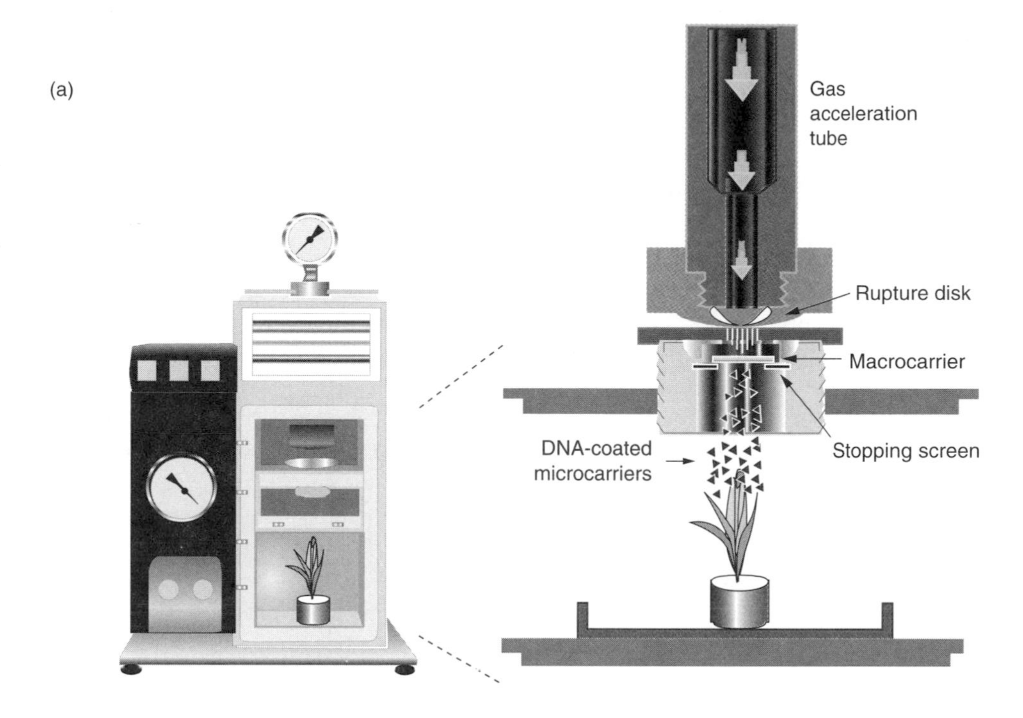

(b)

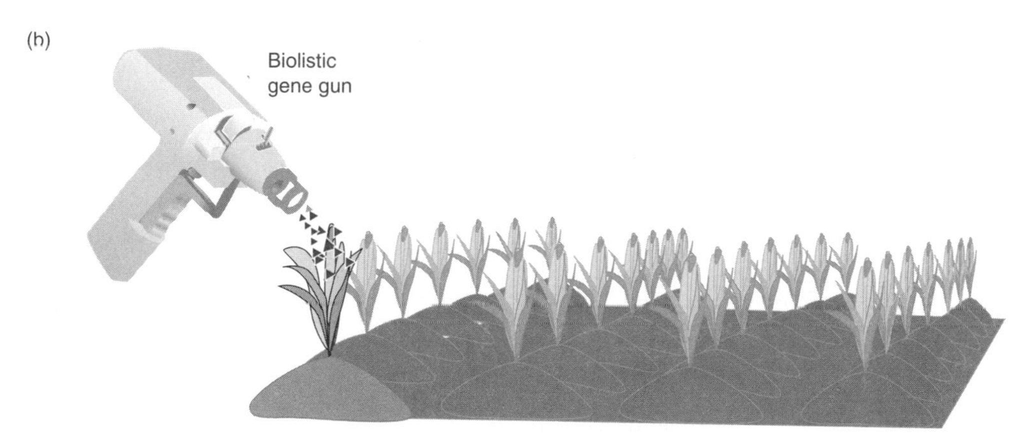

Figure 8.4 Biolistic-mediated DNA transfection of monocot seedlings can be used to generate transgenic plants. (*a*) Schematic drawing of the Bio-Rad Biolistic PDS-1000/HE transformation system. This instrument uses an evacuated sample chamber that positions the target seedling directly in the path of DNA-coated tungsten or gold particles (1–2 μm in diameter) traveling at a velocity of ~500 meters/second (1100 mph!). The particles are released from the macrocarrier when it slams into the stopping screen mounted within the chamber. The initial burst of helium gas that launches the macrocarrier results from a pressure-dependent rupturing of a disk located just above the macrocarrier holding position. (*b*) A recent development in biolistic-mediated plant DNA transfection has been to use a pressurized hand-held gene gun to shoot DNA-coated gold particles directly into the meristematic tissue of plants growing in experimental fields.

against the enzyme polygalacturonase. Because of an inhibition of fruit softening due to overripening, FLAVR SAVR™ tomatoes can be harvested later in the growing season. Transgenic tomatoes, however, have yet to be successful at the grocery store.

MOUSE TRANSGENESIS

The laboratory mouse has traditionally been the mascot of modern biological research. The picture of white mice (or rats) running through mazes in laboratory psychology tests is probably the most common image society has of this furry mammal. However, to molecular geneticists, especially those involved in biomedical research, the laboratory mouse is the mammalian equivalent of *E. coli* because it provides the critical link between observational medical science and mechanistic experimental studies. The emphasis in this section is to describe the two major approaches that molecular geneticists have taken to adopt transgenic mouse models to the study of complex human diseases. The first method involves DNA microinjection into single-cell embryos to produce transgenic mice containing one or more copies of randomly integrated expression vector DNA. A second transgenesis method generates mouse lines that contain specific genetic alterations as a result of homologous recombination events at defined genomic loci. These two molecular genetic advances had their beginnings in the early 1900s when Abbie Lathrop of Granby, Massachusetts decided it would be profitable to breed mice for house pets. Her efforts eventually gave rise to a large number of well-defined laboratory mouse strains that provided geneticists with the necessary tools to develop hundreds of transgenic mouse models for biomedical research.

Generating Transgenic Mice Using Fertilized Egg Cells

Germ-line transmission of microinjected DNA in transgenic mice was accomplished in 1981 by a number of teams of independent researchers, many of whom went on to become leaders in the field of mouse transgenesis. These molecular genetic pioneers include, among others, Frank Ruddle, Elizabeth Lacy, Ralph Brinster, Rudolph Jaenisch, Martin Evans, and Gail Martin. The laboratory methods they refined were based on a working knowledge of early mouse development. Unlike the elegant transposon-based approaches of fruit fly transgenesis, or T-DNA transformation in plants, DNA microinjection of single-cell mouse embryos is an *attack* DNA transfection strategy (Chapter 7). Microinjection is done using a microneedle to inject the male pronuclei of fertilized eggs with about a picoliter of buffer containing several hundred molecules of linearized DNA as shown in Figure 8.5. The surviving intact eggs are implanted into a host female mouse. Three weeks after birth of the pups, genomic DNA from mouse tail snips are analyzed by PCR for the presence of the experimental DNA. Mice that score positive by this test are then bred to establish founder transgenic lines. The frequency of transgenesis is about 10% of the live pups, which represents only about 2% of the injected eggs. Figure 8.6 outlines the general procedure for producing a transgenic mouse by embryo microinjection.

A significant breakthrough in mouse transgenesis, and the key factor in generating cell-specific gene knockouts (laboratory practicum 8), has been the use of experimental DNA expression cassettes that are based on tightly regulated cell-specific promoters. As described in Chapter 7, cell-specific promoters provide a means to direct expression to a limited number of cells owing to the restricted expression of cell-specific transcription

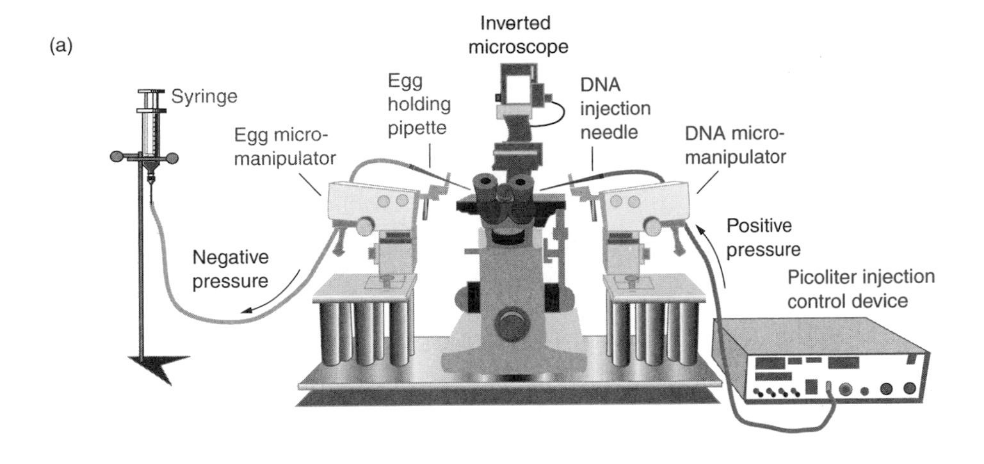

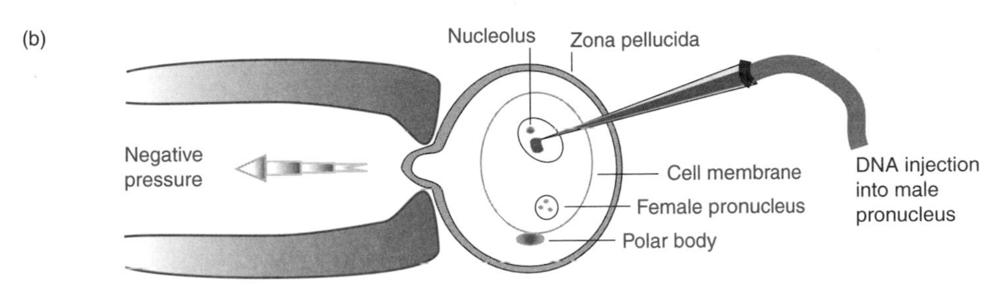

Figure 8.5 Purified experimental DNA is directly injected into the male pronucleus of a fertilized mouse egg using a microscope-mounted manipulator and microneedle. (*a*) The three essential instruments for embryo injection are an inverted microscope with stage magnification of 400×, a micromanipulator-mounted negative pressure holding pipette (left), and a micromanipulator-mounted injection pipette (right). (*b*) The male pronucleus of a fertilized egg is injected with picoliter amounts of purified DNA (1 μg/ml) by stabilizing the egg using the holding pipette. Success depends on the use of properly timed fertilized eggs and accurate injection of the male pronucleus.

factors. For example, founder mice obtained by microinjection of an experimental DNA construct containing the mouse lymphocyte-specific tyrosine kinase (*lck*) gene regulatory region linked to a *CAT* reporter gene display CAT activity only in T lymphocytes, even though the *lck-CAT* expression cassette is contained in every cell of the mouse. This can be a very powerful strategy because it permits the investigation of cell-specific phenotypes throughout mouse development. Figure 8.7 illustrates how a cell-specific promoter could be used to target expression of a dominant oncogene to a single mouse tissue type as a means to generate an animal model of cancer. The transgenic adenocarcinoma mouse prostate model (TRAMP), recently developed by Norm Greenberg and his colleagues, is one example of this approach.

Gene Knockouts by Homologous Recombination in ES Cells

It became clear in the mid 1980s that to take full advantage of molecular genetic approaches, it was necessary to generate transgenic mice with specific genotypes. This

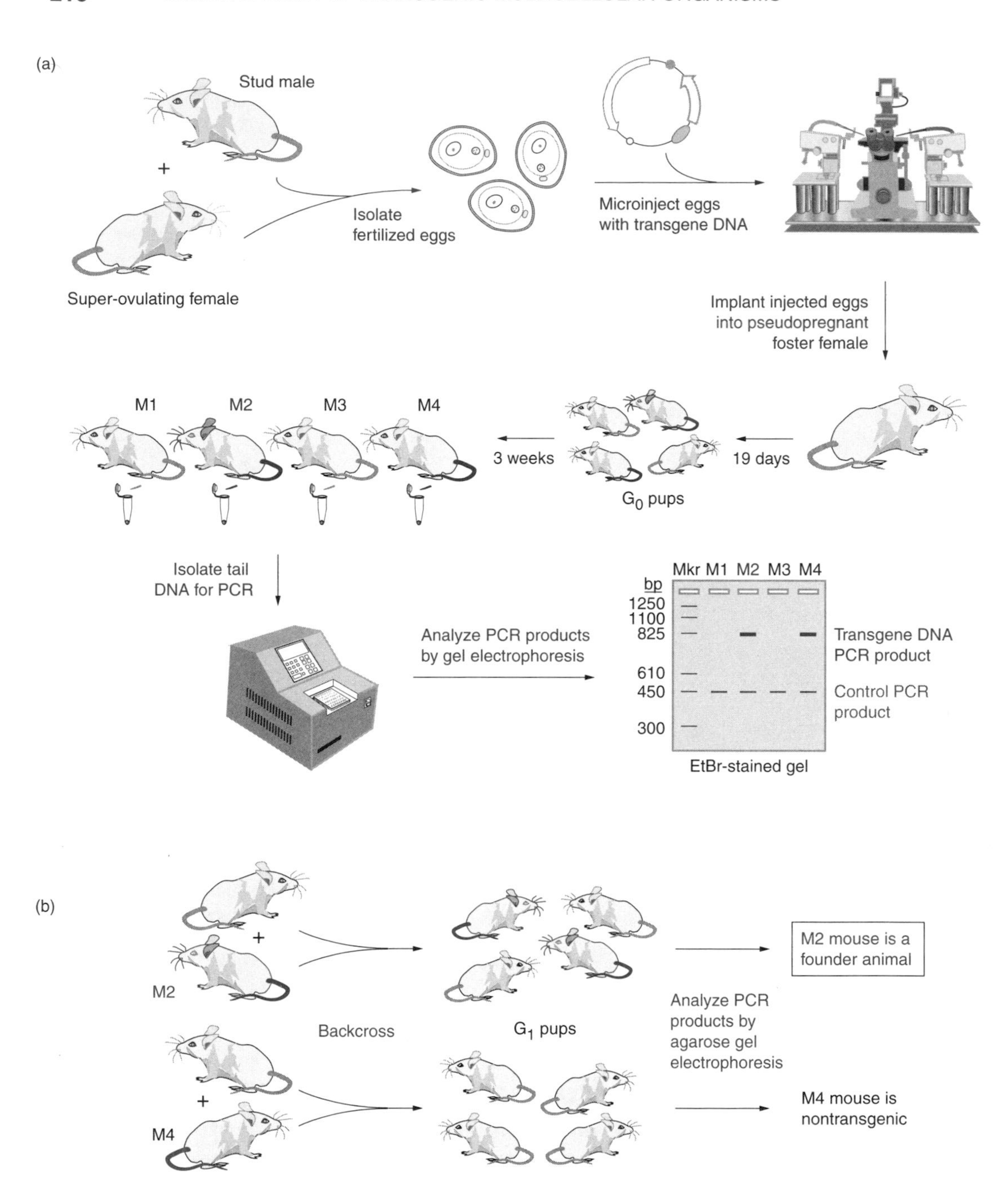

Figure 8.6 Transgenic founder mice are identified by backcrossing PCR-positive first-generation littermates to nontransgenic mates. (*a*) Fertilized one-cell eggs for DNA microinjection are obtained from superovulating female mice that have been mated to a stud male. Following microinjection, 20–30 eggs are implanted into the oviduct of a pseudopregnant surrogate female (recently mated to a vasectomized male) leading to the production of ~5–8 live pups 19 days later. Transgenic littermates are identified by PCR analysis of tail samples performed at 3 weeks of age. (*b*) Mice that score positive by PCR analysis are backcrossed to nontransgenic mates to identify founder animals containing germ-line DNA integration that results in a Mendelian inheritance pattern of the transgene.

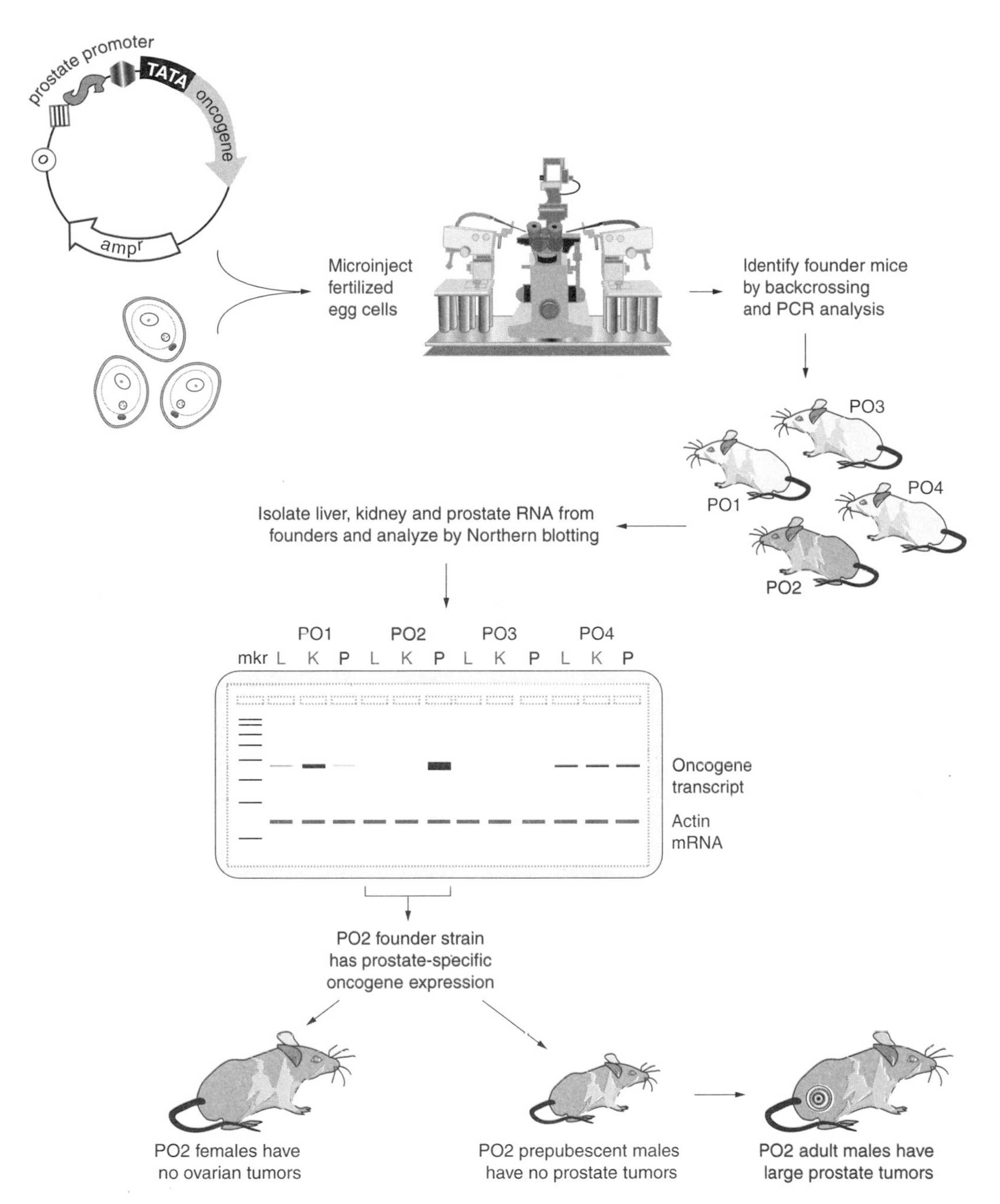

Figure 8.7 Cell-specific promoters can be used to direct expression of a foreign gene to tissue subtypes in transgenic mice. An experimental flow scheme is shown to illustrate how a prostate-specific promoter could be used to generate a transgenic mouse model of prostate cancer. Because DNA integration into regions of the mouse genome influences the cell-specific function of the transgene promoter, Northern blot analysis is needed to identify founder mice with the expected pattern of transgene expression (PO2). Characterization of prostate tissue in the offspring of founder mice would be done to determine how well the transgenic model mimics human prostate cancer.

conclusion came from the realization that random integration of gene expression cassettes often led to variable results that were difficult to interpret. Ideally, mouse molecular geneticists wanted to target integration of the experimental DNA into specific chromosomal locations, in much the same way that yeast geneticists are able to obtain homologous gene replacement in yeast cells. However, because the rate of homologous recombination in mammalian cells is orders of magnitude lower than it is in yeast, researchers needed to screen embryos for rare homologous DNA recombination events.

Two methodological breakthroughs were required before transgenic "knockout" (KO) mice could be created that contained germ-line insertional loss-of-function mutations. The two key discoveries were (1) procedures to cultivate pluripotent mouse embryonic stem (ES) cells that could be combined with nontransgenic mouse blastocysts to produce a chimeric embryo, and (2) the development of DNA cloning vectors and cell line screening strategies that allow the identification of transfected cells that have undergone a homologous recombination event following transfection with experimental DNA. Each of these procedures is described and then several examples are given to illustrate how knockout mice have been used as experimental models of human genetic diseases.

Martin Evans and Gail Martin were the first to report methods to isolate pluripotent ES cells from the inner cell mass (ICM) of a 3 day old mouse embryo. They independently showed that ES cells could be grown in culture under conditions that maintained pluripotential functions. Within a few short years, methods were developed to modify ES cells genetically, by either retroviral infection or stable DNA transfection, in a way that permitted ES cells to be used as embryonic material for generating transgenic mice. This feat was accomplished by exploiting microinjection techniques to deliver ~10 ES cells into freshly isolated 3 day old embryos, which can then be re-implanted into pseudopregnant females to produce chimeric mice 17 days later. Because only a portion of the ES cells injected into the embryo give rise to gametic precursors, the offspring must be bred to nontransgenic mice to identify founders. This screening procedure was facilitated by the use of ES cells and preimplantation embryos from mice with different coat color markers to identify chimeric offspring readily. For example, ES cells isolated from *agouti* mice (yellow-black fur) have a dominant mutation on chromosome 2 that produces mice with a coat color that is distinct from non-*agouti* (black fur) and *albino* (white fur) mice that were derived from donor blastocysts. Figure 8.8 illustrates how stably transfected ES cells can be generated starting with freshly isolated blastocysts, and Figure 8.9 shows how ES-derived transgenic founder mice are identified based on coat color.

The most commonly used method to create KO transgenic mice is to insert the *neo*^{r} gene into the desired target gene by transfecting ES cells with DNA sequences that promote homologous recombination while at the same time permitting genetic selection and/or sensitive screening assays to detect rare targeted integration events. This gene targeting strategy was first described by Oliver Smithies and co-workers. A few years later, a research group led by Mario Cappechi demonstrated the feasibility of using similar homologous recombination strategies to achieve targeted gene integration in cultured ES cells. Numerous improvements in the targeting vectors and screening strategies have led to two basic types of gene targeting strategies. The simplest type is based on a gene insertion process that requires a crossover event between two ends of the targeting vector and the genomic sequence. The net result of gene targeting by insertion is that there is a duplication of the target sequences with a concomitant insertion of a selectable marker gene such as the *neo*^{r} gene (Fig. 8.10a). A second type of targeting vector utilizes a gene replacement strategy. Gene replacement vectors contain two selectable markers that pro-

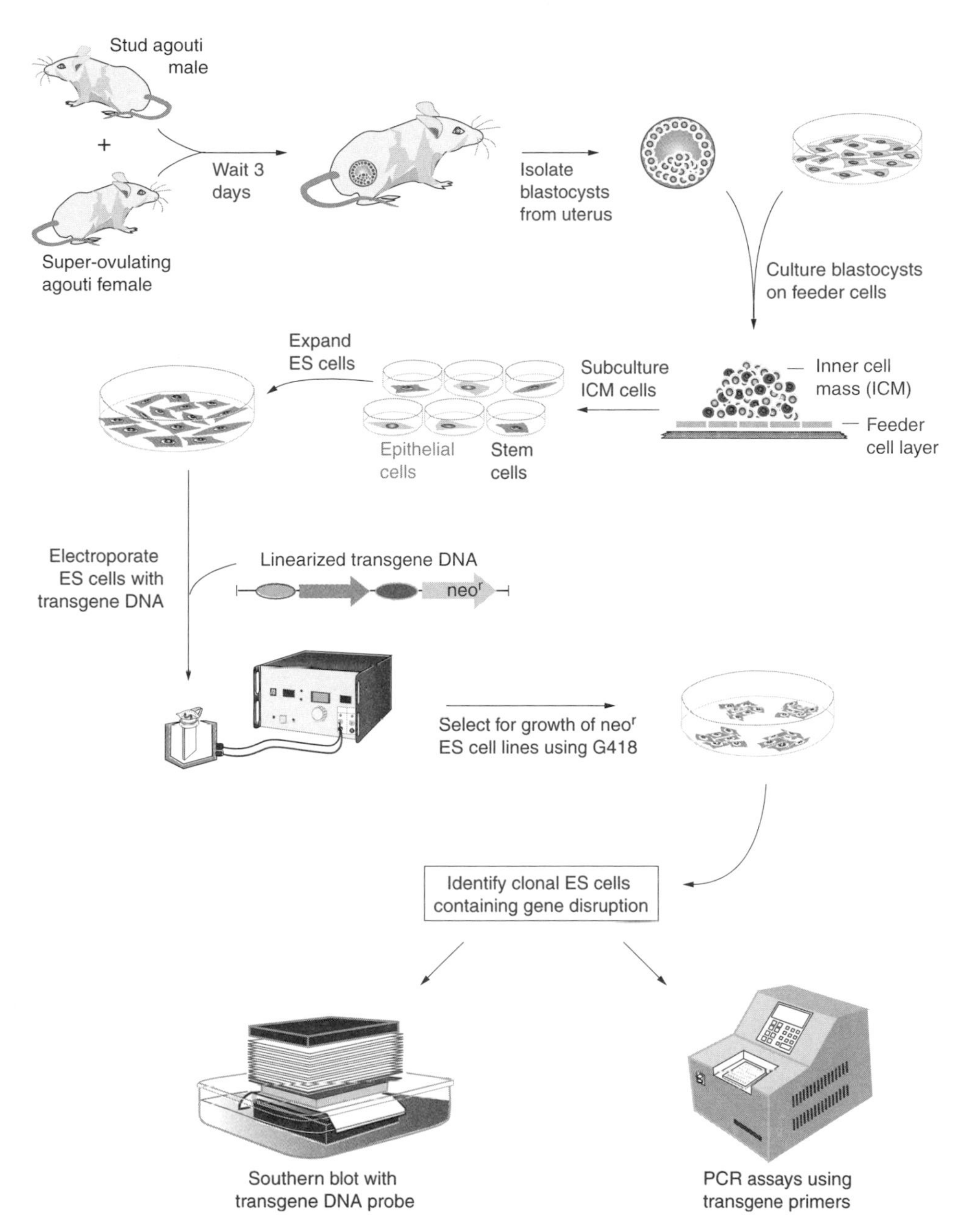

Figure 8.8 Mouse embryonic stem (ES) cells can be manipulated in culture using standard DNA transfection methods. A pluripotent ES cell line is derived from the inner cell mass (ICM) of blastocyst embryos obtained from the uterus of a female mouse 3 days post-fertilization. ES cells are cultured on plates containing nondividing embryonic fibroblasts called feeder cells. Maintenance of the pluripotent state of ES cell cultures depends on the presence of factors such as LIF (leukemia inhibitory factor), which are secreted from the feeder cells or added directly to the media. Experimental DNA is stably transfected into ES cells by electroporation or lipofection using vectors containing a selectable marker such as *neo*^{r}. PCR screening strategies and Southern blotting are used to identify clonal ES sublines containing targeted gene knockouts.

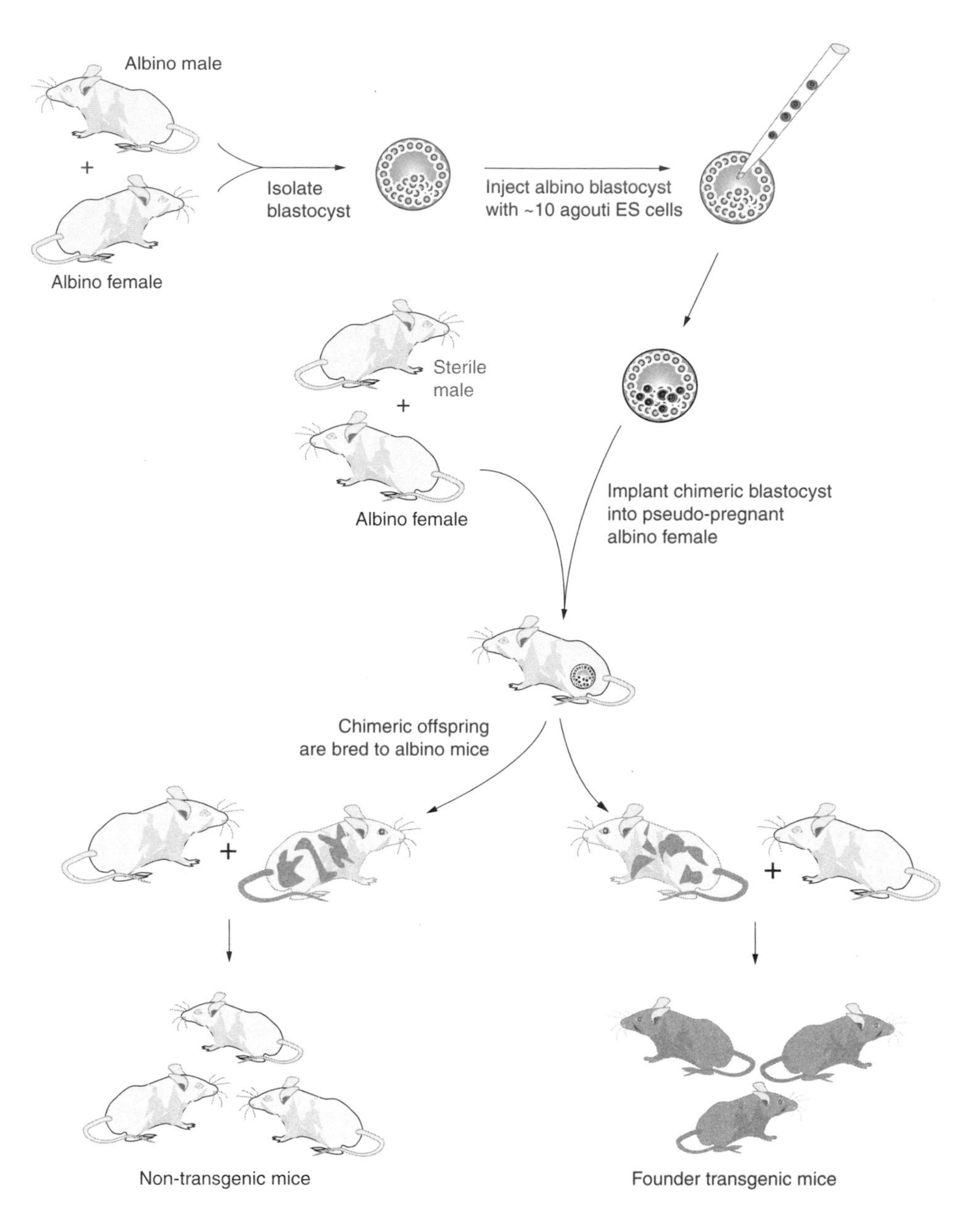

Figure 8.9 Differences in the coat color of ES donor mice and blastocysts host mice are used to identify transgenic founder animals. In this example, chimeric embryos are constructed by injecting *agouti* ES cells into blastocysts from an *albino* donor mouse. The offspring have a mosaic coat color reflecting their chimeric genotype. Transgenic founder lines are identified as *agouti* offspring that arise from crossing chimeric and *albino* mice.

vide a convenient method to identify ES cells that have undergone a positionally defined double-crossover event in the target gene. As shown in Figure 8.10*b*, homologous recombination events using a gene replacement vector produce ES cells that are resistant to both G418 (expression of *neo*^{r} gene) and ganciclovir (loss of the *HSV tk* gene). Ganciclovir (GC) is a suicide nucleotide analogue that is phosphorylated by the *HSV tk* gene product, but is not a substrate for mammalian thymidine kinases.

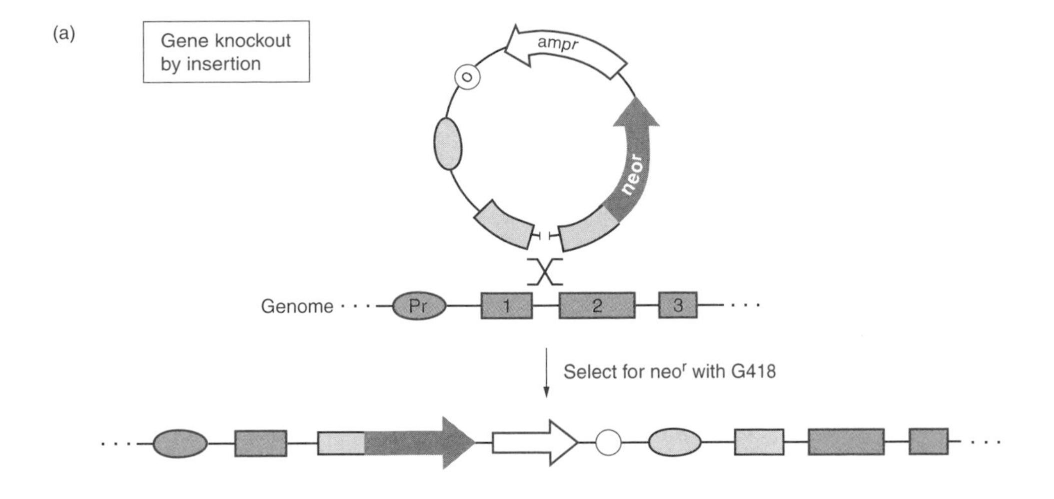

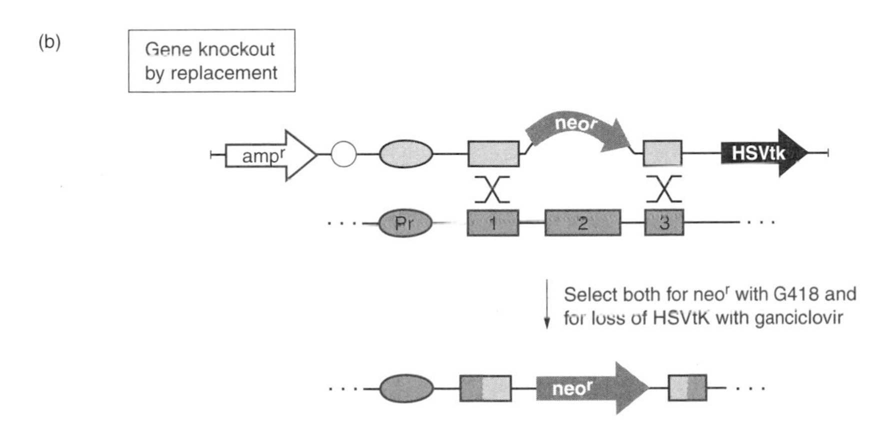

Figure 8.10 Targeting vectors for gene KO strategies result in either gene insertion or gene replacement depending on the arrangement of targeting sequences and marker genes within the vector. (*a*) Gene insertion leads to disruption of the genomic gene copy owing to a duplication of the targeting sequences and incorporation of the *neo^r* coding sequence. (*b*) Gene replacement vectors contain both positive (*neo^r*) and negative (*HSV tk*) marker genes that are used to identify drug-resistant ES cells containing a predicted double-crossover event in the target locus. Gene targeting events are identified in G418-resistant ES cells by PCR screening and Southern blotting using strategies that are designed to distinguish between homologous recombination and random DNA integration.

The frequency of homologous recombination is ~1%, based on comparing the number of G418-resistant colonies to the number of G418- and ganciclovir-resistant colonies. This corresponds to an overall homologous recombination frequency of ~10^{-5}, when considering the total number of ES cells transfected. Factors influencing the frequency of homologous recombination are (1) the length and arrangement of homologous sequences in the targeting construct, (2) the extent of homology between genomic sequences contained in the targeting vector and the target gene, and (3) the chromosomal location of the gene target. Because only one copy of the target gene is disrupted in the ES cell, it is necessary to perform a heterozygous cross between the first generation of transgenic mice to obtain a homozygous line, which, depending on the gene KO, could produce an embryonic lethal phenotype. Interestingly, in more cases than initially anticipated, homozygous KO mice have been constructed that have no detectable mutant phenotype. This result can often be attributed to redundant gene functions that are present in the mouse.

Transgenic Mouse Models of Human Diseases

Mice have 19 autosomal chromosomes and one sex-linked chromosome, many of which contain large segments of DNA that are highly conserved between mouse and humans. The field of mouse genetics is well-developed, and over the past century, a large number of genetic mutations have been mapped that disrupt important metabolic, immunological, and neurological pathways common to all mammals. One of the goals of mouse transgenesis is to use molecular genetic approaches to create better mouse models of human diseases that are based on known genetic lesions. In the past 5 years, most of the transgenic mouse lines have been derived from gene KO strategies, but more recently, methods have been developed that use regulated expression systems or cell-specific gene deletions, to mimic human disease pathophysiology better. Multigenic disorders may also be amenable to biomedical studies as more transgenic founder lines become

TABLE 8.1 Examples of transgenic mouse models of human diseases

Type of disorder	Human disease	Altered mouse gene	Gene locus	Transgenic model
Metabolic diseases	Familial hyper-cholesterolemia	Low-density lipoprotein receptor	*Ldlr*	KO, ectopic
	Hyperlipoproteinemia	Apolipoprotein E	*Apoe*	KO, ectopic
	Tay-Sachs disease	Hexokinase A	*HexA*	KO
	Gaucher disease	Glucocerebrosidase	*Gba*	KO
	Gout	Urate oxidase	*Uox*	KO
Hematological diseases	Hemophilia A	Coagulation factor VIII	*C/8*	KO
	α-Thalassemia	α-Globin gene cluster	*Hba*	KO
	Sickle cell anemia	α-Globin and β-globin	*Hba*, *Hbb*	KO, ectopic
	Chronic granulomatous	Cytochrome b-245	*Cybb*	KO
Neurological diseases	Huntington disease	Huntington disease gene homologue	*Hdh*	KO
	Spinocerebellar ataxia	Spinocerebellar ataxia I	*Sca1*	Ectopic
	Alzheimer's disease	Amyloid β precursor protein	*App*	KO, ectopic
	Ataxia telangiectasia	Ataxia telangiectasia	*Atm*	KO
Oncological diseases	Familial colon cancer, nonpolyposis types	DNA repair genes	*Msh2*, *Mlh1*, *Pms2*	KO
	Breast cancer	Breast cancer 1, 2	*Brca1*, *Brca2*	KO
	Li-Fraumeni syndrome	Transformation-related protein 53	*Trp53*	KO, ectopic
	Familial retinoblastoma	Retinoblastoma-1	*Rb1*	KO

established, characterized, and widely available. Table 8.1 lists examples of mouse transgenic models that have been developed using gain-of-function ectopic gene expression systems and/or loss-of-function transgenic KO mice.

DEVELOPMENT OF TRANSGENIC LIVESTOCK

In this last section, we examine the use of transgenic animals as bioreactors for the production of recombinant proteins that have pharmaceutical applications. There is nothing particularly novel about the science of large animal transgenesis, for it is based on methods first used to generate transgenic mice. Currently, transgenic sheep, goats, pigs, and cattle are primarily produced by pronuclear DNA microinjection of fertilized egg cells, rather than by embryonic stem cell transfection, which has not been successfully applied to livestock animals. Although there have been numerous research reports demonstrating the feasibility of animal "pharming," a term referring to the use of farm animals to produce pharmaceutical products, to date relatively few pharmaceutical products derived from transgenic livestock are ready to be approved for human medical use. This is far fewer than would have been predicted based on the number of proof-of-principle large animal transgenesis experiments that were reported in the mid 1980s.

There are a number of reasons for the slow development of commercial transgenic livestock. First, unlike transgenic crop plants, which have short generation times and produce thousands of progeny (seeds) per individual plant, large farm animals propagate at a comparatively slow rate and generate only small numbers of transgenic offspring. In addition, the inherent variability of gene expression levels, as a result of random DNA integration events arising from egg cell microinjection, leads to an unpredictable outcome that is only further exacerbated by low reproduction rates. Second, there has been public concern that food products derived from transgenic animals, for example, low-fat meat products or humanized cows' milk, may not be widely accepted by consumers, owing to a perception that unknown risk factors could be associated with the production of cloned gene products. Third, animal rights activists have questioned the bioethics of animal pharming on the basis of its impact on decreased genetic diversity, and have questioned the animal safety of proposed large-scale production procedures.

Because of these formidable problems, the established agricultural and pharmaceutical companies have not invested significant resources in large animal transgenesis. However, this has begun to change as a result of two breakthroughs in this branch of the biotechnology industry. First, recent human clinical testing of milk-borne pharmaceuticals, such as antithrombin III and α-1-antitrypsin, have shown high biological activity of these products with no overt side effects. Second, Ian Wilmut and colleagues at the Roslin Institute in Edinburgh reported in the spring of 1997 that they were able to clone a sheep using nuclear transfer methods. Because of this renewed economic interest in the molecular genetic methodologies of milk-borne animal pharming, and in the application of nuclear transfer to animal cloning, we need to take a closer look at how these techniques are currently being performed.

Animal Pharming Using Transgenic Livestock

Several early attempts at producing genetically improved livestock animals by transgenesis indicated that regulated cell-specific promoters were going to be important components of protein expression vectors in animals. For example, high level expression of the porcine growth hormone gene in transgenic pigs was accomplished in 1985 by several groups that used a constitutive promoter to drive expression. Although it was found that these transgenic animals grew faster and had lower fat deposits than nontransgenic controls, the unregulated expression of growth hormone led to serious health problems in the pigs, which included joint disease, infertility, and metabolic disorders. Because of this unexpected outcome, and the uncertainty that transgenesis could be used to improve livestock-derived food products, researchers instead began to develop methods to use animals as bioreactors to synthesize pharmaceutically important proteins. The most successful of these approaches has been to exploit the properties of milk as a renewable resource to produce functional enzymes, antibodies, and structural proteins using existing dairy manufacturing processes. The strategy has been to use the transcriptional promoters of mammary-specific genes to direct the expression of soluble transgenic proteins. Because milk-borne proteins do not cross into the blood or lymphatic fluid of the transgenic animal, there is less danger of producing undesirable side effects in the animal owing to the species origin or biochemical function of the transgenic protein.

In addition to the use of transgenic animals to synthesize pharmaceutical proteins in milk, cell-specific mammary gland expression can also be exploited to produce nutrient enriched dairy products. One such example has been to use transgenic cows to produce milk that contains human lactoferrin, an iron-transporting protein that is present in high levels in human breast milk. It should also be possible to produce specialized cheeses and cream products more efficiently using milk from transgenic animals that contains proteins normally added later in the manufacturing process. Table 8.2 lists some of the transgenic animals that have been engineered to produce human proteins in milk.

Cloning Animals by Nuclear Transfer

It is has been estimated that the cost of developing a single founder transgenic cow using pronuclear injection methodology can be as much as $500,000 owing to the associated cost of inefficient transgenesis (>97% failure rate) and variable gene expression levels.

TABLE 8.2 Some human proteins that have been expressed in the milk of transgenic "pharm" animals

Human gene product	Pharmaceutical use	Mammary gland-specific promoter	Transgenic animal
Factor IX	Blood clotting protein, treatment of hemophilia B	Sheep β-lactoglobin	Sheep
α-1-Antitrypsin	Protease inhibitor, treatment of emphysema and cystic fibrosis	Sheep β-lactoglobin	Sheep
Antithrombin III	Blood clotting protein, treatment of ATIII deficiency disease and use in open heart surgery	Cow casein	Goat
Tissue plasminogen activator	Dissolves blood clots, used as an acute treatment of heart attacks	Mouse whey acidic protein	Goat
Lactoferrin	Iron transport protein, infant formula additive	Cow α-S-casein	Cow
Protein C	Anticoagulant, treatment of hemophilia and used for surgery	Mouse whey acid protein	Pig

Moreover, extensive breeding of the founder line is required to produce sufficient numbers of homozygous transgenic animals to constitute a cost-effective herd, and this could take up to 10 years.

To overcome the economic limitations of using livestock transgenesis for commercial applications, developmental biologists in agrobiotechnology have been trying for years to find conditions under which nuclear material from somatic diploid cells could be used as a pluripotent source of genetic information. In this way, founder animals could be quickly reproduced without the associated costs of breeding. Moreover, if the donor cell could be genetically manipulated in cell culture using standard stable DNA transfection protocols, or better yet, subjected to site-directed homologous recombination, then transgenic pharm animals would be as common as transgenic mice. As a first step toward this goal, Ian Wilmut and Keith Campbell, along with their colleagues at the Edinburgh biotechnology company PPL Therapeutics, used a modified nuclear transfer method to produce a viable sheep using genetic material from a somatic cell that had been isolated from the mammary gland of a pregnant 6 year old ewe. Using electroporation, the donor cell was fused to an unfertilized sheep egg cell that had been enucleated by micromanipulation. These reconstituted egg cells gave rise to 29 multicellular embryos that were implanted into 13 surrogate female sheep. A single lamb was born 148 days later that the Roslin Institute researchers named "Dolly," in honor of their favorite country singer, Dolly Parton. Subsequent molecular genotyping proved that Dolly's DNA was indeed derived from the donor cell nuclei. Dolly was later shown to be fertile (the old-fashioned way!), further substantiating that the regeneration process initiated by nuclear transfer was genetically complete. Figure 8.11 outlines Wilmut's animal cloning method, which is based on maintaining the nuclear donor cell in low concentrations of serum for several days prior to electroporation to induce cell cycle arrest in G_0. However, G_0 quiescence may not be required for transgenesis in all large animals, because nuclear transfer experiments using bovine fetal fibroblasts have shown that actively dividing cells can also be used successfully.

In the first few months following Wilmut's scientific publication, pictures of Dolly could be found on the covers of magazines and newspaper throughout the world, usually with an accompanying story about the possibility of cloning humans. By mid-1997, animal cloning results using genetic material from fetal-derived donor cells were announced by several other biotechnology companies involved in bioproduction and pharming using transgenic livestock. One of these animals was "Gene," a bull that had been cloned by ABS Global of DeForest, Wisconsin, using donor cells from a 30-day old fetus.

The culmination of almost half a century of DNA-based biochemical, developmental, and cell biological research came in December 1997 when the Roslin Institute published its latest results describing the birth of two more transgenic lambs named Molly and Polly. What made these transgenic pharm animals special was that they were generated from fetal donor cells that had been stably transfected with a mammary-specific β-lactoglobulin expression vector containing the human Factor IX gene. Molly and Polly were not only genetically identical to each other with respect to sheep genes, but they also each possess the potential to produce large quantities of the human blood clotting protein Factor IX in their milk. Figure 8.12 outlines how the Roslin Institute team combined a standard DNA co-transfection protocol with nuclear transfer technology to produce cloned transgenic sheep from transfected fetal fibroblasts.

By mid-1998, at a time when people began to think that the generation of Dolly the sheep from *adult* somatic cells may be the exception rather than the rule in animal cloning, Ryuzo Yanagimachi and his colleagues at the University of Hawaii reported

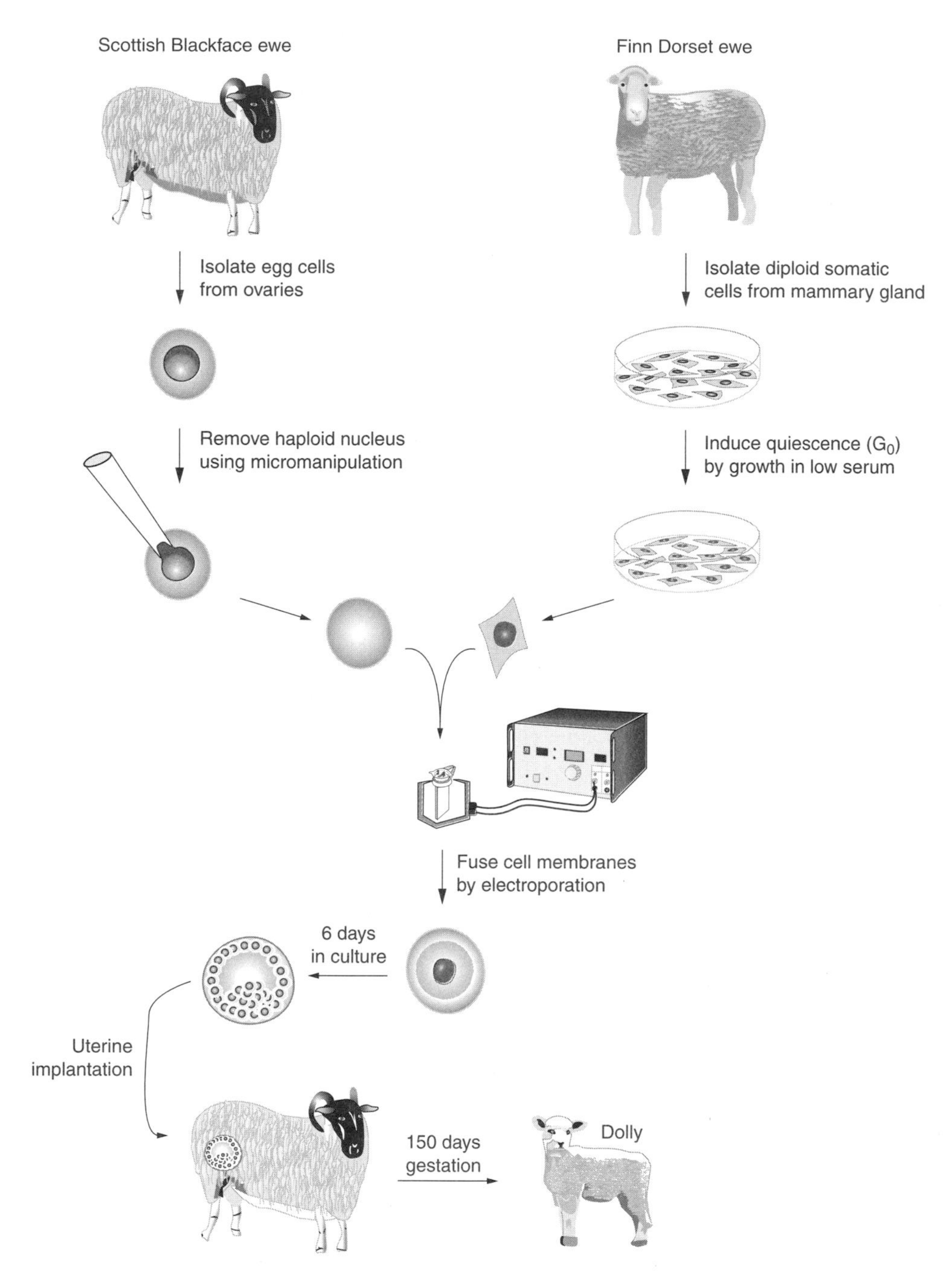

Figure 8.11 Nuclear transfer can be used to clone livestock animals from enucleated egg cells that have been fused to a donor somatic cell by electroporation. Flow scheme depicting how Dolly the sheep was cloned from a reconstituted egg cell of a Scottish Blackface ewe, using genetic material derived from cells that had been isolated from the mammary gland of an adult Finn Dorset ewe. Dolly was the only one of 277 cell fusions to result in a live birth.

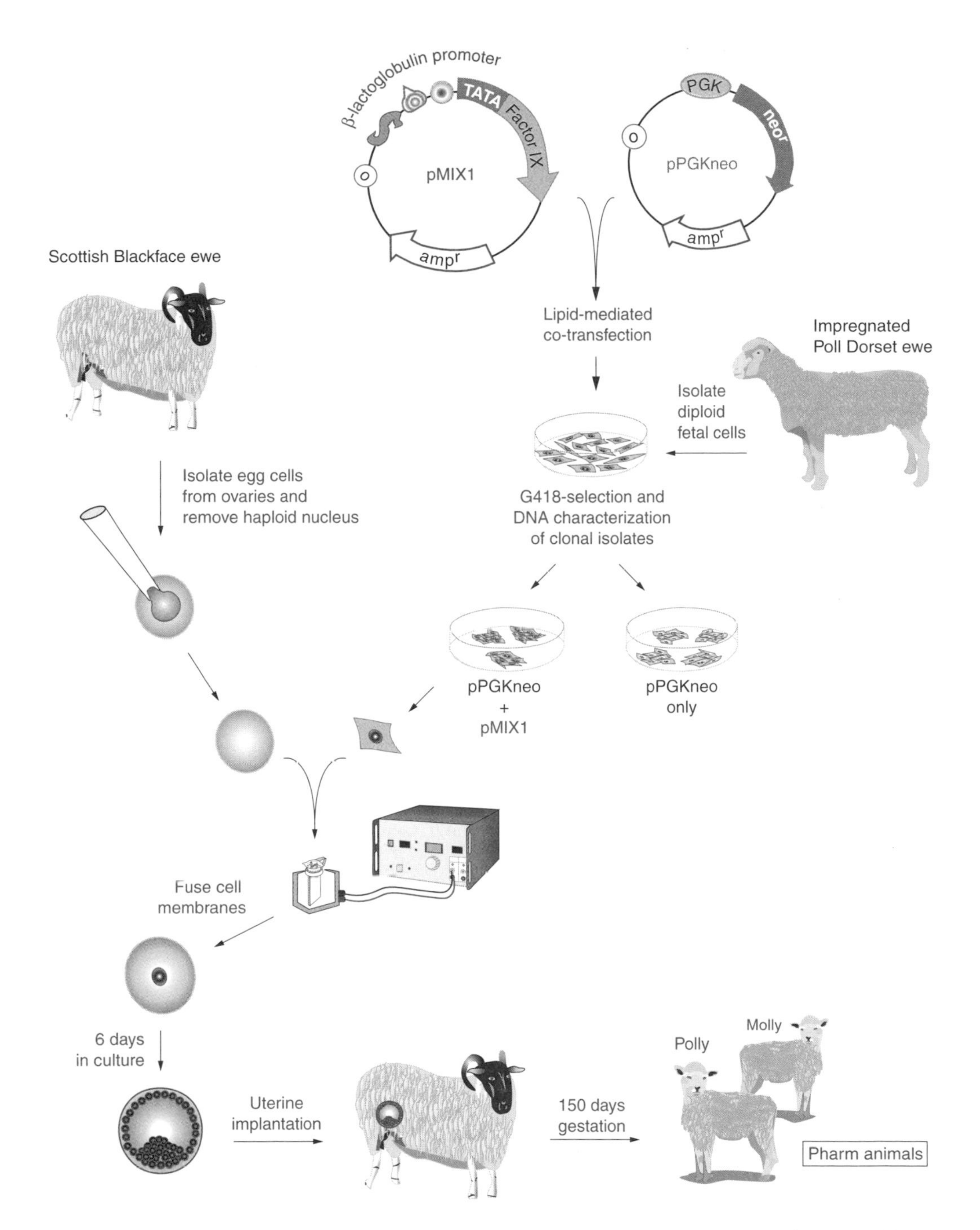

Figure 8.12 Pharm animals can be generated by combining the molecular genetic techniques of DNA co-transfection and nuclear transfer. Flow scheme illustrating how the Roslin Institute researchers used lipofection-mediated gene transfer to transfect diploid fetal donor cells, isolated from a Poll Dorset sheep embryo, stably with a mammary gland-specific expression vector containing human Factor IX cDNA (pMIX1). The ovine β-lactoglobulin (BLG) promoter upstream of the Factor IX coding sequence on pMIX1 had previously been shown to direct high level expression of a heterologous gene in sheep mammary glands. In this co-transfection strategy, a second plasmid was included that encoded the *neo*^{r} gene expressed from the constitutive phosphoglycerate kinase promoter (pPGKneo), which provided a selectable marker for stable transfectants with G418. Molly and Polly represent the first two cloned transgenic pharm animals shown to contain a human gene of pharmaceutical importance.

another breakthrough. Figure 8.13 shows the procedure they pioneered called the "Honolulu technique," that was initially developed to clone laboratory mice. The most significant difference between the Honolulu technique, and the membrane electrofusion method developed by Ian Wilmut (Fig. 8.11), is Yanagimachi's use of direct nuclear injection into enucleated eggs. The first transgenic animal made with this improved

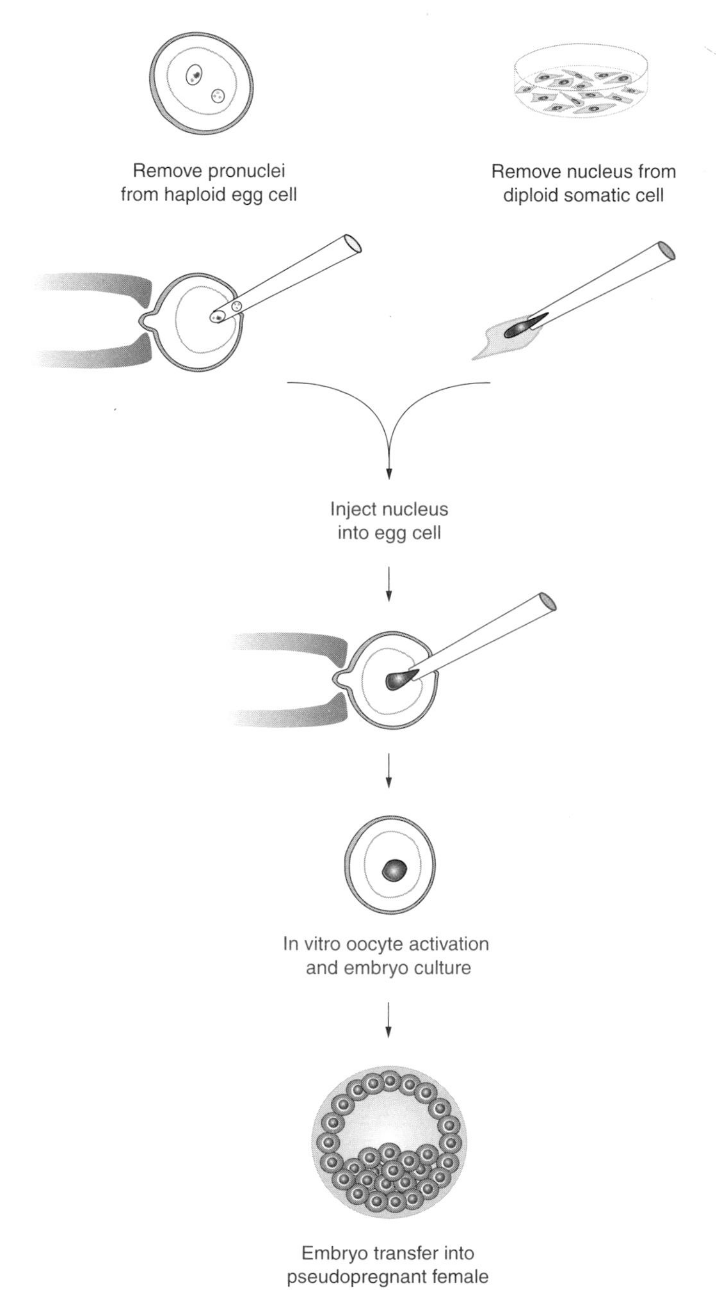

Figure 8.13 The Honolulu nuclear transfer technique utilizes microinjection to generate diploid egg cells for animal cloning. In vitro oocyte activation is done by placing the injected eggs into calcium-free media containing strontium. Cytochalasin B is included in the oocyte activation media to prevent polar body formation which results in chromosome loss.

nuclear transfer technique was, "Cumulina" the mouse, so named because of the adult cumulus cells that were used as the source of her genetic material.

Laboratory Practicum 8. *Creating cell-specific gene knockouts in transgenic mice*

Research Objective

Based on the recent isolation of the hypothetical mouse gene *bean pole* (*bpl*), a physiology graduate student proposes experiments to test her hypothesis that the Bpl protein is required for hormonal signaling in the brain to induce normal feeding behaviors. The mouse *bpl* gene had been positionally cloned as a recessive mutation that is associated with extreme undernourishment owing to an 80% decrease in feeding activity which can be readily measured in homozygous *bpl –/–* mice. Because another lab reported that *bpl* knockout mice are embryonic lethal, the graduate student intends to use the *Cre-loxP* recombination system to target a *bpl* exon deletion to the hippocampus region of the mouse brain. She predicts that a brain-specific loss of *bpl* gene function will result in a quantitatively different feeding behavior in wild-type heterozygous and homozygous *Cre/flox-bpl* mice, while avoiding the embryonic lethality observed with *bpl* KO mice.

Available Information and Reagents

1. A *Cre* recombinase transgenic mouse strain has been obtained that specifically expresses Cre in the forebrain as a result of developmental activation of the calcium-calmodulin-dependent kinase II (*CaMKII*) promoter linked to the *Cre* expression cassette. This mouse strain has already been shown to direct hippocampus-specific deletion of a *loxP*-flanked target gene.
2. A *bpl* targeting vector has been constructed using portions of a 10 kb genomic segment of the mouse *bpl* gene. This replacement vector contains a *lox P* site in intron 1 and another downstream of exon 3. In addition, *neo*^{r} and *HSVtk* genes are included in the vector to facilitate genetic selection of ES cells containing homologous recombination at the *bpl* locus. Transgenic mice will be generated and a homozygous *flox-bpl* offspring will be mated to the *CaMKII-Cre* mouse.
3. An eating behavior assay has been developed that measures the amount of food eaten, the frequency of eating, and water consumption over a 24-hour period. These quantitative measurements can be combined to give a single value, called the "Blimpie" quotient (BQ).

Basic Strategy

Molecular genetic strategies based on loss-of-function phenotypes can often be used to infer the normal function of a gene product. However, this can be difficult to do if the gene disruption results in premature death of the transgenic animal at a time precluding full characterization of a differentiation-dependent phenotype. To circumvent this problem, Brian Sauer developed a method that uses recombination components from the bacteriophage P1 to facilitate site-specific deletion of genomic segments in mammalian cells. The only two functional units required for in vivo targeted DNA deletion with the *Cre-loxP* system are (1) expression of the P1 *Cre* recombinase gene, oftentimes by a cell-specific or regulated promoter, and (2) an integrated DNA segment that is flanked by direct repeat copies of the 34 bp P1 DNA sequence called *loxP; loxP*-targeted DNA is said to be "floxed." The usefulness of the *Cre-loxP* system, as a means to direct cell-specific gene deletions in transgenic mice, has been demonstrated by numerous labs, most notably by Klaus Rajewsky and his colleagues. Figure 8.14 illustrates how the *bpl-loxP* targeting vector could be used to create a homozygous transgenic *flox-bpl* founder mouse that can be bred with the *CaMKII-Cre* mouse to produce offspring with one or both *bpl* genes inactivated in pyramidal cells of the CA1 hippocampal region.

Comments

The *Cre-loxP* system provides a useful method to perform strategic genetic alterations in a spatial (cell-specific) or temporal (acute activation of *Cre* expression using a regu-

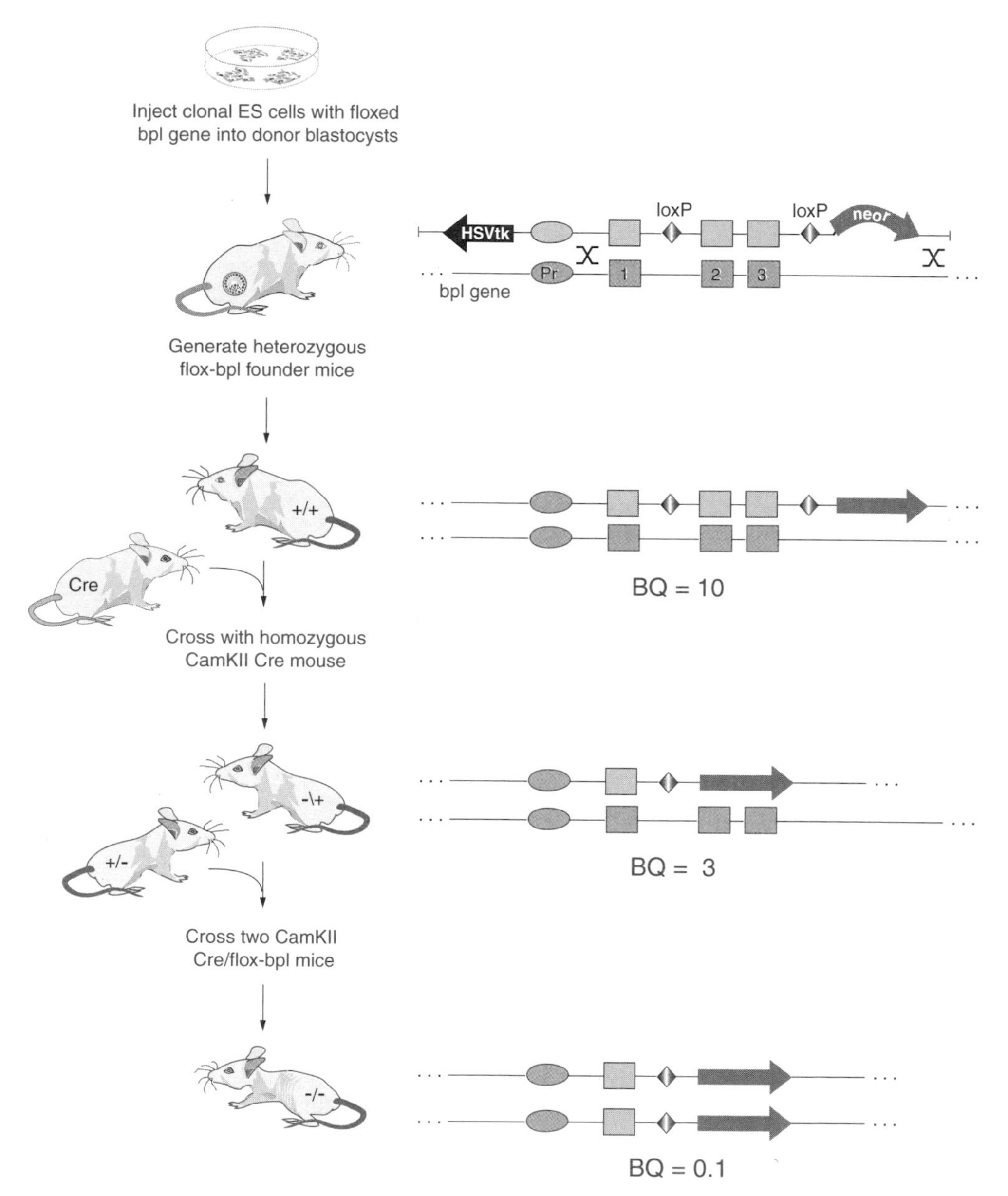

Figure 8.14 The *Cre-loxP* recombination system can be used to create transgenic mice with cell-specific deletions of the hypothetical *bpl* gene. Gene targeting of the *bpl* gene in ES cells with a *loxP* replacement vector will be used to generate a heterozygous *flox-bpl* founder mouse that can be crossed to a homozygous *CaMKII-Cre* mouse. The *bpl* genotypes (+/+, +/–, –/–) and phenotypes (fat, thin, starving) are depicted of the *Cre/flox-bpl* heterozygous offspring and of offspring arising from a cross between *Cre/flox-bpl* heterozygous siblings. These hypothetical BQ data would confirm that bpl function is required for normal feeding behavior.

lated promoter system) context. In this example, the cell specificity of the *CaMKII* promoter had been previously demonstrated, therefore, the next step required only the construction of the *bpl floxed* transgenic mice. Transgenic phenotypes in *Cre-loxP* mice can be difficult to interpret in cases where no suitable promoters have been characterized for a given target tissue, or if a promoter is used that does not express Cre at sufficiently high levels to obtain complete deletion penetrance (some cells in the tissue contain undeleted *floxed* targets). An alternative strategy would be to use *Cre* mice that contain an integrated copy of a regulated *Cre* expression cassette, for example, a tetracycline receptor-activated cell-specific promoter (Chapter 7), which would provide Cre-mediated recombination following acute administration of tetracycline to the animal.

Prospective

Based on what is known about other signaling pathways in the brain, and the sequence homology of *bpl* with other membrane receptors, one future research direction would be to design a binding assay that could be used in a functional expression cDNA cloning scheme to isolate transcripts encoding Bpl binding proteins. This could be done using a modified yeast two-hybrid assay to study peptide hormone–receptor interactions, similar to what has been done with insulin-like growth factor 1 (IGF-1) and its receptor (Chapter 5). Another approach would be to try in vitro expression cloning (IVEC) as a means to identify proteins that bind to recombinant bpl protein using an affinity labeling strategy and pooled plasmids from a normalized cDNA library (Chapter 5). Other research directions to take include finding the human *bpl* (h*bpl*) homologue using either a low stringency cDNA library screening approach (Chapter 5) or a GenBank database homology search (Chapter 9). With the h*bpl* gene in hand, it would be important to determine if any of the previously described human eating disorders are due to defects in hBpl signaling. This could be done by using molecular genetic probes, for example, PCR genotyping assays (Chapter 6), or by using Bpl antibodies as immunocytological probes.

REFERENCES

Molecular genetics of Drosophila development

Ashburner, M. *Drosophila*: A laboratory handbook. Cold Spring Harbor Press, Cold Spring Harbor, NY, 1334 pp., 1989.

*Brand, A.H., Perrimon, N. Targeted gene expression as a means of altering cell fates and generating dominant phenotypes. Development 118:401–415, 1993.

Calleja, M. Moreno, E., Pelaz, S., Morata, G. Visualization of gene expression in living adult Drosophila. Science 274:252–255, 1996.

*Lawrence, P. The making of a fly. Blackwell Scientific Publications, Oxford, 228 pp., 1992.

*Nüsslein-Volhard, C., Wieschaus, E. Mutations affecting segment number and polarity in Drosophila. Nature 287:795–801, 1980.

Rio, D., Molecular mechanisms regulating *Drosophila* P element transposition, Ann. Rev. Gen. 24:543–578, 1990.

Roberts, D.B. (ed.). *Drosophila*: A practical approach, IRL Press, Oxford, 231 pp., 1986.

Rubin, G. M., Kidwell, M. Bingham, P.M. The molecular basis of P-M hybrid dysgenesis: the nature of induced mutations. Cell 29:987–994, 1982.

*Rubin, G.M., Spradling, A.C. Genetic transformation of *Drosophila* with transposable element vectors. Science 218:348–353, 1982.

* Landmark papers in applied molecular genetics

*Wilson, C., Pearson, R., Bellen, H., O'Kane, C., Grossniklaus, U., Gehring, W.J. P-element-mediated enhancer detection: an efficient method for isolating and characterizing developmentally regulated genes in Drosophila. Genes Devel. 3:1301–1313, 1989.

Yeh, E., Gustafson, K., Boulianne, G. Green fluorescent protein as a vital marker and reporter gene expression in *Drosophila*. Proc. Natl. Acad. Sci. USA 92:7036–7040, 1995.

Construction of transgenic crop plants

Babiychuk, E., Fuangthong, M., Montagu, M., Inze, D., Kushner, S. Efficient gene tagging in Arabidopsis thaliana using a gene trap approach. Proc. Natl. Acad. Sci. USA 94:12722–12727, 1997.

Baron, C., Zambryski, P. Plant transformation: a pilus in Agrobacterium T-DNA transfer. Curr. Biol. 12:1567–1569, 1996.

Casas, A., Kononowicz, A., Zehr, U. et al. Transgenic sorghum plants via micropojectile bombardment. Proc. Natl. Acad. Sci. USA 90:11212–11216, 1993.

Chevre, A-M., Eber, F., Baranger, A., Renard, M. Gene flow from transgenic crops. Nature 389: 924, 1997.

Christou, P. Genetic transformation of crop plants using microprojectile bombardment. Plant J. 2:275–281, 1992.

Daniell, H., Datta, R., Varma, S., Gray, S., Lee, S-B. Containment of herbicide resistance through genetic engineering of the chloroplast genome. Nature Biotec. 16:345–348, 1998.

*Gelvin, S., Thomashow, M., McPherson, J., Gordon, M., Nester, E. Sizes and map positions of several plasmid-DNA-encoded transcripts in octopine-type crown gall tumors. Proc. Natl. Acad. Sci. USA 79:76–80.

*Feldman, K., Marks, M., Christianson, M., Quatrano, R. A dwarf mutant of Arabidopsis generated by T-DNA insertion mutagenesis. Science 243:1351–1354, 1989.

*Jefferson, R., Kavanuagh, T., Bevan, M. GIS fusions: glucuronidase as a sensitive and versatile marker in higher plants. EMBO J. 6:3901–3907, 1987.

*Joos, H., Inze, D., Caplan, A., Sormann, M. Van Montagu, M., Schell, J. Genetic analysis of T-DNA transcripts in nopaline crown galls. Cell 32:1057–1067, 1983.

Jorgensen, R., Atkinson, R., Forster, R., Lucas, W. An RNA-based information superhighway in plants. Science 279:1486–1487, 1998.

*Kempion, S., Liljegren, S., Block, L., Rounsley, S., Yanofsky, M., Lam, E. Targeted disruption in Arabidopsis. Nature 389:802–803, 1997.

Kohli, A., Leech, M., Vain, P., Laurie, D., Christou, P. Transgene organization in rice engineered through direct DNA transfer supports a two-phase integration mechanism mediated by the establishment of integration hot spots. Proc. Natl. Acad. Sci. USA 95:7203–7208, 1998.

Koncz, C., Nemeth, K., Redei, G., Schell, J. T-DNA insertional mutagenesis in Arabidopsis. Plant Mol. Biol. 20:963–976, 1992.

Kramer, M.G., Redenbaugh, K. Commercialization of a tomato with an antisense polygalacturonase gene: the FLAVR SAVR™ tomato story. Euphytica 79:293–297, 1994.

Meyerowitz, E. M. Arabidopsis, a useful weed. Cell 56:263–269, 1989.

Moore, I., Galweiller, L., Grosskopf, D., Schell, J., Palme, K. A transcription activation system for regulated expression in transgenic plants. Proc. Natl. Acad. Sci. USA 95:376–381, 1998.

Ronald, P. Making rice disease-resistance. Sci. Am. 277:100–105, 1997.

Rotino, G., Perri, E., Zottini, M., Sommer, H., Spena, A. Genetic engineering of parthenocarpic plants. Nature Biotech. 15:1398–1401, 1997.

*Sanford, J., Klein, T., Wolf, E., Allen, N. Delivery of substances into cells and tissues using a particle bombardment process. J. Part. Sci. Tech. 6:559–563, 1987.

Sanford, J., Smith, F., Russell, J. Optimizing the biolistic process for different biological applications. Methods Enzmol. 217:483–509, 1993.

Shah, D., Rommens, C., Beachy, R. Resistance to diseases and insects in transgenic plants: progress and applications to agriculture. Trends Biotech. 13:362–368, 1995.

*Smith, C., Watson, C., Ray, J., Bird, C., Morris, P., Schuch, W., Grieson, D. Antisense RNA inhibition of polygalacturonase gene expression in transgenic tomatoes. Nature 334:724–726, 1988.

Walden, R., Wingender, R. Gene-transfer and plant regeneration techniques. Trends Biotech. 13:324–331, 1995.

*Zambryski, P., Tempe, J., Schell, J. Transfer and function of T-DNA genes from Agrobacterium Ti and Ri plasmids in plants. Cell 56:193–201, 1989.

Williams, N. Agricultural biotech faces backlash in Europe. Science 281:768–771, 1998.

Mouse transgenesis

Barlow, C., Hirotsune, S., Paylor, R., Liyanage, M., Eckhaus, M., Collins, F., Shiloh, Y., Crawley, J.N., Ried, T., Tagle, D., Wynshaw-Boris, A. *Atm*-deficient mice: A paradigm of ataxia telangiectasia. *Cell* 86:159–171, 1996.

Bedell, M.A., Jenkins, N.A., Copeland, N.G. Mouse models of human disease .1. Techniques and resources for genetic analysis in mice. *Genes Dev.* 11:1–10, 1997.

Bedell, M.A., Largaespada, D.A., Jenkins, N.A., Copeland, N.G. Mouse models of human disease, 2. Recent progress and future directions. *Genes Dev.* 11:11–43, 1997.

*Constatini, F., Lacy, E. Introduction of a rabbit β-globin gene into the mouse germ line. Nature 294:92–94, 1981.

Copp, A.J. Death before birth: clues from gene knockouts and mutations. *TIG* 11:87–93, 1995.

*Evans, M., Kaufman, M. Establishment in culture of pluripotential cells from mouse embryos. Nature 292:154–156, 1981.

*Gordon, J., Scangos, D., Plotkin, J., Barbosa, J., Ruddle, F.H. Genetic transformation of mouse embryos by microinjection of purified DNA. Proc. Natl. Acad. Sci. USA 77:7380–7384, 1980.

Greenberg, N.M., DeMayo, F., Finegold, M.J., Medina, D., Tilley, W.D., Aspinall, J.O., Cunha, G.R., Donjacour, A.A., Matusik, R.J., Rosen, J.M. Prostate cancer in a transgenic mouse. Proc. Natl. Acad. Sci. USA 92:3439–3443, 1995.

*Hanahan, D. Heritable formation of pancreatic β-cell tumors in transgenic mice expressing recombinant insulin/simian virus 40 oncogenes. Nature 135:115–122, 1985.

Hogan, B., Beddington, R., Constantini, F., Lacy, E. Manipulating the mouse embryo: A laboratory manual, 2nd ed. Cold Spring Harbor Press, Cold Spring Harbor, NY, 497 pp., 1994.

*Jaenisch, R., Mintz, B. Simian virus 40 DNA sequences in DNA of healthy adult mice derived from preimplantation blastocysts injected with viral DNA. Proc. Natl. Acad. Sci. USA 71:1250–1254, 1974.

Jasin, M., Moynahan, M.E., Richardson, C. Targeted transgenesis. Proc. Natl. Acad. Sci. USA 93:8804–8808, 1996.

*Koller, B., Smithies, O. Inactivating the β2-microglobulin locus in mouse embryonic stem cells by homologous recombination. Proc. Natl. Acad. Sci. USA 86:8932–8935, 1989.

Lewis, J., Yang, B.L., Detloff, P., Smithies, O. Perspectives series: Molecular medicine in genetically engineered animals - Gene modification via "plug and socket" gene targeting. J. Clin. Invest. 97:3–5, 1996.

Liu, Y., Suzuki, K., Reed, J. et al. Mice with type 2 and 3 Gaucher disease point mutations generated by a single insertion mutagenesis procedure (SIMP). Proc. Natl. Acad. Sci. USA 95:2503–2508, 1998.

Ludwig, T., Chapman, D., Papaioannou, V., Efstratiadis, A. Targeted mutations in breast cancer susceptibility gene homologs in mice. Genes Dev. 11:1226–1241, 1997.

*Mansour, S., Thomas, K., Capecchi, M.R. Disruption of the proto-oncogene int-2 in mouse embryo derived stem cells: a general strategy for targeting mutations to nonselectable genes. Nature 336:348–352, 1988.

*Martin, G.M. Isolation of a pluripotent cell line from early mouse embryos cultured in medium conditioned by teratocarcinoma cells. Proc. Natl. Acad. Sci. USA 78:7634–7638, 1981.

Monastersky, G., Robl, J. (eds.). Strategies in transgenic animal science. ASM Press, Washington, D.C., 357 pp., 1995.

*Palmiter, R., Behringer, R., Quaife, C., Maxwell, F., Maxwell, I., Brinster, R. Cell lineage ablation in transgenic mice by cell-specific expression of a toxin gene. Cell 50:435–443, 1987.

*Ryan, T., Ciavatta, D., Townes, T. Knockout-transgenic mouse model of sickle cell disease. Science 278:873–880, 1997.

*Smithies, O., Gregg, R., Boggs, S., Koralewski, M., Kucherlapati, R. Insertion of DNA sequences into the human chromosomal β-globin locus by homologous recombination. Nature 317:230–235, 1985.

*Swift, G., Hammer, R., MacDonald, R., Brinster, R. Tissue-specific expression of the rat pancreatic elastase I gene in transgenic mice. Cell 38:639–646, 1984.

Takada, T., Lida, K., Awaji, T., Itoh, K., Takahashi, R., Shibui, A., Yoshida, K., Sugano, S., Tsujimoto, G. Selective production of transgenic mice using green fluorescent protein as a marker. Nature Biotech. 15:458–461, 1997.

*Thomas, K., Capecchi, M. Site-directed mutagenesis by gene targeting in mouse embryo-derived stem cells. Cell 51:503–512, 1987.

Whithers, D., Gutierrez, J., Towery, H. et al. Disruption of IRS-2 causes type 2 diabetes in mice. Nature 391:900–904, 1998.

Zambrowicz, B., Friedrich, G., Buxton, E. et al. Disruption and sequence identification of 2,000 genes in mouse embryonic stem cells. Nature 392:608–611, 1998.

Zhou, X-Y, Tomatsu, S., Fleming, R. E. et al. HFE gene knockout produces a mouse model of hereditary hemochromatosis. Proc. Natl. Acad. Sci. USA 95:2492–2497, 1998.

*Zijlstra, M., Li, E., Sajjadi, F., Subramani, S., Jaenisch, R. Germ-line transmission of a disrupted β2-microglobulin gene produced by homologous recombination in embryonic stem cells. Nature 342:435–438, 1989.

Development of transgenic livestock

Ashworth, D., Bishop, M., Campbell, K.H.S. et al. DNA microsatellite analysis of Dolly. Nature 394:329, 1998.

*Campbell, K.H.S., McWhir, J., Ritchie, W.A., Wilmut, I. Sheep cloned by nuclear transfer from a cultured cell line. Nature 380:64–66, 1996.

*Cibelli, J.B., Stice, S., Golueke, P. et al. Cloned transgenic calves produced from nonquiescent fetal fibroblasts. Science 280:1256–1258, 1998.

Echelhard, Y. Recombinant protein production in transgenic animals. Curr. Opin. Biotechnol. 7:536–540, 1996.

*Hammer, R., Pursel, V., Rexroad, C., Wall, R., Bolt, D., Ebert, K., Palmiter, R., Brinster, R. Production of transgenic rabbits, pigs and pigs by microinjection. Nature 315:680–683, 1985.

Kahn, A. Clone mammals . . . Clone man? Nature 386:119, 1997.

MacQuitty, J.J. The real implications of Dolly. Nature Biotech. 15:294, 1997.

*Pittius, C., Hennighausen, L., Lee, E., Westphal, H., Nicols, E., Vitale, J., Gordon, K. A milk protein gene promoter directs the expression of human tissue plasminogen activator cDNA to the mammary gland in transgenic mice. Proc. Natl. Acad. Sci. USA 85:5874–5878, 1988.

Pursel, V., Pinkert, C., Miller, K., Bolt, D., Campbell, R., Palmiter, R., Brinster, R., Hammer, R. Genetic engineering of livestock. Science 244:1281–1281, 1989.

*Schnieke, A., Kind, A., Ritchie, W., Mycock, K., Scott, A., Ritchie, M., Wilmut, I., Colman, A., Campbell, K. Human Factor IX transgenic sheep produced by nuclear transfer of nuclei from transfected fetal fibroblasts. Science 278:2130–2133, 1997.

Signer, E., Dubrova, Y., Jeffreys, A.J. et al. DNA fingerprinting Dolly. Nature 394:329–330, 1998.

Stewart, C. Nuclear Transplantation—An udder way of making lambs. Nature 385:769–771, 1997.

Velander, W.H., Lubon, H., Drohan, W.N. Transgenic livestock as drug factories. Sci. Am. 276:70–74, 1997.

*Wakayama, T., Perry, A., Zuccotti, M., Johnson, K., Yanagimachi, R. Full-term development of mice from enucleated oocytes injected with cumulus cell nuclei. Nature 394:369–374, 1998.

Wall, R., Kerr, D., Bidioli, K. Transgenic dairy cattle: genetic engineering on a large scale. J. Dairy Sci. 80:2213–2224, 1997.

Wall, R.J. A new lease on life for transgenic livestock. Nature Biotech. 15:416–417, 1997.

*Wilmut, I., Schnieke, A.E., McWhir, J., Kind, A.J., Campbell, K.H.S. Viable offspring derived from fetal and adult mammalian cells. Nature 385:810–813, 1997.

*Wright, G., Carver, A., Cottom, D., Reeves, D., Scott, A., Simons, P., Wilmut, I., Garner, I., Coman, A. High level expression of active human alpha-1-antirypsin in the mile of transgenic sheep. Bio/Tech. 9:830–834, 1991.

Young, M., Okita, W., Brown, E., Curling, J. Production of biopharmaceutical proteins in the milk of transgenic dairy animals. Biopharm. 10:34–38, 1997.

Creating cell-specific gene knockouts in transgenic mice

Feil, R., Brocard, J., Mascrez, B., LeMeur, M., Metzger, D., Chambon, P. Ligand-activated site-specific recombination in mice. Proc. Natl. Acad. Sci. USA 93:10887–10890, 1996.

*Gu, H., Marth, J.D., Orban, P.C., Mossmann, H., Rajewsky, K. Deletion of a DNA polymerase β gene segment in T cells using cell type-specific gene targeting. Science 265:103–106, 1994.

Jiang, R.L., Gridley, T. Gene targeting: Things go better with Cre. Curr. Biol. 7:R321–R323, 1997.

King, R.W., Lustig, K.D., Stukenberg, P.T., McGarry, T.J., Kirschner, M.W. Expression cloning in the test tube. Science 277:973–974, 1997.

*Kistner, A., Gossen, M., Zimmermann, F., Jerecic, J., Ullmer, C., Lübbert, H., Bujard, H. Doxycycline-mediated quantitative and tissue-specific control of gene expression in transgenic mice. Proc. Natl. Acad. Sci. USA 93:10933–10938, 1996.

Kühn, R., Schwenk, F., Aguet, M., Rajewsky, K. Inducible gene targeting in mice. Science 269: 1427–1429, 1995.

Lakso, M., Sauer, B., Mosinger, B., Jr., Lee, E.J., Manning, R.W., Yu, S.-H., Mulder, K.L., Westphal, H. Targeted oncogene activation by site-specific recombination in transgenic mice. Proc. Natl. Acad. Sci. USA 89:6232–6236, 1992.

Metzger, D., Clifford, J., Chiba, H., Chambon, P. Conditional site-specific recombination in mammalian cells using a ligand-dependent chimeric Cre recombinase. Proc. Natl. Acad. Sci. USA 92:6991–6995, 1995.

Morris, R.G.M., Morris, R.J. Memory floxed. Nature 385:680–681, 1997.

Reichardt, H.M, Kaestner, K., Tuckerman, J. et al. DNA binding of the glucocorticoid receptor is not essential for survival. Cell 93:531–542, 1998.

*Sauer, B., Henderson, N. Site-specific DNA recombination in mammalian cells by the Cre recombinase of bacteriophage P1. Proc. Natl. Acad. Sci. USA 85:5166–5170, 1988.

*Tsien, J.Z., Chen, D.F., Gerber, D., Tom, C., Mercer, E.H., Anderson, D.J., Mayford, M., Kandel, E.R., Tonegawa, S. Subregion- and cell type-restricted gene knockout in mouse brain. Cell 87:1317–1326, 1996.

Zhu, J., Kahn, C.R. Analysis of a peptide hormone-receptor interaction in the yeast two-hybrid system. Proc. Natl. Acad. Sci. USA 94:13063–13068, 1997.

CHAPTER

9

CONTEMPORARY APPLIED MOLECULAR GENETICS

In this last chapter, we turn our attention toward techniques that are beginning to emerge as the next major advances in applied molecular genetics. First, we examine how automated instrumentation is becoming the method of choice for high-throughput data collection. One example of this technology is automated DNA sequencing, a technique that is quickly replacing manual DNA sequencing methods because of its cost-effectiveness, speed, and accuracy. Second, an overview is given of how the Internet is becoming a dry-lab molecular genetic tool. This has come about primarily because genomic researchers need to access large public DNA sequence databases to conduct nucleotide homology searches. The third section is devoted to a description of recent advances in applied medical molecular genetics, with a focus on DNA forensics, molecular pathophysiology, and somatic cell gene therapy.

AUTOMATED INSTRUMENTATION FOR GENOME ANALYSIS

As is often the case, big government projects, particularly in science, provide the impetus for commercialization of new technologies. The Human Genome Project is a good example of this paradigm. In this section, several examples are profiled in which the development of automated instrumentation for genome analysis has sparked a new generation of service industries for molecular genetic studies. The three most visible developments in this area of applied molecular genetics are (1) high-throughput DNA sequencing, (2) mRNA profile analysis using microarray hybridization assays, and (3) utilization of robot-driven computerized work stations for routine laboratory procedures.

High-Throughput DNA Sequencing

As described in Chapter 1, DNA sequence determination using the dideoxy chain termination method results in the synthesis of reaction products that have the same 5′ terminus, but random 3′ ends due to the incorporation of dideoxynucleotides (ddNTPs). Manual DNA sequencing requires excessive technician time for laborious reaction preparation procedures, processing of gel electrophoresis materials, and data acquisition. By the early 1980s it became clear that efficient whole genome sequencing would depend on the development of an automated process. The leader in this effort was Lee Hood, who with his colleagues pioneered automated DNA sequencing using a fluorescence detection system with an integrated gel electrophoresis component. This DNA sequencing instrument was commercialized by Applied Biosystems Incorporated (ABI) and it soon became the primary workhorse of the Human Genome Project.

The ABI system uses four spectrally distinct fluorescent ddNTPs. This strategy makes it possible to perform the entire sequencing reaction in a single tube and to resolve the chain-terminated products in one lane of a sequencing gel. The ABI DNA detection system permits direct real-time data acquisition by laser-activated dye excitation at a point in the electrophoresis run that maximizes fragment resolution. Using this method, reliable sequence information can be obtained for DNA segments >500 nucleotides long in less than 4 hours. Based on the number of lanes per gel, one ABI instrument can generate ~50,000 nucleotides of sequence information in an 8 hour day. Figure 9.1 illustrates how automated fluorescence-based dideoxy DNA sequencing systems are used for high-throughput data acquisition. An initial drawback to automated DNA sequencing using fluorescence-based detection systems was the reduced signal to noise ratios that occur when small amounts of template are used in the reaction. The solution to this problem was to develop a one-sided PCR amplification strategy, called cycle sequencing, that utilizes a modified *Taq* DNA polymerase and temperature cycling to generate high levels of chain-terminated product from a small amount of template. The use of cycle sequencing has greatly extended the automation capabilities of DNA sequencing because it requires much less starting material for each round of sequencing.

Automated DNA sequencing has not only provided an essential tool for whole genome analysis, but it has also led to the development of new molecular genetic strategies that would have been logistically impossible using manual DNA sequencing methods. One such example is serial analysis of gene expression (SAGE), a PCR-based method developed by Bert Vogelstein and Ken Kinzler to identify differences in steady-state mRNA levels between two RNA samples. SAGE is based on the idea that the relative proportion of gene-specific expressed sequence tags (ESTs) in a cDNA pool reflects the relative abundance of the corresponding mRNA transcripts in the original RNA preparation. By determining the sequence of a statistically significant number of ESTs in one cDNA pool, relative to another, it is possible to identify gene transcripts that are differentially represented. Because only very short segments of cDNA are required to establish identity, the SAGE protocol uses restriction enzyme digestion, adaptor-mediated ligation, and PCR to link large numbers of short ESTs (10–15 bp) into a single array for automated DNA sequencing. In this way, fewer DNA templates have to be sequenced than would be required if each cDNA in the starting pool were treated as a single EST. The SAGE strategy is outlined in Figure 9.2.

Genome Probing Using DNA Microarrays

Technological advances in the computer industry have closely followed the key improvements in microchip design and manufacturing that resulted in more data processing power using less surface area. This same idea is now being applied to genome analysis, and indeed, much of the groundwork for developing DNA microchips has been through a direct application of methods used to miniaturize silicon-based computer

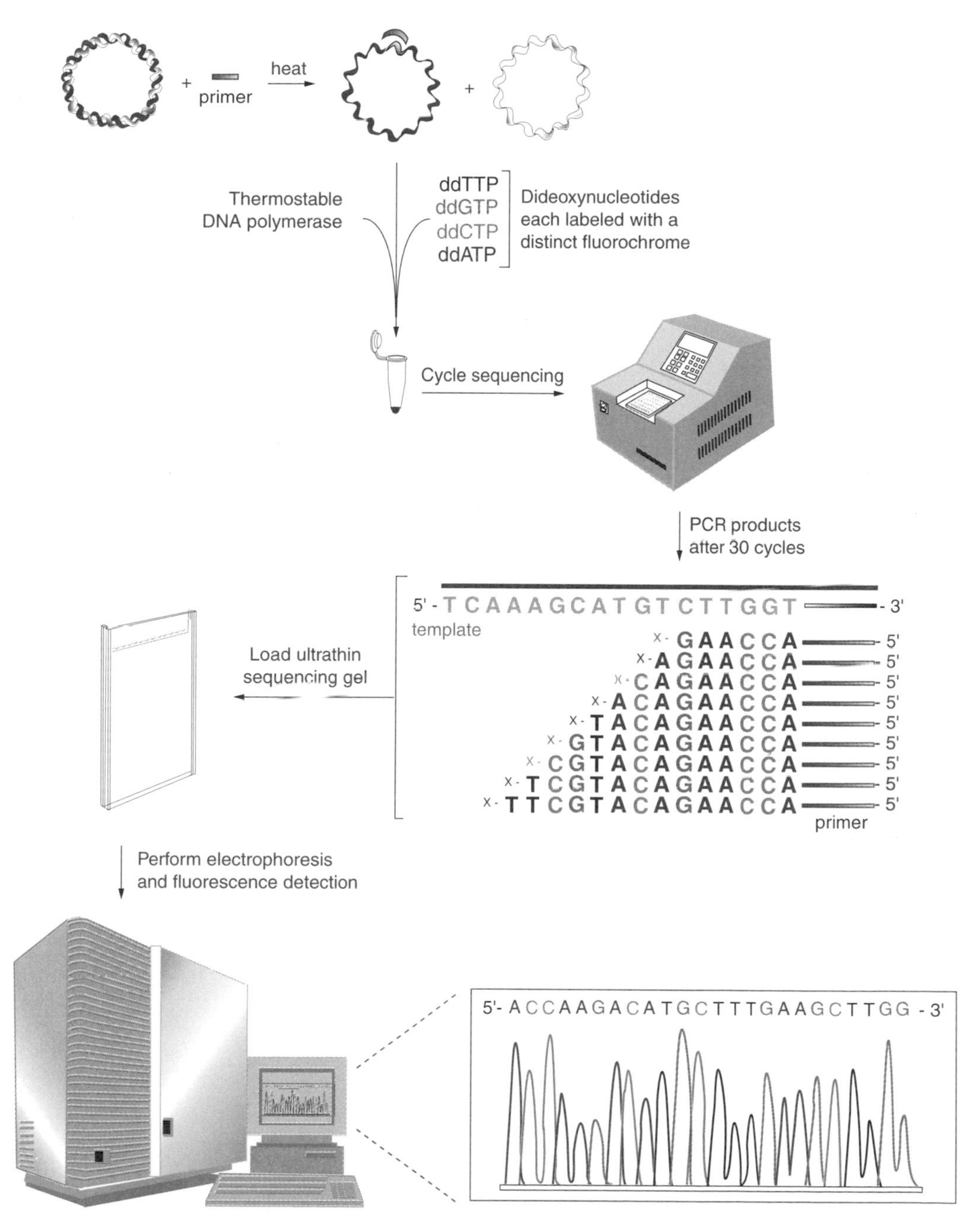

Figure 9.1 Fluorescence-based DNA sequencing systems use dideoxy chain termination reactions to generate labeled reaction products that are size-fractionated by gel electrophoresis. In the example shown, each ddNTP contains a spectrally distinct fluorochrome, which permits the sequencing reaction to be performed in a single tube and the products to be resolved in one lane. Fluorochrome excitation and emission detection occur at a fixed point near the bottom of the gel to permit maximum resolution. Sequence data are recovered as an emission spectra readout for each fluorescent dye as a function of fragment size. A text file of the inferred DNA sequence is output directly to an in-line computer.

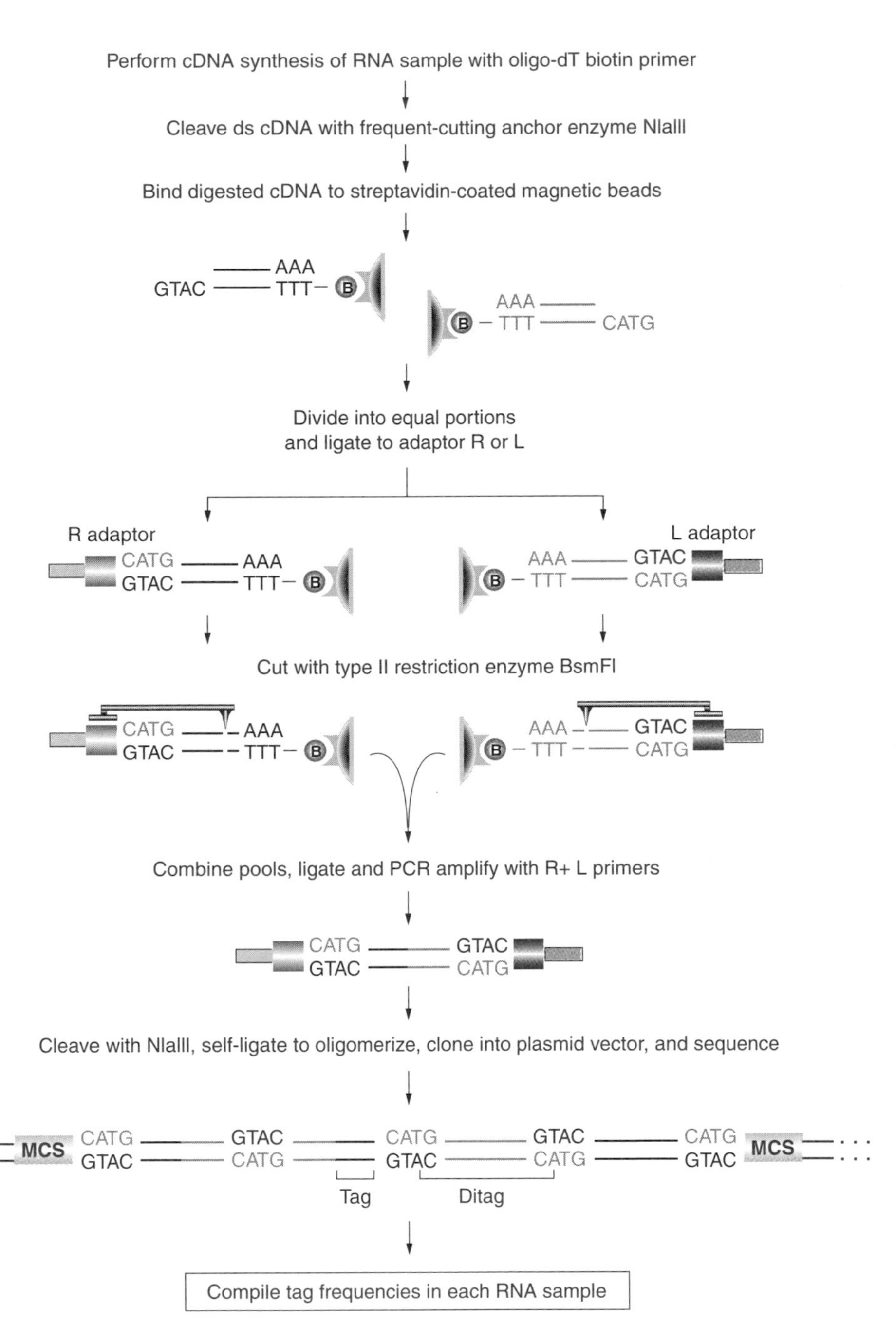

Figure 9.2 SAGE is a DNA sequencing-dependent gene analysis method that has been used to identify differentially expressed gene transcripts. The basic idea of SAGE is to generate very short 3′ cDNA sequence tags using various mRNA samples, and then compile the tag frequencies using automated DNA sequencing information. A flow scheme is shown for the generation of cDNA tags from just one RNA population. In this example, the cDNA synthesis is initiated with a biotin-labeled oligo-dT primer that permits physical separation of 3′ ESTs using magnetic beads coated with streptavidin. The two enzymes used for tag delineation are NlaIII, which recognizes the sequence CTAG, and BsmFI, a type II restriction enzyme that cleaves 10 bp downstream of the sequence GGGAC. Amplification bias during the PCR step is normalized by monitoring nearest neighbor tags in each ditag, with the assumption that ditag formation is a random event for cDNA tags representing low abundance transcripts.

chips. The basic idea has been to attach known DNA sequences covalently onto a solid support surface in a way that will facilitate rapid hybridization to a fluorescently labeled pool of DNA (or cDNA) molecules. Because each DNA microchip contains a standardized set of DNA sequences, it is referred to as the probe, whereas the labeled experimental DNA is called the target. By using laser-scanning and fluorescence detection devices to "read" the chip surface, the hybridization pattern can be quantitatively analyzed to determine the sequence complexity of the DNA target population.

The simplest form of DNA chip technology has been to spot denatured cDNA molecules onto a glass microscope slide using a computer-driven robotic arm to position the probe material precisely. The probes consist of known EST sequences that are catalogued by gene identity in a computer database. This type of cDNA microchip was pioneered by Pat Brown and Ron Davis. One example of how cDNA microchips can be used to monitor global changes in gene expression comes from experiments using a microchip containing all ~6000 ORF sequences known to be present in the *S. cerevisiae* genome. These yeast probe sequences were spotted onto a square grid (80×80 spots) that fit completely within an area less than 4 cm^2. The labeled target cDNAs were generated from mRNA samples prepared from yeast grown under various metabolic conditions. Figure 9.3 shows how a cDNA microchip can be used to determine the mRNA profile of two populations of human cells cultured under different media conditions.

A second type of DNA microchip has been developed by Affymetrix, a California biotechnology company, using a combination of photolithography and solid phase oligonucleotide chemistry to attach short oligonucleotide probes (25-mer oligos) covalently to a solid support surface. The Affymetrix chips contain 100,000 different oligos in a 4 cm^2 area and each probe site has $\sim 10^7$ oligo molecules. Chip design is a critical component of this strategy, and typically, multiple oligos are chosen that represent different regions of a single gene segment to provide a built-in control for variability in the hybridization signal. Moreover, because each oligo is custom-synthesized directly on the chip, it is possible to include oligo probes that have single nucleotide changes that cause a decreased hybridization efficiency owing to nucleotide mismatching. By designing oligos that span an entire exon using a register of 1 nucleotide change between adjacent probe sites, and a window of 25 nucleotides at a time, it is possible to utilize Affymetrix chip technology for DNA resequencing. In this resequencing strategy, the midpoint nucleotide (number 13 in a 25-mer) is synthesized as an A, G, C, or T. Using PCR products and hybridization conditions that discriminate between perfect and single base pair mismatch duplexes, it is possible to read the sequence across the target DNA based on the most intense signal in each set of four oligo probes. This strategy has been used to identify cancer gene mutations present in genomic DNA target pools that were isolated from patient biopsy samples as shown in Figure 9.4. Affymetrix-designed chips are also being produced that contain oligos corresponding to every known SNP in the human genome (see Chapter 4). Hybridization of labeled genomic target DNA to these "SNiP chips" is being used to create human genomic fingerprints for gene mapping studies. Other applications of oligo DNA microchips include the generation of mRNA profile arrays that can be used to measure differences in gene expression levels, similar to what is being done with cDNA microchips.

Robot-Driven Computerized Laboratory Work Stations

Many of the recent advances in high-throughput genome analysis and gene function bioassays have included improvements in automated laboratory robots. An important

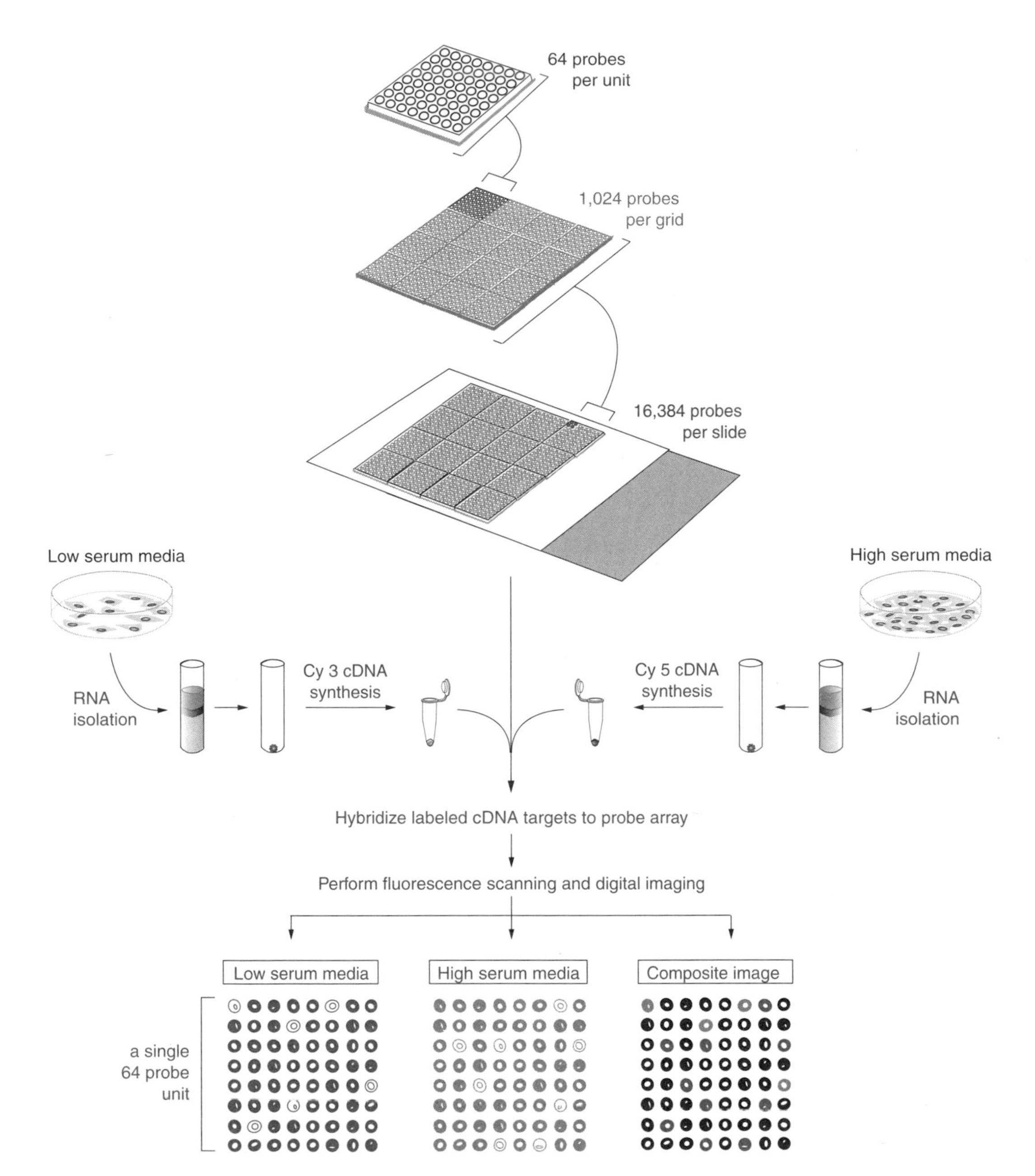

Figure 9.3 Global changes in gene expression can be monitored using cDNA microchips containing known ESTs. In this example, the cDNA microchip contains ~16,000 human ESTs that have been spotted onto a glass slide and hybridized to a target cDNA mixture consisting of fluorescently labeled sequences (Cy3 or Cy5 labeling). The cDNA hybridization pattern can be spectrally separated by fluorescence microscopy and digital imaging scanning. The probe spots on this type of array are not completely uniform owing to physical irregularities introduced by the metal stylus and mechanical arraying device.

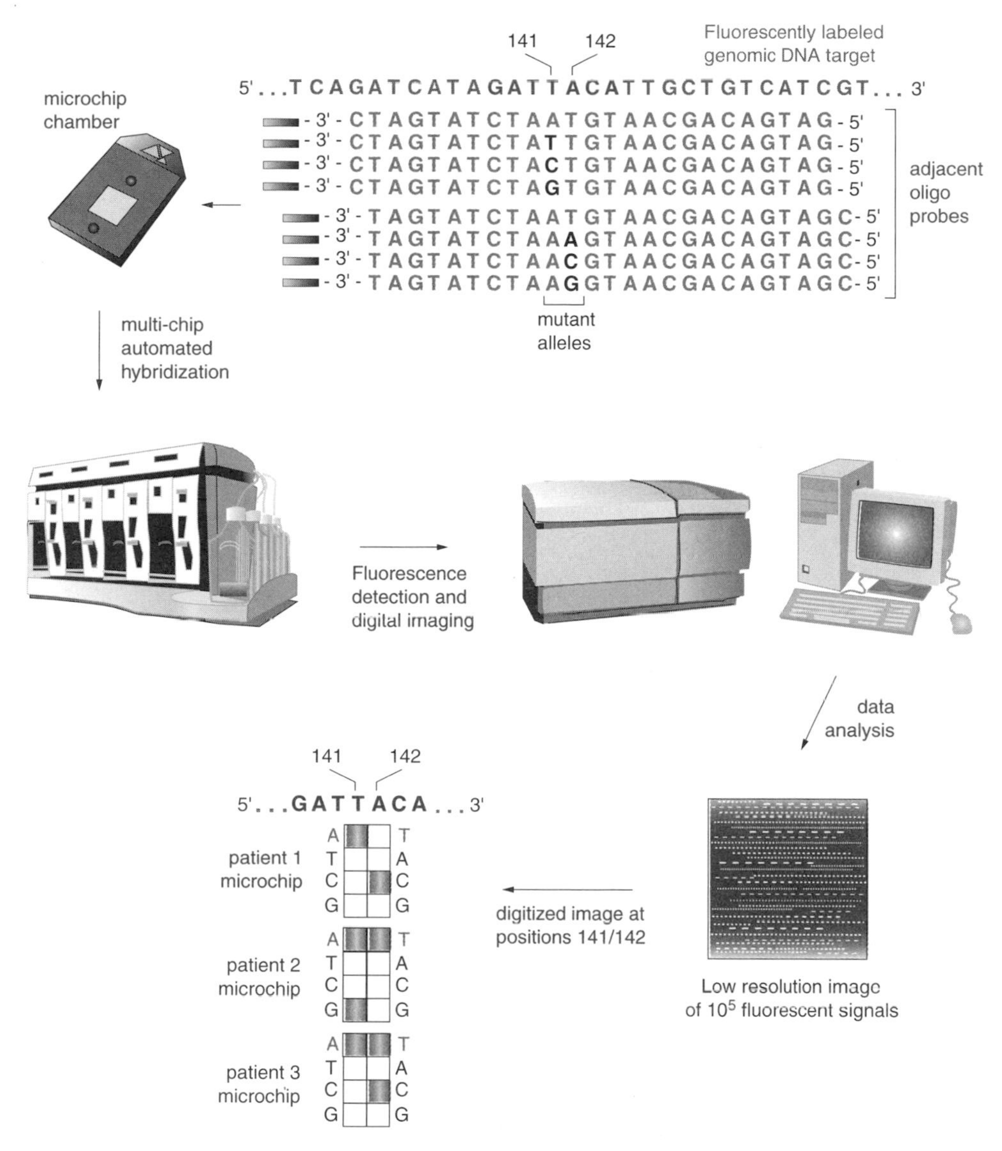

Figure 9.4 High density oligonucleotide DNA microchips have been produced commercially and used to probe genomic DNA targets for single nucleotide alterations. By designing chips that contain each set of four oligos, with a single nucleotide difference at the mid-position (A, G, C, or T), it is possible to resequence known target DNA sequences based on hybridization signal intensity. In this example, three different patient samples were analyzed and compared for sequence variation at two mutant alleles (141, 142) in a hypothetical human disease gene.

development has been the integration of PCR amplification steps into molecular genetic screening strategies. Because PCR conditions are amenable to automation and miniaturization, many types of laboratory work stations have been designed with PCR machines serving a central role. This in turn has led to the development of robotic instruments that are capable of accurate liquid transfer on a large and repetitive scale. An example of such a robotic pipetman is the Beckman Biomek 2000™. This instrument is routinely used for setting up DNA sequencing and PCR reactions, for constructing high density bacterial and yeast colony arrays for library screening, and for array printing of cDNA microchips. Different Biomek models can use a 96-well, 384-well, or 1536-well plate format, which provides flexibility in the design of high-throughput experiments. By combining robotic pipetting devices with available computer- controlled robotic arms and automated detection systems, it is now fairly straightforward, albeit costly, to design a fully automated molecular genetic screening assay. Indeed, whole genome sequencing on the scale of the Human Genome Project would not be *humanly* possible without robot-controlled automation.

ACCESSING MOLECULAR GENETIC INFORMATION THROUGH THE INTERNET

The 1990s ushered in what has been called the Information Age. Molecular geneticists, like many other scientists, have been quick to capitalize on the technological advances brought on by this new era. Initially, the goal was to store and disseminate DNA sequence information through government-sponsored computer links connecting academic research centers. More recently, however, the Internet has become a new type of electronic tool for molecular genetic research. The rapid growth in the number of Internet users (exceeding 50 million in 1997) and the application of information technology to molecular genetics have been due to the introduction of a powerful graphical interface called the World Wide Web (WWW). The WWW protocols were developed in 1992 by Tim Berners-Lee while he was at CERN, the European high energy physics laboratory located near Geneva, Switzerland. The essential WWW software programs developed by the CERN computer scientists are named HTTP (HyperText Transfer Protocol) and HTML (HyperText Markup Language).

The WWW permits the display of files as multimedia objects that are linked together by hypertext addresses. Software programs, called web browsers, allow individual Internet users to interface with the WWW through personal computers that are electronically connected to the Internet. The first web browser program, Mosaic, was developed in 1993 by a student at the University of Illinois named Mark Andreessen, who went on to become one of the founders of Netscape Communications. Individual WWW files are web pages that can be located by unique Internet addresses. A web page contains text, graphics, and marked hypertext links that are encoded in the HTML language. Web page files must be stored on dedicated Internet-connected computers that function as WWW servers. A WWW "home" page is the assigned starting point for a set of related WWW sites. Appendix G contains a representative list of WWW sites that are of particular interest to molecular geneticists.

The Internet network itself was established in 1969 when four universities in the United States were connected to form the first data transfer file exchange network. During the 1970s and 1980s, the Internet grew both in the number of networked computer servers and in the amount of information that could be accessed through text searching programs such as Gopher. The Internet today is a vast collection of linked high-speed computers that serve as regional centers for user access though smaller local area networks, called LANs. Through the use of routing protocols established at regional network servers, a single user can send or receive files by connecting to the Internet with a dial-up telephone link to an Internet service provider (ISP), or by an Ethernet-type connection with a LAN. These two common methods for Internet access are shown in Figure 9.5. Currently, no one governmental or commercial entity manages the world-wide Internet, and therefore, its growth is analogous to a spider web. Although the ever-changing environment of the WWW, coupled with its almost exponential growth rate,

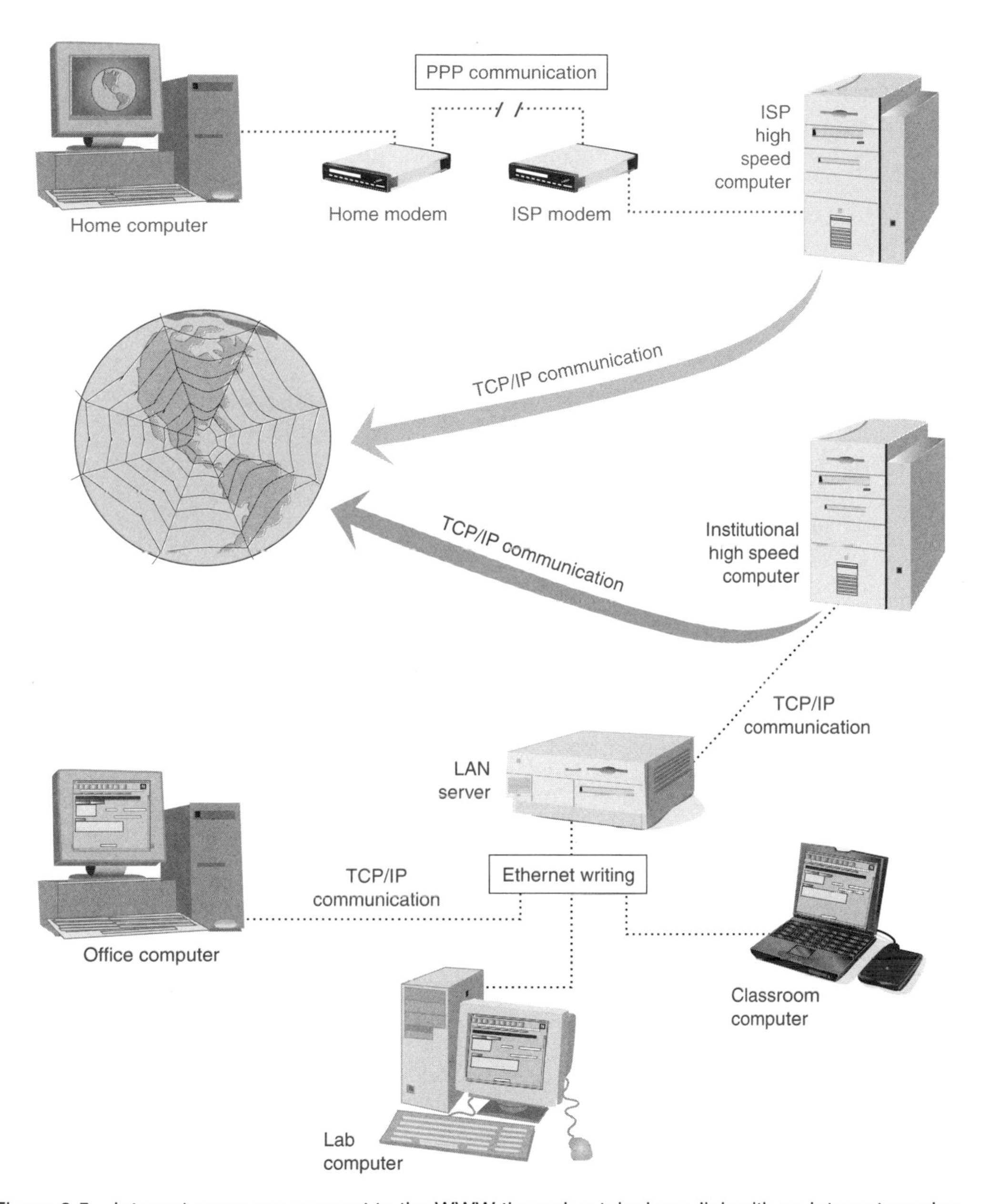

Figure 9.5 Internet users can connect to the WWW through a telephone link with an internet service provider (ISP) or through an institutional local area network (LAN). Dial-up connections to the Internet require telephone lines and modem devices to link users with an ISP which provides regional access to the Internet through a high-speed host computer. LAN connections use on-site hardware servers to funnel users through Ethernet links to organizational Internet host computers (or to an ISP). All Internet connections require a special communication software called TCP/IP (Transmission Control Protocol/Internet Protocol). Dial-up users require PPP (Point to Point Protocol) software for proper modem communication.

precludes giving a complete description of WWW-based molecular genetic applications, it is worthwhile to describe three ways that the WWW is being used for information retrieval by molecular genetic researchers. Aside from the use of E-mail and on-line journal publications, the primary research-related activities are (1) DNA sequence analysis using sequence alignment programs, (2) searching integrated databases to locate biological resources and reagents, and (3) interactive communication with specialized biological sciences user groups. To emphasize further the utility of these molecular genetic tools on the WWW, laboratory practicum 9 describes how searching DNA sequence databases can be used as a dry-lab method to identify functionally related genes.

DNA Sequence Analysis Programs and Databases

Molecular biologists have used the Internet since the early 1980s as a mechanism to search nucleotide databases. Using a file transfer protocol (FTP), and on-site DNA sequence algorithms such as the University of Wisconsin Genetics Computer Group (GCG) programs, researchers were able to perform sophisticated DNA sequence analyses using UNIX computers located at their own institution. However, early versions of GCG, as well as most other text-dependent sequence analysis programs, had two major limitations. First, it was difficult for the average computer novice to master the necessary on-line command routines that were required to utilize the full potential of these cumbersome sequence analysis programs. Second, the analyses had to be performed using data terminals that were directly linked to a large institutional main frame computer. This often meant that in addition to learning on-line commands and becoming familiar with the voluminous GCG documentation manuals, researchers needed to reserve access time to the main frame computer to run the time-consuming and memory-intensive algorithms.

In just a few short years, the graphical interface of the WWW transformed DNA sequence analyses into a simple point and click task that could be performed with any lab, office, or home computer connected to the Internet. There are numerous ways to access public domain DNA sequence databases, but one of the most convenient methods is to use an Internet connection to search GenBank with the Basic Local Alignment Search Tool (BLAST). As of 1998, GenBank contained almost 3 million sequence file entries that together amount to $>10^9$ nucleotides. GenBank is maintained by the National Center for Biotechnology Information (NCBI), a division of the National Library of Medicine located on the NIH campus in Bethesda, Maryland. GenBank is a member of the International Nucleotide Sequence Database Collaboration, an organization that includes the European Molecular Biology Laboratory (EMBL) and the DNA DataBank of Japan (DDBJ). The GenBank, EMBL, and DDBJ databases freely exchange DNA sequence files on a daily basis to maintain a comprehensive set of all known sequences. The number of entries in the collective GenBank databases have been doubling every 2 years for the past decade, and presently, ~50% of the GenBank sequences are from cDNA clones that have been isolated as uncharacterized expressed sequence tags (ESTs).

BLAST is a sequence homology algorithm that uses statistical calculations to identify significant sequence matches between a known sequence provided by the user, called a query, and DNA sequences in the GenBank database. The search parameters for a database query are chosen by the user and ultimately determine the type of matches that will be scored as significant. The BLAST homology search programs were developed by David Lipman and his colleagues at NCBI and can be accessed and executed

TABLE 9.1 Several BLAST homology search programs developed at NCBI[a]

Program	Query sequence type	Database type
blastn	Nucleotide sequence	Nucleotide sequences
blastx	Nucleotide sequence translated in all six reading frames	Protein sequences
blastp	Protein sequence	Protein sequences
tblastn	Protein sequence	Nucleotide sequences translated in all six reading frames
tblastx	Nucleotide sequence translated in all six reading frames	Nucleotide sequence translated in all six reading frames

[a]All of the BLAST programs, with the exception of *tblastx*, which is very computer-intensive, can be executed through the NCBI Internet web site (http://www.ncbi.nlm.nih.gov/BLAST/).

through the WWW. Full explanations of the default BLAST search parameters, as well as statistical considerations used for determining significant homology values, can be found through the NCBI WWW site (see Appendix G). The simplest GenBank sequence comparison uses the BLAST program *blastn* to compare a nucleotide query sequence to all nucleotide sequences in GenBank. Table 9.1 describes five of the BLAST search programs that can be used to compare DNA or protein sequences with any of the available NCBI databases (Table 9.2). Results from BLAST searches are typically used to design follow-up laboratory experiments that are capable of assigning biological relevance to the most significant nucleotide or amino acid homologies (Fig. 9.6). Laboratory practicum 9 at the end of the chapter illustrates how a researcher would use the BLAST programs to perform a sequence homology search.

Integrated Biological Resource Databases

One of the most exciting molecular genetic applications of the WWW has been the ability to exploit hypertext file linking functions using strategies that create Internet entry

TABLE 9.2 Representative NCBI databases that can be queried with BLAST programs using the WWW interface[a]

Type	Name	Description
Nucleotide	nr	All nonredundant GenBank + EMBL + DDBJ + PDB sequences (no ESTs)
	month	All new or revised nr sequences released in the last 30 days
	dbest	All nonredundant GenBank + EMBL + DDBJ EST nucleotide sequences
	dbsts	All nonredundant GenBank + EMBL + DDBJ STS sequences
	htgs	High-throughput genomic sequences (direct nucleotide sequences)
	epd	Eukaryotic promoter database
Protein	Swissprot	Recent major release of the SWISS-PROT protein sequence database
	pdb	Sequences from protein structures in Brookhaven Protein Data Bank
	kabat	Kabat's database of sequences of immunological interest
	yeast	*S. cerevisiae* protein sequences

[a]By interconverting between nucleic acid and inferred amino acid sequence data, it is possible to query these databases with either nucleotide or protein sequence files.

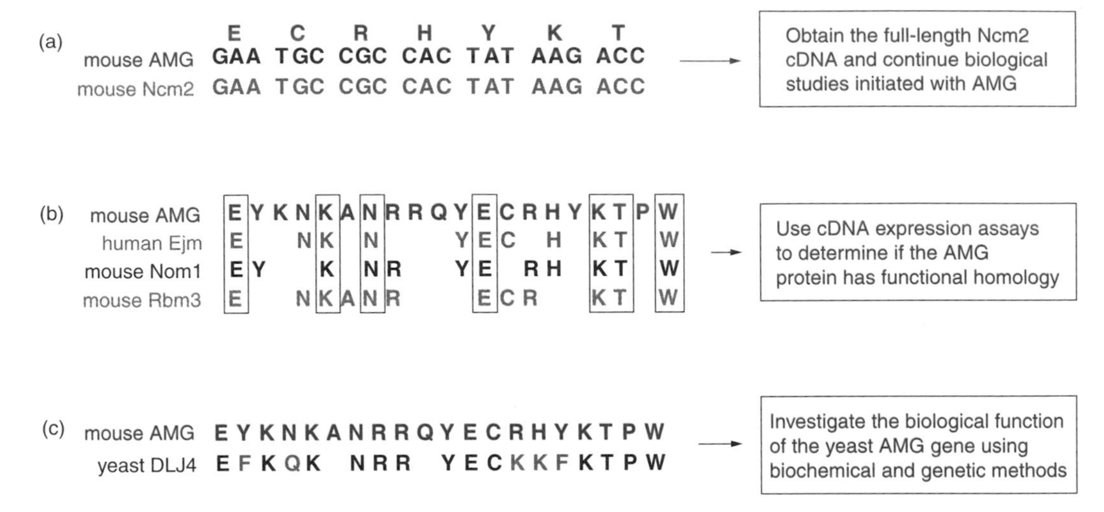

Figure 9.6 The results of BLAST search queries are used to design laboratory experiments that test the biological relevance of DNA and protein sequence homologies. For example, a BLAST search using coding sequences from the hypothetical mouse *AMG* gene could be done to determine if there are any homologous or orthologous (evolutionarily conserved) gene sequences in GenBank. Three possible outcomes are illustrated: (*a*) *AMG* identity with a previously cloned mouse cDNA; (*b*) limited, but highly significant homology, with several mouse and human genes encoding a well-characterized protein function; and (*c*) a yeast gene that shares the same region of homology as the mouse and human homologues, as well as additional identical and conserved residues that indicate it is the yeast *AMG* homologue. The examples shown in (*b*) and (*c*) would be derived from the inferred amino acid sequence of *AMG* based on codon assignments. A fourth possible outcome not depicted is a search result in which no significant homologies with GenBank would be found. See Figure 9.12 at the end of the chapter for a depiction of the GenBank WWW interface.

points for integrated resource databases. The basic idea of these linked web sites has been to enable researchers to place DNA sequence information into a proper biological context. By linking gene names, GenBank sequence files, MedLine abstracts, and investigator-generated research databases into common biological themes using extensively annotated web pages, it is now possible for scientists, clinicians, and interested WWW visitors to find valuable experimental data through the Internet.

One example of an integrated database resource is the GenomeNet project in Japan, which is operated jointly by Kyoto University and the University of Tokyo. GenomeNet is a basic science resource consisting of database retrieval tools and hypertext-linked databases that have been organized so that researchers can better exploit database information. GenomeNet represents one of several new types of molecular genetic research tool that utilizes complex search algorithms to "mine" existing nucleic acid and protein databases for potential novel relationships of biological relevance. Figure 9.7 shows the database relationships between various resources that have been linked through the GenomeNet DBGET retrieval system on the WWW.

Bionet News Groups for Molecular Genetic Research

So far we have only described ways that molecular genetic researchers use the Internet to obtain information, but it is also an important means of communication between individuals (E-mail) and between groups of users that have a common interest (news groups

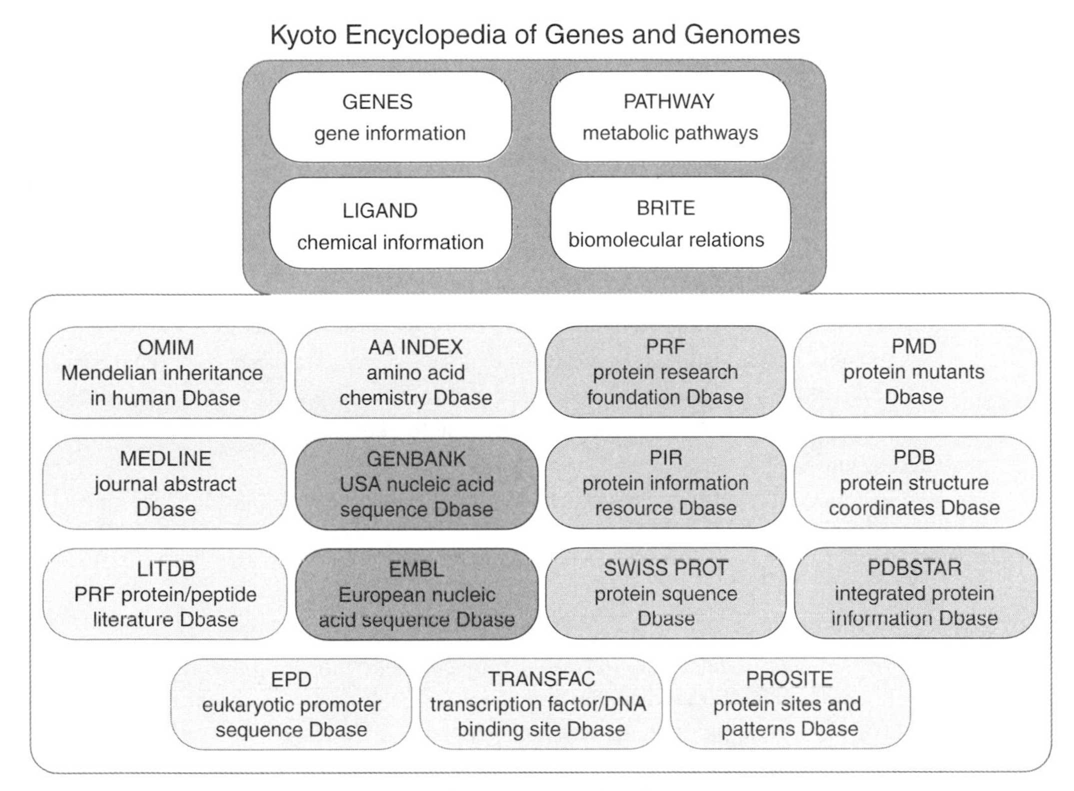

Figure 9.7 GenomeNet is a WWW-accessible network of computational services and databases that were initially developed as a research tool for molecular and cell biological studies in Japan (http://www.genome.ad.jp/). A graphical representation of the GenomeNet DBGET databases are shown with color codes to outline the different types of resources (text information, nucleic acid sequences, and protein sequences) that are available. Each of these resources is linked together in a way that provides powerful on-line search strategies for researchers using a WWW browser interface.

or forums). Internet news groups function as electronic bulletin boards and have been around since the inception of the Internet network. News groups were originally created by computer scientists as an efficient mechanism to exchange technical information. Participants in Internet news groups can read current announcements, post questions, answer questions, or search the archives for specific topics. The news-group postings are organized as threads of related subtopics which permit users to carry on an electronic conversation in the context of the entire forum. The benefit of participating in an Internet news group is that it provides a forum for global communication on related topics. Importantly, however, the usefulness of a news group to its members depends completely on the quality and diversity of the postings, and therefore some newsgroups use monitoring protocols to maintain proper "netiquette" behavior.

News groups specifically dedicated to biological research are administered by an organization called BIOSCI, presently located at Stanford University. BIOSCI was initiated through grants by the Department of Energy and the National Science Foundation, but since 1996, has depended on WWW advertising revenue and company sponsorship to cover overhead expenses. The BIOSCI home page provides a convenient entry point for news group users to begin participating in the forum(s) of their choice.

TABLE 9.3 Representative BIOSCI news groups that can be accessed through the BIOSCI WWW home page[a]

News Group	Topic
Amyloid	Forum for Alzheimer's disease disorders including prion diseases
Arabidopsis	News group for the Arabidopsis Genome Project
Bioforum	Discussions about biological topics lacking a dedicated news group
Bionews	General announcements of widespread interest to biologists
Bio-Matrix	Applications of computers to biological databases
Biotechniques	Discussions of articles in the journal Biotechniques
Cell-Biology	Discussions about cell biology and cancer research at the cellular level
EMBL-Databank	Messages to and from the EMBL database staff
GenBank-Bb	Messages to and from the GenBank database staff
Genstructure	Genome and chromatin structure and function
Protein-Analysis	Discussions about research on proteins and SWISS-PROT data bank

[a]BIOSCI is an organization that maintains biological science-related news groups on the Internet (http://www.bio.net/).

One of the most popular BIOSCI newsgroups, called "bionet.molbio.methds-reagnts," attracts more than 1000 molecular biological research postings a month. Table 9.3 list some of the other topics that are represented by BIOSCI newsgroups on the WWW.

APPLIED MEDICAL MOLECULAR GENETICS

In this last section of the book, we examine three areas in which applied medical molecular genetic methods are considered essential alternatives to existing protein- and chemical-based approaches. The first two, DNA forensics and molecular genetic pathophysiology, represent fields in which molecular genetic approaches are already in widespread use. In contrast, the third area, somatic cell gene therapy, is still considered by many to be only in the research and development phase. All three of these applications have met with substantial resistance on ethical and legal grounds. Nevertheless it now appears, as is historically true with most forms of new technologies in modern society, that applications offering a safer, healthier, and more prosperous life-style quickly earn mainstream public approval.

DNA Forensics

Forensic medicine has been around for more than a century and has historically been the realm of pathologists and criminologists who developed biochemical methods to deduce probable scenarios based on criminal evidence. In this context establishing human identity, of either the victim or the suspect, has been the most definitive and relevant to criminal justice (*who* did it is usually more important to the police than *why* or *how*). Two examples of forensic tests commonly used to establish human identity are the dermal-derived fingerprint pattern and ABO/Rh blood groups. In cases where high-quality fingerprints can be obtained from the crime scene, and a suspect with identical fingerprints has been apprehended, it is possible to make a sound statistical argument that the suspect was at the scene of the crime. Blood group identity is not very informative with regard to human identity because of the reduced complexity of human variation in erythrocytic allotypes; however, unlike fingerprints, the very presence of blood at a scene is reason to be suspicious!

All humans, with the exception of monozygotic siblings, have numerous differences in their genomic DNA sequence, and therefore are genotypically distinct. Based on this principle, Alec Jeffreys of the University of Leicester reasoned that it should be possible to develop **DNA fingerprinting** methods that exploit polymorphic genetic markers as identity determinants. Beginning in the 1980s, Jeffreys and others began to compile a set of DNA markers that could be used to distinguish between individuals (or human samples) on the basis of restriction fragment length polymorphisms (RFLPs) at specific loci. The presence or absence of these RFLPs was used in conjunction with rigorous statistical calculations to determine the probability that two samples were the same or different. A landmark example demonstrating the utility of DNA fingerprinting methods came in 1985 when Jeffreys used RFLP markers to determine that a man who had confessed to a murder could not, on the basis of a statistical argument, be the person matching the genotype profile found in the crime scene sample. The police then used Jeffreys' forensic test to survey a large number of possible suspects in the local village.

Over the past decade, legal arguments against the use of DNA fingerprinting as admissible courtroom evidence in criminal cases, and the frustrating inability of jury members to comprehend the statistical significance of DNA data, have led to increased quality control and sample analysis methodologies on the part of the forensic scientists. One of the biggest improvements has been to include PCR assays, in addition to RFLP analyses, to amplify polymorphic regions of human genomic DNA. PCR is a much more sensitive technique than RFLP, and importantly, PCR assays performed properly are more amenable to standardized quality control (QC) protocols than is Southern blotting. By using a minimum of three single locus genetic markers, each of which is highly polymorphic (>90%) and stably inherited (<0.2% mutation rate per generation), it is possible, for example, in paternity cases to assign statistically significant values of inclusion or exclusion. Figure 9.8 illustrates how PCR can be used to detect polymorphic SNPs and STRPs in human DNA samples as a forensic test to assign paternity in a hypothetical criminal case involving a suspected serial rapist. It will soon be possible to use microarray oligonucleotide chips to determine human identity on the basis of >10,000 human SNPs quickly and accurately. The application of SNiP chip genotyping to human forensics should provide an irrefutable statistical argument connecting a crime scene sample to a suspect or a victim.

Molecular Genetic Pathophysiology

The application of molecular genetics to the development of improved disease diagnosis and treatment strategies is an emerging area of medicine that has been described as molecular genetic pathophysiology. Molecular genetic pathophysiology falls somewhere between genetic identity testing, which is a reliable scientifically based forensic method, and somatic cell gene therapy, a very exciting but still unproved molecular genetic approach to disease treatment. In this section we look at two such examples: (1) molecular genetic diagnostics for the purposes of embryo preimplantation decisions, a technology that is already available in private clinics on a fee-for-service basis, and (2) personalized disease treatment strategies, sometimes called pharmacogenetics, that will be tailored to molecular genetic information obtained from DNA microchips.

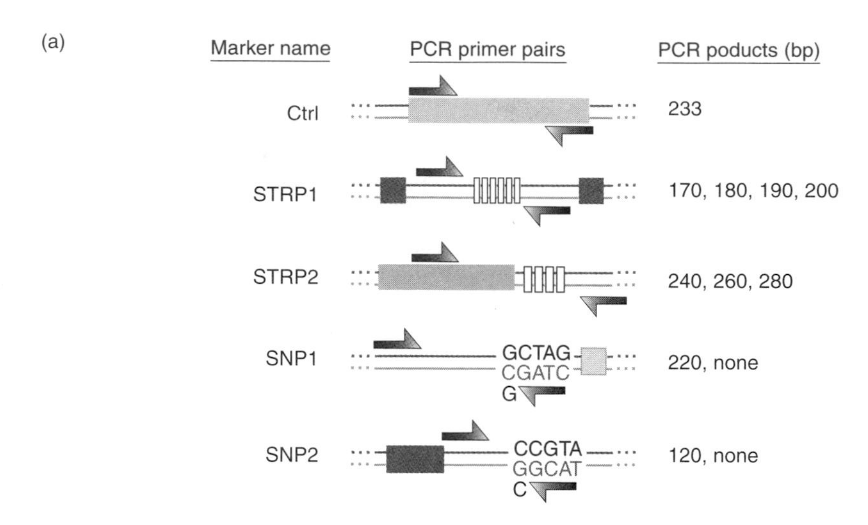

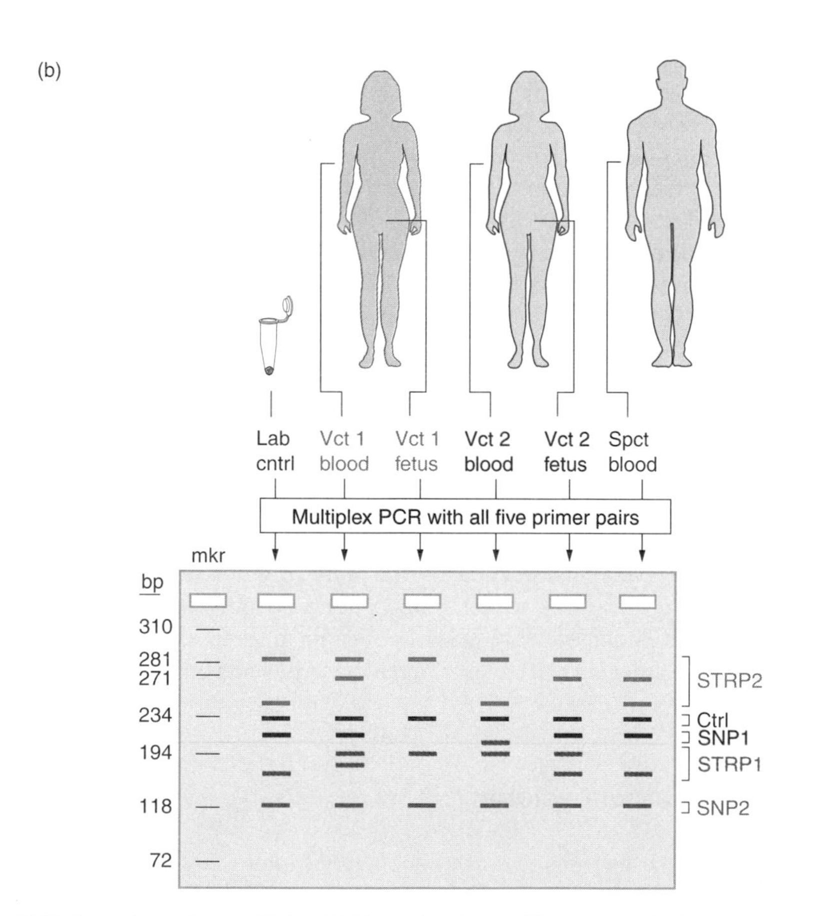

Figure 9.8 DNA forensics using multiplex PCR can be done with comparative human samples to assign identity in criminal cases. This example shows results from forensic DNA testing in a hypothetical crime case involving two women who were sexually assaulted and impregnated. The police have one man in custody who may be responsible for both rapes. DNA forensics was used to determine if identity could be assigned to the suspect using cell samples obtained from the victims, the fetuses, and the suspect. (*a*) Schematic representation of four polymorphic markers, and one nonpolymorphic marker control, that could be used in a multiplex PCR assay. (*b*) Results of agarose gel electrophoresis comparing the five human samples and one laboratory standard using the designated PCR primer pairs. These data would indicate high probability that the suspect sexually assaulted victim 2.

There is no doubt that as the Human Genome Project nears completion in the middle of the next decade, medical diagnostic procedures will expand dramatically. Moreover, with the recent advances in in vitro fertilization procedures, it is likely that couples seeking reproductive services will face choices that were not available just a few years ago. One way to illustrate this concept is to examine a technique called preimplantation genetic diagnosis (PGD) that was developed in the early 1990s by fertility researchers including, among others, Mark Hughes and Alan Handyside. PGD has been used as a means to screen human embryos for known genetic mutations present in the maternal and paternal gene pools. This PCR-based diagnostic assay involves the removal of a single blastomere from an in vitro eight-cell embryo as the source of genomic DNA for a single-cell PCR technique, called primer-extension preamplification (PEP). The PEP method was first described by Norm Arnheim, a molecular geneticist who had earlier played a key role in developing the Southern blot (Chapter 1) and PCR (Chapter 6) for research applications. Figure 9.9 illustrates how PGD could be used to genotype embryos derived from in vitro fertilization methods in a case where the mother was a carrier for chromosome X-linked Factor VIII hemophilia.

A second way that molecular genetic applications are having an impact on medical science is in the area of personalized medical treatment. One example of this approach is in the area of cancer diagnosis and treatment where PCR-based genotyping of various patient cell samples (sputum, blood, and tumor biopsies) are compared prior to, during, and following therapy. In this way, the efficacy of the chosen chemotherapeutic or radiation treatment can be monitored by determining the relative numbers of cancer cells and normal cells in a patient sample. A potential extension of this strategy would be to use microchip genotyping to create a genomic DNA profile of individuals who have a family history of cancer. If known cancer gene alleles could be fingerprinted early in life, then it may be feasible to advise patients about life-style choices that may lessen their risk of developing cancer (Fig. 9.10). This type of integrated molecular genetic approach to cancer prevention, diagnosis, and treatment is a type of "pharmacogenetics," which refers to the use of genetic information for therapeutic purposes.

A more futuristic form of pharmacogenetic medicine is being pursued by genome-based biotechnology companies that aim to develop personalized approaches to disease treatment that are based on a patient's microchip genotype. In this approach, large pilot studies would be conducted using patients that have a complex disease that can be treated, such as adult-onset diabetes. The diabetic patient's genotype would be fingerprinted using oligonucleotide microchips and then the individual would be tested to determine responsiveness to different pharmaceutical compounds that are known to be efficacious in some diabetic patients. Because differences in drug metabolizing pathways would affect pharmacokinetic parameters, it has been proposed that pharmacogenetics could be used to predict which therapies would be the most efficacious for a given patient. These genotype-based approaches would be especially valuable for the treatment of diabetes, cardiovascular disease, inflammatory diseases, and mental illness.

Somatic Cell Gene Therapy

Human somatic cell gene therapy is a form of medical treatment that is being developed as a genetic approach to disease management. This anticipated clinical application of molecular genetics has resulted in more than 200 human gene therapy clinical trials having been conducted since 1990. These various phase I and phase II studies have enrolled thousands of patients and have provided clinical researchers with a clear view of the

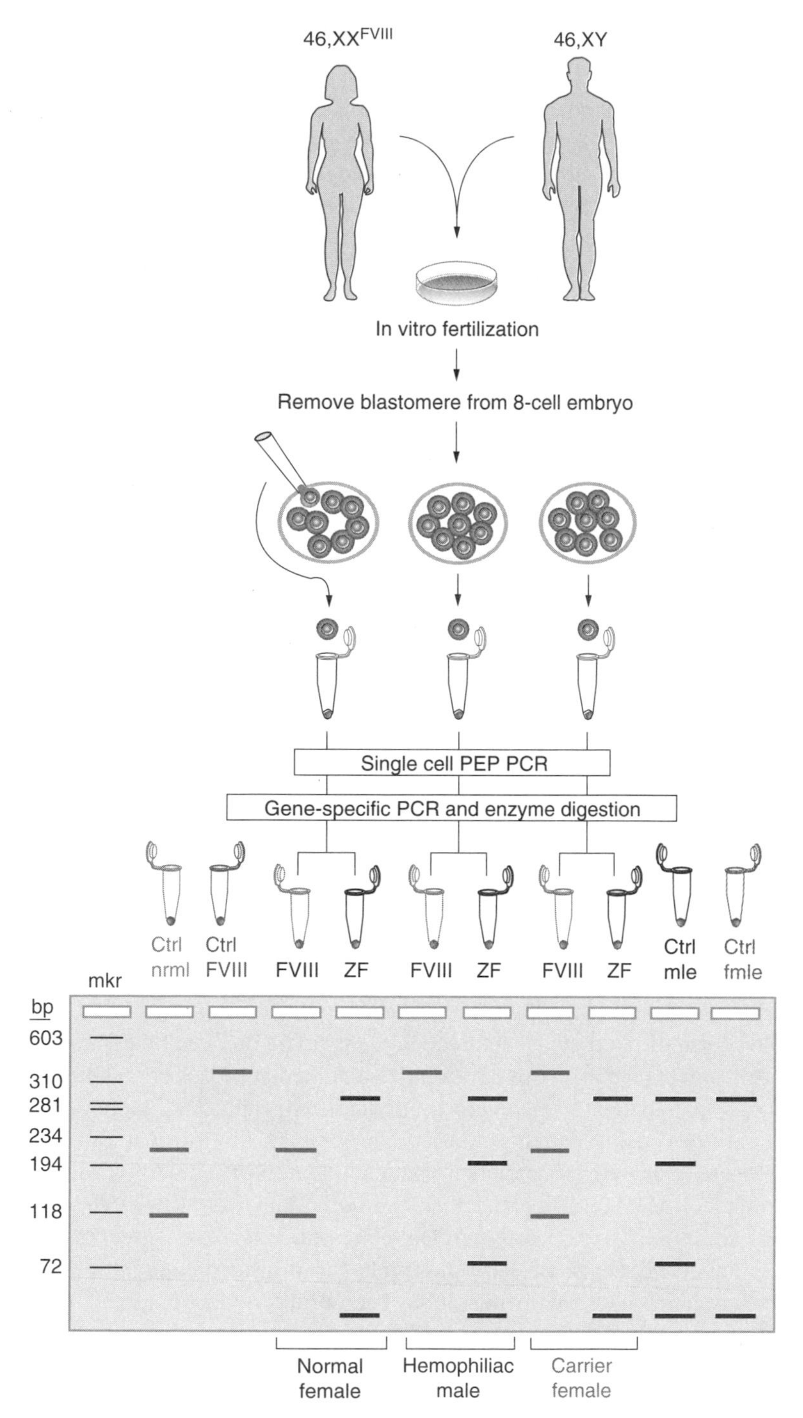

Figure 9.9 Preimplantation genetic diagnosis (PGD) is a method used to genotype human embryos for known genetic defects prior to uterine implantation. In this hypothetical example, the Factor VIII gene defect is carried on one of the two X chromosomes on the maternal side. PEP/PCR amplification is used to amplify the entire genome randomly to increase the sensitivity of the assay and provide material for additional tests. A second PCR reaction is performed using the amplified genomic DNA and Factor VIII specific primers that will produce a 339 bp product. Lack of a diagnostic Bcl1 site within the target segment is associated with the Factor VIII defect. Amplification of the ZFX and ZFY regions of the X and Y chromosomes, respectively, followed by HaeIII enzyme digestion of the PCR product, provides a diagnostic test for embryo gender. Agarose gel electrophoresis, including appropriate control PCR reactions, is used to resolve the restriction enzyme fragments.

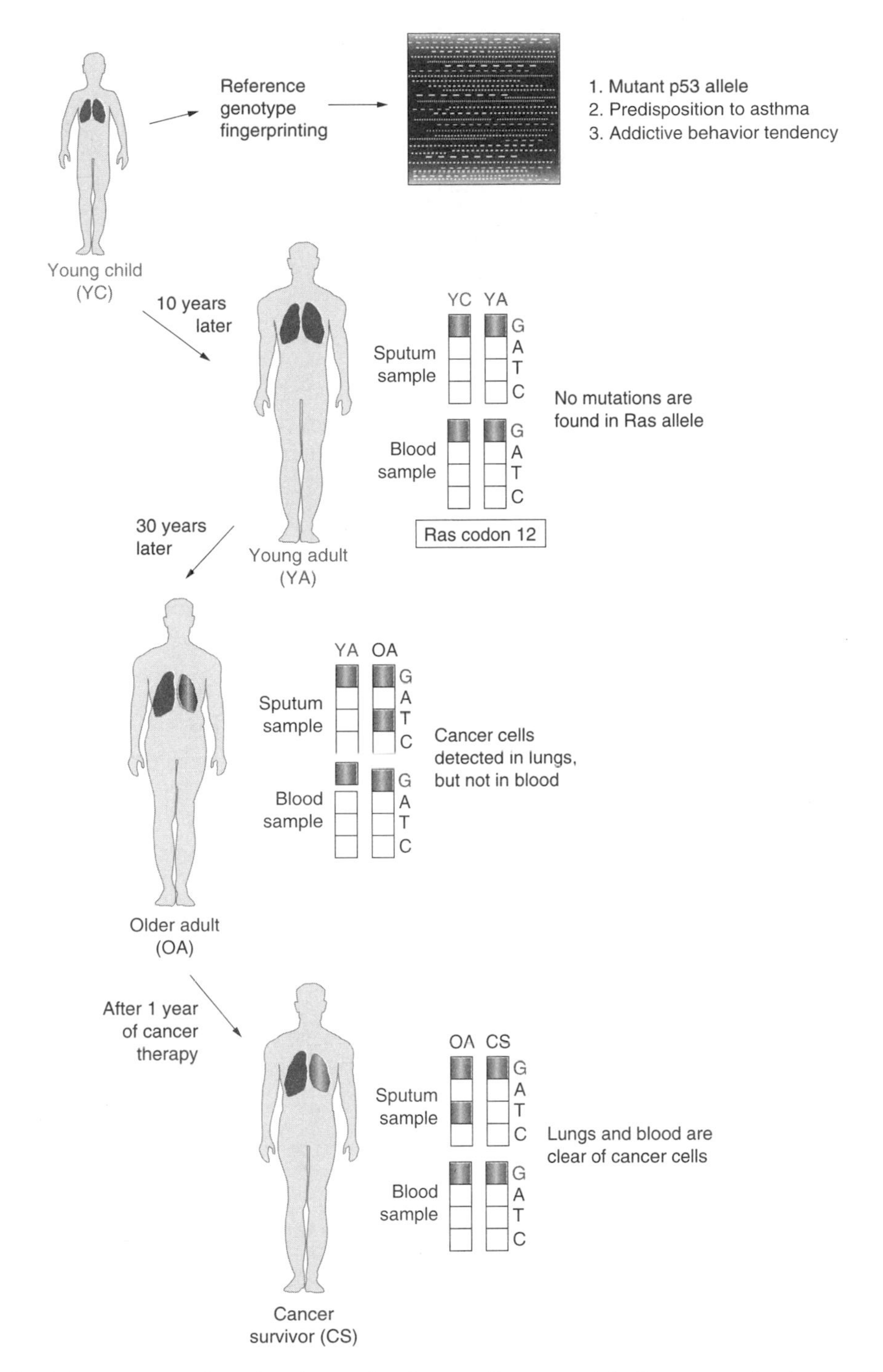

Figure 9.10 Pharmacogenetics refers to a molecular genetic approach to the personalized prevention, diagnosis, and treatment of disease. In this scenario, genotype fingerprinting would be performed early in life to identify gene mutations known to be associated with an increased risk of lung cancer. For example, inherited mutations in the *p53* gene would be predictive of a predisposition to somatic mutations in known oncogenes. Codon 12 of the oncogenic *ras* gene represents a mutational hot spot that could easily be monitored using a high resolution DNA microchip assay. Early detection methods for lung cancer would include routine PCR-based DNA microchip analysis of lung fluid (sputum) to detect rare cancer cells that would indicate disease, and blood samples would be analyzed to detect tumor metastasis. Other uses of DNA-based assays include tissue biopsies to determine tumor grade, in situ PCR as a tool for pathologists, treatment efficacy profiles, and long-term follow-up.

major obstacles that need to be addressed before human gene therapy will be considered a viable alternative to traditional small molecule therapies. Three key improvements appear to be needed for gene therapy to move ahead: (1) more efficient DNA delivery systems to target cells, (2) increased levels of long-term gene expression, and (3) abrogation of host immune reactions directed against transfected cells.

This section summarizes current approaches researchers are using to develop effectual human somatic cell gene therapies. Figure 9.11 shows the two basic gene therapy strategies that have been investigated. In vivo gene therapy requires that the gene transfer vector be delivered in a cell-type selective manner, either through direct tissue injection or, perhaps someday, by receptor-mediated processes. Ex vivo gene therapy involves removing tissue from the patient, transfecting (or virally infecting) the cells in culture, and then reimplanting the genetically altered cells to the patient. Systemic infusion of cell-targeted gene transfer vectors would be ideal because this method presents the least number of complications. Note that DNA transfection of human embryonic stem cells to correct genetic mutations prior to uterine implantation, or as a method to alter the germ-line genotype of an early human embryo in any way, including by nuclear transfer methods, is considered by many to be unethical and has consequently been banned by professional medical organizations.

Monogenic loss-of-function diseases have been used as paradigms to develop gene therapy strategies using animal models. Cancer is another disease target for potential gene therapy applications. Gene therapy could be used to activate anti-tumor immune responses in the patient, inhibit cell proliferation by reintroduction of functional tumor suppressor genes, or induce the apoptotic cell death program in cancer cells. Table 9.4 lists representative human diseases that have been proposed as possible candidates for directed gene therapy. In this context, gene therapy refers to genotypic alterations (chromosomal or episomal) of somatic cells, resulting in sustained gene expression. This is to be distinguished from transient DNA transfection approaches, such as drug therapies utilizing ribozymes or antisense oligonucleotides.

One of the biggest problems to solve in somatic cell gene therapy is finding an efficient gene delivery system. Two basic strategies have been investigated, viral-based vectors and nonviral transfection methods. There are currently four viral vectors and two nonviral transfection methods being tested for in vivo and ex vivo gene therapy approaches. Each of these gene delivery strategies has its own advantages and disadvantages, as listed in Table 9.5. In addition to being safe and cost-effective, the most

TABLE 9.4 Human diseases that are target candidates for somatic cell gene therapy

Disease	Target cells	Transfected gene(s)
Hemophilia A	Liver, muscle, bone marrow cells	Factor VIII
Hemophilia B	Fibroblasts	Factor IX
Familial hypercholesterolemia	Liver	Low-density lipoprotein receptor
Severe combined immunodeficiency	Bone marrow cells, T cells	Adenosine deaminase (ADA)
Hemoglobinopathies	Red blood precursor cells	α-Globin, β-globin
Cystic fibrosis	Lung airway cells	Cystic fibrosis gene (CFTR)
Gaucher's	Bone marrow cells, macrophages	Glucocerebrosidase
Cancer	Tumor cells	p53, Rb, interleukins, growth-inhibitory genes, apoptosis genes

(a)

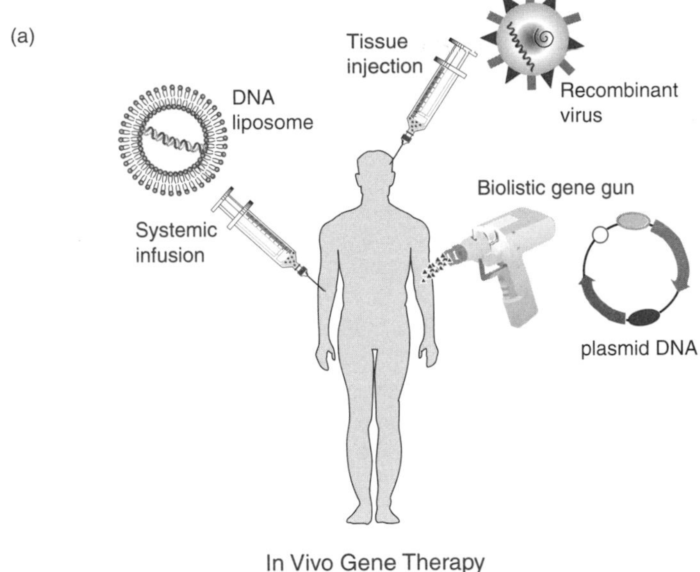

In Vivo Gene Therapy

(b)

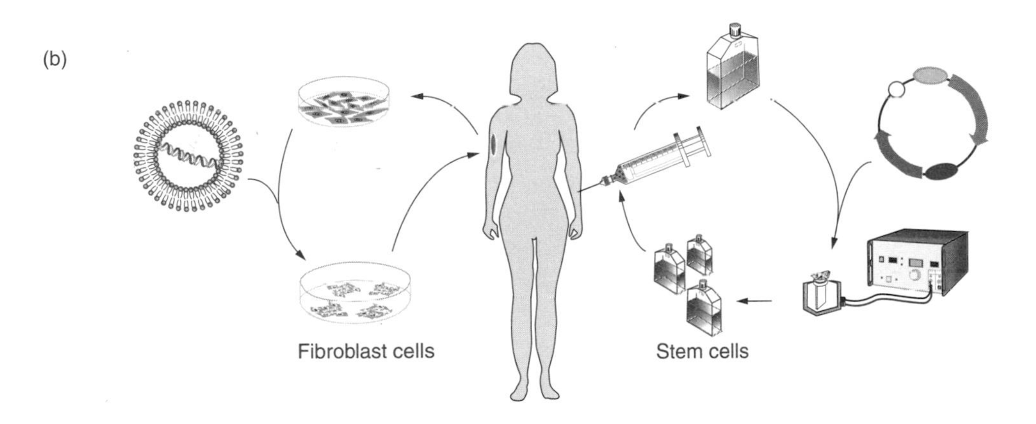

Ex Vivo Gene Therapy

Figure 9.11 Somatic cell gene therapy can be performed using in vivo or ex vivo gene delivery strategies. (*a*) In vivo gene therapy is done by targeting the gene delivery system to the desired cell type in the patient either using physical means such as tissue injection (brain tumor) or biolistics (dermal DNA vaccination), or using, potentially in the systemic infusion of cell-specific receptor-mediated DNA carriers (reconstructed liposomes or viruses). (*b*) Ex vivo gene therapy is performed by transfecting or infecting patient-derived cells in culture with vector DNA and then reimplanting the transfected cells into the patient. Two types of ex vivo gene therapies under development are those directed at fibroblasts and hematopoietic stem cells.

TABLE 9.5 Potential advantages and disadvantages of various gene delivery systems

Gene delivery system	Advantages	Disadvantages
Retrovirus	Integrates into host cell genome providing stable gene expression, ontains no viral genes, few immunity problems	Infects only dividing cells, random integration may cause insertional mutations, relatively low titer stocks
Adenovirus	Can contain >30 kb of nonviral DNA, very high titer stocks, infects both nondividing and dividing cells	Does not provide long-term gene expression due to lack of integration, very immunogenic, can be toxic
Adeno-associated virus	Can integrate into genome providing stable expression, contains no viral genes, nonpathogenic, no immunity problems, infects nondividing cells	Small capacity for foreign gene sequences, difficult to obtain large amounts of viral stock
Liposomes	No viral components, nonpathogenic, no immunity problems, no limit to size of foreign gene	Low transfection efficiency, low rate of stable integration and long-term gene expression
Biolistics	Same as liposome-mediated transfer, promising as a vaccination method	Limited to dermal tissue, low rate of stable integration, difficult to QC

important properties of an efficacious gene transfer system are that it is, (1) target cell selective, (2) transcriptionally competent for the desired length of time, (3) available in a highly concentrated active form, and (4) immunologically neutral. Although it is still too early to know how useful somatic cell gene therapy will be as a routine treatment for human genetic diseases, most biomedical researchers agree that it will represent just one of the *many* new molecular genetic tools that 21st-century physicians will have at their disposal to prevent and diagnose disease and treat their patients better.

Laboratory Practicum 9. *Identifying orthologous genes by searching Internet DNA databases*

Research Objective

A pharmacogenetic-based biotechnology company, Ocard Inc., has developed a new treatment for asthma it calls PulmaKine. This inhaler-delivered compound was discovered using a combinatorial chemistry approach directed against known cytokine receptors expressed on lung epithelial cells. During the clinical trial phase, it was observed that PulmaKine was highly effective in blocking respiratory impairment in 65% of asthmatics who used the Ocard inhaler during an acute asthma attack. However, the other 35% of the test subjects showed no measurable improvement with the drug, and in fact, many of these PulmaKine nonresponders required emergency medical treatment to stop the asthma attack.

Peripheral blood lymphocytes (PBL) and bronchial lavage fluid (BLF) had been collected from all of the subjects in the PulmaKine drug trial. DNA and RT-PCR microchip fingerprinting was performed using the PBL and BLF cell samples, respectively. This characterization led to the identification of a single genomic DNA marker (STS.1030) and cDNA transcript (EST.4591) that differentiated the PulmaKine responders (STS.1030–/EST.4591+) from the nonresponders (STS.1030+/EST.4591–). The molecular genetic research group at Ocard intend to use the microchip fingerprinting data to identify the candidate gene that positively correlates with PulmaKine responsiveness. Their long-term goal is to develop a second-generation therapeutic compound that would be used specifically to treat PulmaKine nonresponders.

Available Information and Reagents

1. The DNA marker STS.1030 maps to human chromosome 11 and is within 100 kb of a putative gene encoding EST.4591. The sequence of EST.4591 has been obtained from the dbest database.
2. The molecular genetic researchers have a LAN-internet connection and use the Netscape Communications browser software to access the NCBI BLAST home page through the WWW.
3. Ocard Inc. has purchased an on-site license for the MacVector™ DNA sequence analysis software package to perform advanced DNA sequence characterizations and to produce graphical images of the data.

Basic Strategy

The human EST.4591 cDNA sequence was identified as the only probe on the ~16,000 member microchip cDNA array that gave a >98% association coefficient with PulmaKine responsiveness. Because the STS.1030 DNA marker was inversely correlated with expression of the EST.4591 transcript in lung epithelial cells, the researchers reasoned that PulmaKine nonresponders had a germ-line mutation that prevented expression of the EST.4591-encoded protein, which they named AMG (asthma modulating gene). In order to investigate the mechanisms of PulmaKine resistance, the researchers needed to obtain a full-length cDNA containing the *AMG* ORF, and then determine if any orthologous genes (evolutionarily related) existed in model organisms for which a gene function had already been assigned.

Using the NCBI BLAST Search WWW site, the *AMG* sequence from EST.4591 was used in *blastn* query against the dbest database to find other human ESTs that might be overlapping. The sequence of an assembled *AMG* EST contig was then analyzed in the Ocard computer lab using the MacVector™ programs to determine if an extended ORF was present. Figure 9.12 shows the results of the *AMG* sequence analysis using both the BLAST and MacVector™ programs. The identified *AMG* encoded protein sequence obtained from the initial sequence analysis was used to query GenBank using the *tblastn* search strategy as a means to identify *AMG* orthologous genes in other eukaryotes. The results of this BLAST search are shown in Figure 9.13. It can be seen that the hypothetical AMG protein shares homology with both the human interleukin-3 receptor and the insulin receptor, as well as a *C. elegans* receptor gene called *daf-2*, which may represent an orthologous tyrosine receptor kinase gene involved in signal transduction.

Comments

The cDNA microchip used in this study contained EST probes that had not yet been linked to known human protein coding sequences. This is also true of the first-generation cDNA microchips that are currently available; however, this lack of identity will eventually be corrected once the Human Genome Project is completed. The *S. cerevisiae* ORF microchip, for example, provides a comprehensive analysis of all predicted protein coding sequences in the yeast genome; a human EST equivalent should be available in the not too distant future. The BLAST homology searches run through the NCBI WWW site provide information regarding evolutionary similarities; however, most of the detailed DNA sequence analysis has to be performed using on-site software such as the MacVector™ programs. This EST "cloning by modem" strategy depends on the fact

(a)

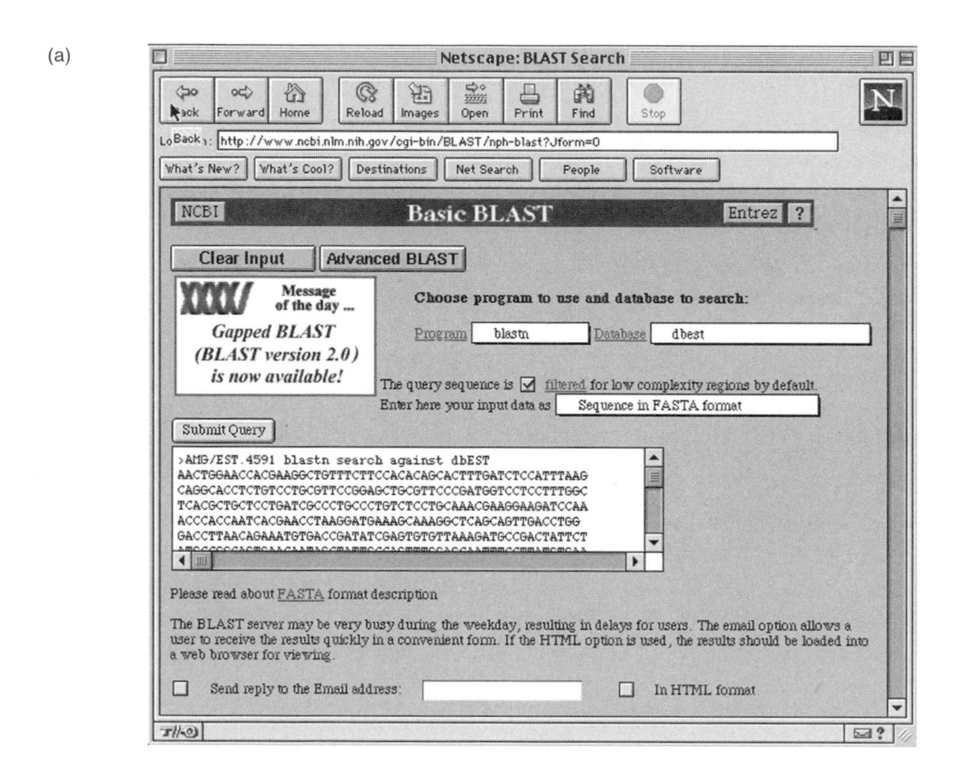

(b)

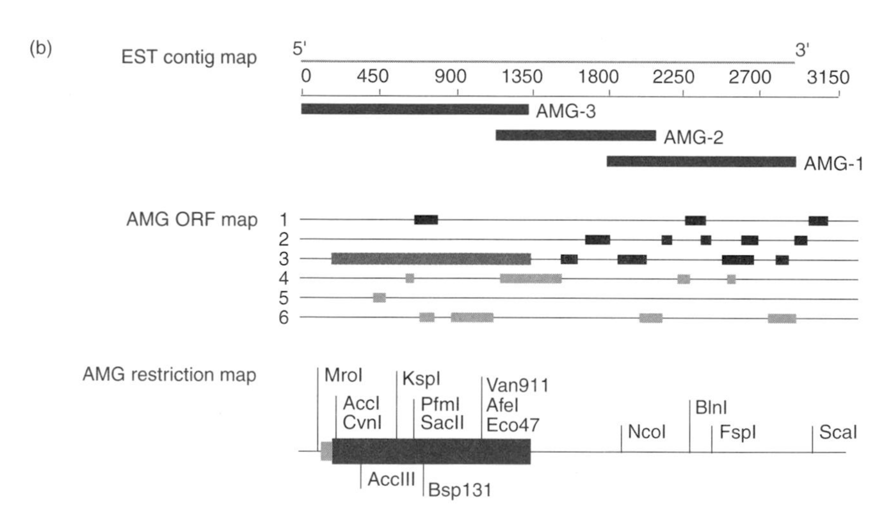

Figure 9.12 Sequence homology searches with the *AMG* sequence can be used to identify related genes in the GenBank dbEST database. (*a*) Screen capture of the NCBI WWW browser window used to enter the BLAST search program using the hypothetical *AMG* cDNA sequence. (*b*) The MacVector™ DNA sequence analysis programs are run on the laboratory computer to find overlapping EST contigs and predicted ORFs. The *AMG-3* sequence corresponds to the original EST.4591 which was used in the BLAST search to identify *AMG-2*. A subsequent BLAST search using *AMG-2* sequences as the query identified *AMG-1*. In this example, the combined *AMG* sequence from the EST contig map was used to identify a predicted 453 amino acid AMG protein using in reading frame 3 using an ORF mapping algorithm in MacVector™. A limited restriction enzyme map of the *AMG* contig was also generated by the MacVector™ sequence analysis programs.

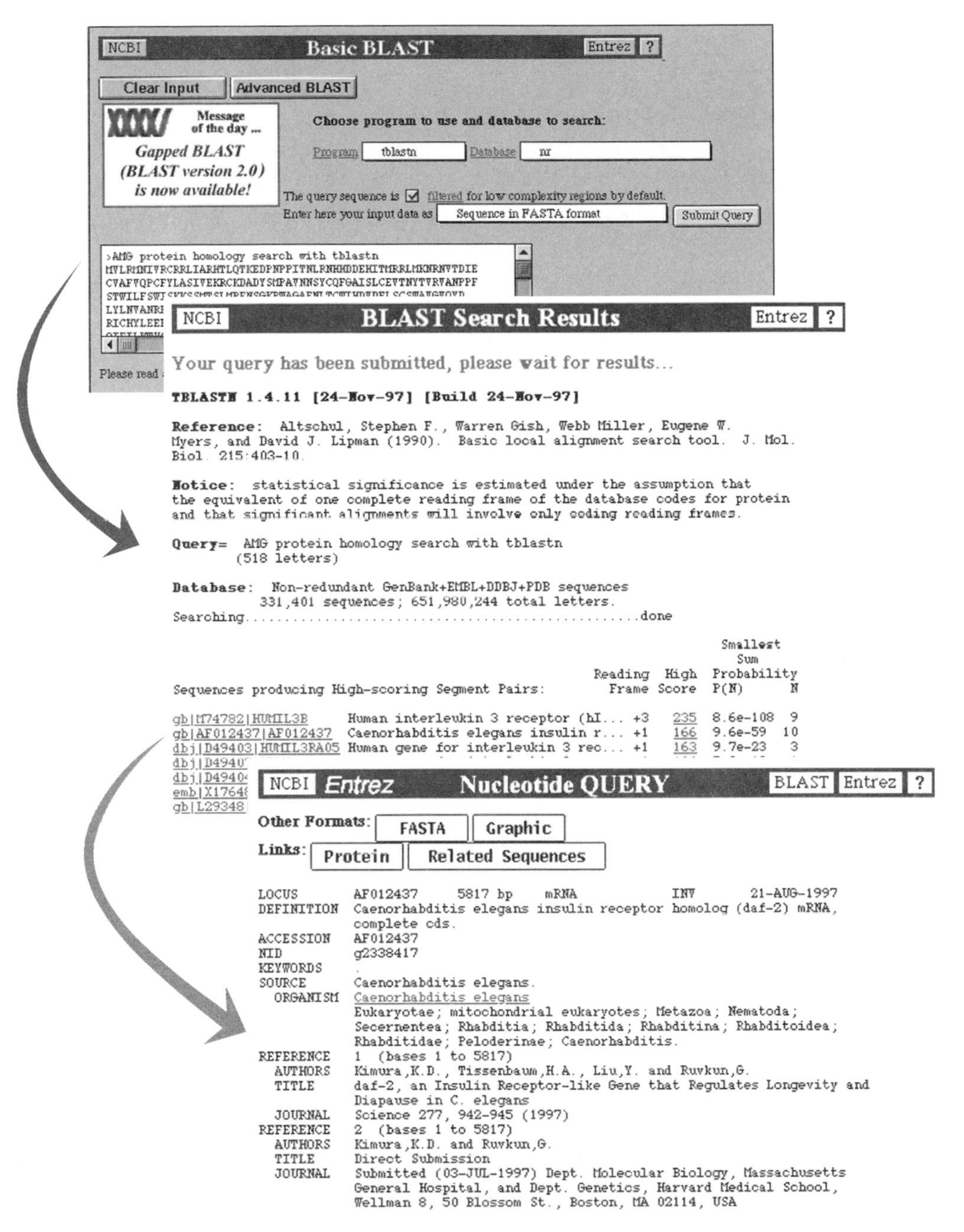

Figure 9.13 The *tblastn* search program is used to find orthologous proteins by comparing query amino acid sequences with the translated GenBank database. The results of a *tblastn* search using the hypothetical AMG protein sequence indicate that this gene is most related to the human interleukin-3 receptor (probability of match due to random chance is 8.6×10^{-108}) and a recently cloned insulin receptor-like gene from *C. elegans* called *daf-2*. These three screen captures demonstrate how the WWW can be used quickly to find investigative leads to the function of a newly cloned gene.

that all of the EST clones used to assemble the dbest database are readily available from a government sponsored group called the I.M.A.G.E. (Integrated Molecular Analysis of Genomes and their Expression) consortium, which provides recombinant DNA samples at cost for research purposes.

Prospective

The next logical set of laboratory experiments would be to determine what role the *AMG* receptor has in mediating PulmaKine efficacy. Future laboratory experiments would require a cell line or animal model, such as a transgenic mouse or even a *C. elegans*-based model. Once this were developed, then it should be possible to use standard mutagenesis experiments, in the context of a bioassay, to determine the requirement of various conserved AMG protein motifs in mediating PulmaKine function. However, it would also be very important from a medical diagnostic point of view, and economically advantageous to Ocard Inc., to develop a modified microchip assay to identify candidates for PulmaKine treatment on the basis of genotype information. Because the STS.1030 is a genomic marker that can be used to identify individuals with a genetic defect in *AMG* gene expression, an STS-based oligonucleotide microchip could be used easily to obtain *AMG* genotype information using DNA isolated from patient PBL samples.

The application of genotype chips for selective medical diagnostic purposes, however, raises the complex bioethical issue of genetic privacy, because disease markers other than just STS.1030 would also be identified. The potential misuse of personal DNA information is currently the subject of much public debate. The National Human Genome Research Institute (NHGRI) at the NIH, in conjunction with a panel organized to oversee bioethical issues related to the Human Genome Project called ELSI (ethical, legal and social issues), has been involved in developing guidelines for legislation that would prevent organizations such as insurance companies or employers from using DNA sequence data to deny access to individuals with perceived "risky" genotypes. Questions about protecting genetic privacy, as well as other more complex bioethical issues relating to nuclear transfer applications and cloning, will require careful thought to maximize benefits and minimize moral negligence.

REFERENCES

Automated instrumentation for genome analysis

Beugelsdijk, T. (ed.). Automation technologies for genome characterization. John Wiley & Sons, New York, 303 pp., 1997.

Chee, M., Yang, R., Hubbell, E., Berno, A., Huang, X.C., Stern, D., Winkler, J., Lockhart, D.J., Morris, M.S., Fodor, S.P.A. Accessing genetic information with high-density DNA arrays. Science 274:610–614, 1996.

Cho, R., Fromont-Racine, M., Wodicka, L., Feierbach, B., Stearns, T., Legrain, P., Lockart, D., Davis, R. W. Parallel analysis of genetic selections using whole genome oligonucleotide arrays. Proc. Natl. Acad. Sci. USA 95:3752–3757, 1998.

Chuan, T-C., Sparks, A., Rago, C. et al. Identification of a c-Myc as a target of the APC pathway. Science 281:150–154, 1998.

DeRisi, J., Iyer, V., Brown, P. Exploring the metabolic and genetic control of gene expression on a genomic scale. Science 278:680–686, 1997.

* Landmark papers in applied molecular genetics.

DeRisi, J., Penland, L., Brown, P.O., Bittner, M.L., Meltzer, P.S., Ray, M., Chen, Y., Su, Y.A., Trent, J.M. Use of a cDNA microarray to analyze gene expression patterns in human cancer. Nature Genet. 14:457–460, 1996.

*Hacia, J., Brody, L., Chee, M., Fodor, S., Collins, F. Detection of heterozygous mutations in BRCA1 using high density arrays and two colour fluorescence analysis. Nature Genet. 14:441–447, 1996.

Johnston, M. Gene chips: Array of hope for understanding gene regulation. Curr. Biol. 8:R171–174, 1998.

*Lashkari, D., DeRisi, J., McCusker, J., et al. Yeast microarrays for genome wide parallel genetic and gene expression analysis. Proc. Natl. Acad. Sci. USA 94:13057–13062, 1997.

*Lee, L., Connell, C., Woo, S., et al. DNA sequencing with dye-labeled terminators and T7 DNA polymerase: effect of dyes and dNTPs on incorporation of dye-terminators and probability analysis of termination fragments. Nucl. Acids Res. 20:2471–2483, 1992.

Marshall, A., Hodgson, J. DNA chips: an array of possibilities. Nature Biotech. 16:27–31, 1998.

Metzker, M.L., Lu, J., Gibbs, R.A. Electrophoretically uniform fluorescent dyes for automated DNA sequencing. Science 271:1420–1422, 1996.

Pardee, A. Complete genome expression monitoring: the human race. Nature Biotech. 15: 1343–1344, 1997.

Rowen, L., Mahairas, G., Hood, L. Sequencing the human genome. Science 278:605–607, 1997.

*Sanger, F., Nicklen, S., Coulson, A. DNA sequencing with chain termination inhibitors. Proc. Natl. Acad. Sci. USA 74:5463–5467, 1977.

Schena, M., Shalon, D., Heller, R., Chai, A., Brown, P.O., Davis, R.W. Parallel human genome analysis: Microarray-based expression monitoring of 1000 genes. Proc. Natl. Acad. Sci. USA 93:10614–10619, 1996.

*Smith, L.M., Sanders, J., Kaiser, R., Hughes, P., Dodd, C., Connel, C., Heriner, C., Kent, S., Hood, L.E. Fluorescence detection in automated DNA sequence analysis. Nature 321: 674–679, 1986.

*Velculescu, V.E., Zhang, L., Vogelstein, B., Kinzler, K.W. Serial analysis of gene expression. Science 270:484–487, 1995.

Venter, J.C., Adams, M., Sutton, G., Kerlavage, A., Smith, H.O., Hunkapiller, M. Shotgun sequencing of the human genome. Science 280:1540–1542, 1998.

*Wang, D., Fan, J., Siao, C., et al. Large-scale identification, mapping, and genotyping of single-nucleotide polymorphisms in the human genome. Science 280:1077–1082, 1998.

*Winzeler, E., Richards, D., Conway, A. et al. Direct allelic variation scanning of the yeast genome. Science 281:1194–1197, 1998.

*Wodicka, L., Dong, H., Mittmann, M., Ho, M-H, Lockart, D. Genome-wide expression monitoring in *S. cerevisiae*. Nature Biotech. 15:1359–1367, 1997.

Zhang, L., Zhou, W., Velculescu, V.E., Kern, S.E., Hruban, R.H., Hamilton, S.R., Vogelstein, B., Kinzler, K.W. Gene expression profiles in normal and cancer cells. Science 276:1268–1272, 1997.

Accessing molecular genetic information through the internet

*Altschul, S.F., Gish, W., Miller, W., Myers, E.W., Lipman, D.J. Basic local alignment search tool. J. Mol. Biol. 215:403–410, 1990.

*Andreessen, M. NCSA Mosaic technical summary, revision 2.1. National Center for Supercomputing Applications, University of Illinois, Champaign, 1993.

Benson, D., Boguski, M., Lipman, D., Ostell, J. GenBank. Nucl. Acids Res. 25:1–6, 1997.

Botstein, D., Cherry, J.M. Molecular linguistics: Extracting information from gene and protein sequences. Proc. Natl. Acad. Sci. USA 94:5506–5507, 1997.

Heiter, P., Boguski, M. Functional genomics: it's all how you read it. Science 278:601–602, 1997.

Kanehisa, M. Linking databases and organisms: GenomeNet resources in Japan. Trends Bioch. Sci. 22:442–444, 1997.

*Karlin, S., Altschul, S. Applications and statistics for multiple high-scoring segments in molecular sequences. Proc. Natl. Acad. Sci. USA 90:5873–5877, 1993.

Karp, P.D. Metabolic databases. Trends Biochem. Sci. 23:114–116, 1998.

Lawrence, S., Giles, C. L. Searching the world wide web. Science 280:98–100, 1998.

Mann, M. A shortcut to interesting human genes: Peptide sequence tags, expressed-sequence tags and computers. Trends Biochem. Sci. 21:494–495, 1996.

Pennisi, E. Genomics—A catalog of cancer genes at the click of a mouse. Science 276: 1023–1024, 1997.

Schatz, B. Information retrieval in digital libraries: bringing search to the net. Science 275: 327–334, 1997.

Schultz, J., Milpetz, F., Bork, P., Ponting, C. SMART, a simple modular architecture research tool: Identification of signaling domains. Proc. Natl. Acad. Sci. USA 95:5857–5864, 1998.

Smith, C., Shindyalov, I. Veretnik, S., Gribskov, M., Taylor, S., Eyck, L., Bourne, P. The protein kinase resource. Trends Biochem. Sci. 22:444–446, 1997.

Smith, R. Perspectives: sequence database searching in a the era of large-scale genomic sequencing. Genome Res. 6:653–660, 1996.

Sommerville, C., Flanders, D., Cherry, J. Plant biology in the post-Gutenber era: everything you wanted to know and more on the world wide web. Plant Physiol. 113:1015–1022, 1997.

Strausberg, R.L., Dahl, C.A., Klausner, R.D. New opportunities for uncovering the molecular basis of cancer. Nat Genet 16 suppl., 415–516, 1997.

Vasmatzis, G., Essand, M., Brinkmann, U., Lee, B., Pastan, I. Discovery of three genes specifically expressed in human prostate by expressed sequence tag database analysis. Proc. Natl. Acad. Sci. USA 95:300–304, 1998.

Walsh, L. (ed.). Guide to the Internet 1997, Trends Journals supplement. Elsevier Science Publications, Cambridge, UK, 1997.

Applied medical molecular genetics

Cantor, C. How will the Human Genome Project improve our quality of life? Nature Biotech. 16:212–213, 1998.

Doll, J. The patenting of DNA. Science 280:689—690, 1998.

Eng, C. From bench to bedside . . . but when? Genome Res. 7:669–672, 1997.

Gjertson, D., Mickey, M., Hopfield, J., Takenouchi, T., Terasaki, P. Calculation of of the probability of paternity using DNA sequences. Am. J. Human Genet. 43:860–869, 1988.

Glaser, V., Hodgson, J. Before anyone knew the future nature of biotechnology. Nature Biotech. 16:239–242, 1998.

*Handyside, A., Kontogianni, E., Hardy, K., Winston, R. Pregnancies from biopsied human preimplantation embryos sexed by Y-specific DNA amplification. Nature 340:768–773, 1990.

Felgner, P.L. DNA vaccines. Curr. Biol. 18:R551–R553, 1998.

Jeffreys, A. J., MacLeod, A., Tamaki, K., Neil, D., Monckton, D. Minisatellite repeat coding as a digital approach to DNA typing. Nature 354:204–209, 1991.

*Jeffreys, A.J., Wilson, V., Neumann, R., Keyte, J. Amplification of human minisatellites by the polymerase chain reaction: towards DNA fingerprinting of single cells. Nucl. Acids Res. 16: 10953–10971, 1988.

Kay, M., Liu, D., Hoogerbrugge, P. Gene therapy. Proc. Natl. Acad. Sci. USA 94:12744–12746, 1997.

Lander, E. DNA fingerprinting on trial. Nature 339:501–505, 1989.

Marshall, A. Laying the foundations for personalized medicines. Nature Biotech. 15:954–957, 1997.

Sidransky, D. Nucleic acid-based methods for the detection of cancer. Science 278:1054–1058, 1997.

Siver, L. M. Remaking of eden: Cloning and beyond in a brave new world. Avon/Weiderfeld and Nicolson Publishers, pp. 315, 1998.

*Snabes, M., Chong, S., Subramanian, S., Kristjansson, K., DiSepio, D., Hughes, M. Preimplantation single-cell analysis of multiple genetic loci by whole genome amplification. Proc. Natl. Acad. Sci. USA 91:6181–6185, 1994.

The evaluation of forensic DNA evidence. Proc. Natl. Acad. Sci. USA 94:5498–5500, 1997.

Verma, I., Somia, N. Gene therapy: promises, problems and prospects. Nature 389:239–242, 1997.

Vile, R. Gene therapy: Primer. Curr. Biol. 8:R73–R75, 1998.

Weinstein, J.N., Myers, T.G., O'Connor, P.M., et al. An information-intensive approach to the molecular pharmacology of cancer. Science 275:343–349, 1997.

Word, C., Sawosik, T., Bing, D. Summary of validation studies from twenty-six forensic laboratories in the United States and Canada on the use of AmpliType PM PCR amplification and typing kit. J. Forensic Sci. 42:39–48, 1997.

*Zhang, L., Cui, X., Schmitt, K., Hubert, R., Navdi, W., Arnheim, N. Whole genome amplification from a single cell: implications for genetic analysis. Proc. Natl. Acad. Sci. USA 89: 5847–5851, 1992.

Identifying orthologous genes by searching internet DNA databases

Cohen, J. Medicine—Developing prescriptions with a personal touch. Science 275:776, 1997.

Harris, J. Is cloning an attack on human dignity? Nature 387:754, 1997.

Henikoff, S., Greene, E., Pietrokovoski, Bork, P., Attwood, T., Hood, L. Gene families: the taxonomy of protein paralogs and chimeras. Science 278:609–614, 1997.

Hudson, K.L., Rothenberg, K.H., Andrews, L.B., Kahn, M.J.E., Collins, F.S. Genetic discrimination and health insurance: An urgent need for reform. Science 270:391–393, 1995.

Inohara, Y, Koseki, T., Hu, Y., Chen, S., Nunez, G. CLARP, a death effector domain-containing protein interacts with caspase-8 and regulates apoptosis. Proc. Natl. Acad. Sci. USA 94: 10717–10722, 1997.

Kimura, K., Tissenbaum, H., Liu, Y., Ruvken, G. *daf-2*, an insulin receptor-like gene that regulates longetivity and diapause in *Caenorhabditis elegans*. Science 277:942–946, 1997.

Makalowski, W., Boguski, M. Evolutionary parameters of the transcribed mammalian genome: An analysis of 2,820 orthologous rodent and human sequences. Proc. Natl. Acad. Sci. USA 95:9407–9412, 1998.

Marshall, E. ELSI's cystic fibrosis experiment. Science 274:489, 1996.

Tatusov, R., Koonin, E., Lipman, D. A genomic perspective on protein families. Science 278: 631–637, 1997.

Vogel, G. From science fiction to ethics quandary. Science 277:1753–1754, 1997.

Wadman, M. Business booms for guides to biology's moral maze. Nature 389:658–659, 1997.

SECTION

4

APPENDICES

APPENDIX

A

STANDARD ABBREVIATIONS

A260	absorbance at 260 nm
Ab	antibody
ABI	Applied Biosystems Incorporated
acetyl CoA	acetyl coenzyme A
AD	activation domain (transcriptional)
ADA	adenosine deaminase
ADP	adenosine 5′-diphosphate
Ag	antigen
AIDS	acquired immune deficiency syndrome
Ala	alanine
AMH	artificial mismatch hybridization
AMP	adenosine 5′monophosphate
amp^r	ampicillin-resistance gene
AMV	avian myeloblastosis virus
APH	aminoglycoside phosphotransferase
Arg	arginine
ARS	autonomous replication sequences
ASCII	American standard code for information interchange
Asn	asparagine
Asp	aspartate
ATCC	American Type Culture Collection
ATP	adenosine 5′–triphosphate
β-*gal*	β-galactosidase gene
BAC	bacterial artificial chromosome
BAP	bacterial alkaline phosphatase
BESS	base excision sequence scanning
BIOSCI	bionet biological sciences newsgroup
bis	bisacrylamide *N,N′*-methylenebisacrylamide
BLAST	basic local alignment research tool
bp	base pair
BPB	bromophenol blue
BrdU	5-bromodeoxyuridine
BSA	bovine serum albumin
BUdR	5-bromodeoxyuridine
CAD	carbamoylphosphate synthetase
cal	calorie
CaMKII	calcium-calmodulin-dependent kinase II
cAMP	adenosine 3′,5′-cyclic-monophosphate
CAP	catabolite activator protein
CAT	chloramphenicol acetyltransferase
CDC	Centers for Disease Control
cDNA	complementary deoxyribonucleic acid
CDP	cytidine 5′-diphosphate
CEN	yeast centromeric sequence
cfu	colony-forming units
CGAP	Cancer Genome Anatomy Project
CHEF	contour-clamped homogeneous electric field
CHO	Chinese hamster ovary (cells)
Ci	curie
CIP	calf intestinal phosphatase
cM	centimorgan
CMP	cytidine 5′-monophosphate
CMV	cytomegalovirus
CNBr	cyanogen bromide
CoA	Coenzyme A
Con A	concanavalin A

C_0t (concentration) x (time)
cpm counts per minute
CTL cytotoxic T lymphocytes
CTP cytidine 5′-triphosphate
Cys cysteine
Da dalton
DAG diacylglycerol
dAMP deoxyadenosine monophosphate
DAPI 4′,6-diamidino-2-phenylindole
dATP deoxyadenosine triphosphate
DBD DNA binding domain
DBM diazobenzyloxymethyl
dCMP deoxycytidine monophosphate
dCTP deoxycytidine triphosphate
ddATP dideoxyadenosine triphosphate
DDBJ DNA Data Bank of Japan
ddCTP dideoxycytidine triphosphate
ddGTP dideoxyguanosine triphosphate
ddH_2O deionized distilled water
ddNTP dideoxynucleoside triphosphate
DDRT-PCR differential display RT-PCR
ddTTP dideoxythymidine triphosphate
DEAE diethylaminoethyl
DEPC diethyl pyrocarbonate
DGGE denaturing gradient gel electrophoresis
dGTP deoxyguanosine triphosphate
DHFR dihydrofolate reductase
dITP deoxyinosine 5′-triphosphate
DMEM Dulbeccos modified Eagle medium
DMF dimethylformamide
DMS dimethyl sulfate
DMSO dimethyl sulfoxide
DMT dimethoxytrityl
DNA deoxyribonucleic acid
DNase deoxyribonuclease
DNP dinitrophenol
dNTP deoxynucleoside triphosphate
DOPA L-dihydroxyphenylalanine
dpm disintegrations per minute
ds double-strand
DTT dithiothreitol
dTTP deoxythymidine triphosphate
dUMP deoxyuridine monophosphate
dUTP deoxyuridine triphosphate
EBP enhancer binding protein
EBV Epstein–Barr virus
EcR ecdysone receptor
EDTA ethylenediaminetetraacetic acid
ELISA enzyme-linked immunosorbent assay
ELSI ethical, legal, and social issues (committee)
EMBL European Molecular Biology Laboratory
EMS ethyl methanesulfonate
EMSA electrophoretic mobility shift assay
ER endoplasmic reticulum
ES embryonic stem (cells)
EST expressed sequence tag
EtBr ethidium bromide
EtOH ethanol
FACS fluorescence activated cell sorting
FAQ frequently asked question
FBS fetal bovine serum
FCS fetal calf serum
FISH fluorescence in situ hybridization
FITC fluorescein isothiocyanate
fMet *N*-formylmethionine
FOA fluoroorotic acid
FPLC fast protein liquid chromatography
FSH follicle-stimulating hormone
FTP File Transfer Protocol
GABA γ -aminobutyric acid
GAPDH glyceraldehyde-3-phosphate dehydrogenase
GCG Genetics Computer Group
GDP guanosine 5′-diphosphate
GFP green fluorescent protein
GIF graphics interchange format
Glc glucose
GlcNAc *N*-acetylglucosamine
Gln Glutamine
Glu Glutamate
Gly Glycine
GMP guanosine monophosphate
gpt guanine phosphoribosyl transferase gene
GR glucocorticoid receptor
GRAIL gene recognition and analysis Internet link
GRE glucocorticoid response element
GST glutathione S-transferase
GTP guanosine 5′-triphosphate
GUS β-glucuronidase

HAT hypoxanthine/aminopterin/thymidine media
Hb hemoglobin
HBS HEPES-buffered saline
HBSS Hanks buffered saline solution
HCL hydrochloric acid
HEK human embryonic kidney (cells)
HEPES *N*-2-hydroxyethylpiperazine-*N*′-2-ethanesulfonate
hGH human growth hormone (gene)
HGPRT hypoxanthine-guanine phosphoribosyltransferase
His histidine
HIV human immunodeficiency virus
hnRNA heterogeneous nuclear RNA
HPH hygromycin-B-phosphotransferase
HPLC high-performance liquid chromatography
HRPO horseradish peroxidase
HSV herpes simplex virus
HTML hypertext markup language
HTTP hypertext transfer protocol
IAA indole-3-acetic acid
ICM inner cell mass (embryo)
IE immediate early (viral gene)
IEF isoelectric focusing
Ig immunoglobulin
Ile isoleucine
IMAGE integrated molecular analysis genomes expression
IP_3 inositol-1,4,5-triphosphate
IPTG isopropyl-1-thio-β-D-galactoside
ISP Internet service provider
ISPCR in situ PCR
IVEC in vitro expression cloning
JIPID Japan International Protein Information Database
JPEG joint photographic expert group
kb kilobase
kcal kilocalories
KCI potassium chloride
kDa kilodalton
KLH keyhole limpet hemocyanin
KO knockout (gene disruption)
LA-PCR ligation-anchored PCR
lacZ β-galactosidase gene
LAN local area network
lck lymphocyte-specific tyrosine kinase gene
Leu leucine
LMPCR ligation-mediated PCR
LOD likelihood of the odds
LOH loss of heterozygosity
LTR long terminal repeat
luc luciferase (gene)
Lys lysine
mA milliampere
MAb monoclonal antibody
MBP maltose-binding protein
MCS multiple cloning site
MEF mouse embryo fibroblasts
MEM minimal essential medium
Met methionine
MLV murine leukemia virus
MMLV Moloney murine leukemia virus
MMTV mouse mammary tumor virus
MOI multiplicity of infection
mol mole
MOPS 3-(*N*-morpholino)propane sulfonate
MPA mycophenolic acid
mRNA messenger RNA
MTX methotrexate
MUP methylumbelliferyl phosphate
MW molecular weight
NAD nicotinamide adenine dinucleotide
NBRF National Biomedical Research Foundation
NBT nitroblue tetrazolium
NCBI National Center for Biotechnology Information
NCI National Cancer Institute
NCSA National Center for Supercomputing Applications
neo^r neomycin (G418)-resistance gene
NHGRI National Human Genome Research Institute
NIH National Institutes of Health
NMU *N*-nitroso-*N*-methyl urea
NP-40 nonidet P-40 (detergent)
OD_{260} optical density at 260 nm
ODN oligodeoxynucleotide
oligo oligonucleotide
oligo(dT) oligodeoxythymidylic acid
ONPG *O*-nitrophenyl-β-D-galactosidase

ORF open reading frame
ori origin of replication
p plasmid designation
PAC P1 bacteriophage-derived artificial chromosome
PAGE polyacrylamide gel electrophoresis
PAP peroxidase-antiperoxidase
par partition loci on plasmid DNA
PBL peripheral blood lymphocytes
PBS phosphate-buffered saline
PCR polymerase chain reaction
PDF portable document format
PEG polyethylene glycol
PEP-PCR primer-extension preamplification PCR
PFA paraformaldehyde
PFGE pulsed-field gel electrophoresis
pfu plaque-forming units
PGD preimplantation genetic diagnosis
PGK phosphoglycerate kinase
Phe phenylalanine
pI isoelectric point
PI phosphatidylinositol
PIPES piperazine-*N*,*N*′-bis(2-ethanesulfonic acid)
PMSF phenylmethylsulfonyl fluoride
PNA peanut agglutinin
poly(A)+ polyadenylated mRNA
POP post office protocol
PP_I inorganic pyrophosphate
PPP point to point protocol
Pu purine
PVC polyvinyl chloride
PVP polyvinylpyrrolidone
Py pyrimidine
QC quality control
RACE rapid amplification of cDNA ends
RAP-PCR RNA arbitrarily primed PCR
RDA representational difference analysis
RF replicative form
RFLP restriction-fragment-length polymorphism
RH radiation hybrid
RIA radioimmunoassay
RLM-RACE RNA ligase-mediated RACE
RNA ribonucleic acid
RNase ribonuclease
rRNA ribosomol RNA
RSV Rous sarcoma virus
RT-PCR reverse transcriptase-mediated PCR
RTase reverse transcriptase
RXR retinoid X receptor
SAGE serial analysis of gene expression
SAM *S*-adenosylmethionine
SDS sodium dodecyl sulfate
SEAP secreted alkaline phosphatase
Ser serine
Sf9 cell line used in baculovirus studies
SM suspension medium
SNP single nucleotide polymorphism
SpCCM solid-phase chemical cleavage
ss single-strand
SSB single-strand DNA-binding protein
SSC sodium chloride/sodium citrate (buffer)
SSCP single-strand conformational polymorphism
SSH suppression subtraction hybridization
STRP short tandem repeat polymorphism
STS sequence tagged site
$t^1/_2$ half-life
TAE Tris/acetate (buffer)
Taq *Thermus aquaticus* DNA polymerase
TBE Tris/borate/EDTA (buffer)
TBP TATA box binding protein
TBS Tris-buffered saline
TCA trichloroacetic acid
TCP/IP transmission control protocol/internet protocol
TdT terminal deoxynucleotidyl transferase
TE Tris/EDTA (buffer)
TEA triethanolamine acetate
TEAE triethylaminoethyl
TEMED *N*,*N*,*N*′,*N*′-tetramethylethylenediamine
TetR tetracycline receptor
tet^r tetracycline-resistance gene
Thr threonine
Tk thymidine kinase gene
TLC thin-layer chromatography
T_m melting temperature (midpoint denaturation)
TMAC tetramethylammonium chloride
TMB 3,3′,5,5′-tetramethylbenzidine

TMP thymidine monophosphate
tPA tissue plasminogen activator
TRAMP transgenic adenocarcinoma mouse prostate
Tris tris(hydroxymethyl)aminomethane
Tris-Cl Tris hydrochloride
tRNA transfer RNA
Trp Tryptophan
ts temperature-sensitive (mutation)
TTP thymidine 5′-triphosphate
UAS upstream activating sequence
UDP uridine 5′-diphosphate
UMP uridine 5′-monophosphate
URL uniform resource locator
UTP uridine 5′-triphosphate
UTR untranslated region (5′ or 3′)
UV ultraviolet
UWGCG University of Wisconsin Genetics Computer Group
V volts
Val valine
VNTR variable number tandem repeat
VRC vanadyl-ribonucleoside complex
VSG variant surface glycoprotein
WAIS Wide Area Information Service
WWW World Wide Web
XC xylene cyanol
X-gal 5-bromo-4-chloro-3-indolyl-β-D-galactoside
XGPRT xanthine-guanine phosphoribosyl transferase
Xyl xylose
YAC yeast artificial chromosome
YCp yeast centromeric plasmid
YEp yeast episomal plasmid
YIp yeast integrating plasmid
YPD yeast/peptone/dextrose (media)
YRp yeast replicating plasmid

APPENDIX

B

PROPERTIES OF NUCLEIC ACIDS

TABLE B.1 Properties of Nucleotides

Purines			*Pyrimidines*		
Abbreviation	Chemical name	MW	Abbreviation	Chemical name	MW
ATP	Adenosine triphosphate	507.2	CTP	Cytidine triphosphate	483.2
ADP	Adenosine diphosphate	427.2	CDP	Cytidine diphosphate	403.2
AMP	Adenosine monophosphate	347.2	CMP	Cytidine monophosphate	323.2
dATP	Deoxyadenosine triphosphate	491.2	dCTP	Deoxycytidine triphosphate	467.2
dAMP	Deoxyadenosine monophosphate	331.2	dCMP	Deoxycytidine monophosphate	307.2
GTP	Guanosine triphosphate	523.2	UTP	Uridine triphosphate	484.2
GDP	Guanosine diphosphate	443.2	UDP	Uridine diphosphate	404.2
GMP	Guanosine monophosphate	363.2	UMP	Uridine monophosphate	324.2
dGTP	Deoxyguanosine triphosphate	507.2			
dGMP	Deoxyguanosine monophosphate	427.2	TTP	Deoxythymidine triphosphate	482.2
			TMP	Deoxythymidine monophosphate	322.2

TABLE B.2 Genome Sizes of Representative Organisms

Organism	Haploid genome size in base pairs	Organism	Haploid genome size in base pairs
SV40	5,243	*Caenorabditis elegans*	8.0×10^7
φX174	5,386	*Mus musculus* (mouse)	2.7×10^9
Adenovirus 2	35,937	*Rattus norvigeticus* (rat)	3.0×10^9
Lambda	48,502	*Xenopus laevis*	3.1×10^9
Escherichia coli	4.64×10^6	*Homo sapiens*	3.3×10^9
Saccharomyces	1.21×10^7	*Zea mays*	3.9×10^9
Dictyostelium	5.4×10^7	*Nicotiana tabacum*	4.8×10^9
Arabidopsis	7.0×10^7		

TABLE B.3 RNA Facts

~1 g of eukaryotic cells yields → ~2 mg of total RNA yields → ~60 μg poly A+ mRNA (~5 × 10^9 cells)

Total RNA consists of approximately 2% poly A+ RNA, 80% ribosomal RNA (rRNA), and ~18% transfer RNA (tRNA) and small nuclear RNAs (snRNA)

The ratio of the absorbance of purified RNA at 260 nm to its absorbance at 280 nm should be in the range of 1.7–2.1. Lower values may indicate protein contamination of the RNA sample

The ratio of the absorbance of purified RNA at 260 nm to its absorbance at 230 nm should be about 2.0. Lower values may indicate carbohydrate contamination of RNA sample

TABLE B.4 Nucleotide Length of Small and Large rRNAs in Several Organisms

Organism	Small rRNA nucleotides	Large rRNA nucleotides
E. coli	1540 (16S)	2900 (23S)
S. cerevisiae	1800 (18S)	3390 (25S)
D. melanogaster	1980 (18S)	3900 (28S)
Mouse	1870 (18S)	4710 (28S)
Human	1870 (18S)	5020 (28S)

TABLE B.5 Nucleic Acid Conversion Factors

Average MW of a DNA base pair = 660 Da

1 pmol of 1000 bp DNA = 0.66 μg

1 μg of 1000 bp DNA fragment = 1.5 pmol = 3.0 pmol ends

To calculate picomoles of ends per microgram of linear double strand DNA:

$(2 \times 10^6)/(660 \text{ x number of bases})$ = pmol ends/μg DNA

In solution:

1 A_{260} unit = ~50 μg/ml of double strand DNA = 0.15 m*M*

1 A_{260} unit = ~40 μg/ml of single strand RNA = 0.11 m*M*

1 A_{260} unit = ~33 μg/ml of single strand DNA = 0.10 m*M*

To calculate the concentration of plasmid DNA in solution using absorbance at 260 nm:

(observed A_{260}) × (dilution factor) × (0.050) = DNA concentration in μg/μl

1000 bp DNA open reading frame = 333 amino acids = 37,000 Da protein

APPENDIX

C

PROPERTIES OF AMINO ACIDS

TABLE C.1 Properties of Amino Acids

Amino acid	Single letter code	Triple letter code	MW (g/mol) at pH 7	Classification
Alanine	A	Ala	89	Aliphatic
Arginine	R	Arg	174	Basic
Asparagine	N	Asn	132	Amide
Aspartic acid	D	Asp	133	Acidic
Cysteine	C	Cys	121	Sulfur
Glutamic acid	E	Glu	147	Acidic
Glutamine	Q	Gln	146	Amide
Glycine	G	Gly	75	Aliphatic
Histidine	H	His	155	Basic
Isoleucine	I	Ile	131	Aliphatic
Leucine	L	Leu	131	Aliphatic
Lysine	K	Lys	146	Basic
Methionine	M	Met	149	Sulfur
Phenylalanine	F	Phe	165	Aromatic
Proline	P	Pro	115	Imino
Serine	S	Ser	105	Hydroxyl
Threonine	T	Thr	119	Hydroxyl
Tryptophan	W	Trp	204	Aromatic
Tyrosine	Y	Tyr	181	Aromatic
Valine	V	Val	117	Aliphatic

TABLE C.2 Chemical Grouping for Conservative Amino Acid Substitution Analysis

Chemical properties of side group	Amino acids included
Neutral, weakly hydrophobic	P, A, G, S, T
Acidic, hydrophilic	Q, N, E, B, D, Z
Basic, hydrophilic	H, K, R
Hydrophobic, aliphatic	L, I, V, M
Hydrophobic, aromatic	F, Y, W
Cross-link forming	C

APPENDIX D

PROPERTIES OF COMMON RESTRICTION ENZYMES

Enzyme name	Recognition site	Reaction buffers[a]	Sites in λ	Comments
AatII	GACGT^C	A	10	Activity is pH-sensitive
AluI	AG^CT	A	143	Identifies Alu repeat sequences in humans
ApaI	GGGCC^C	A	1	Inhibited by high salt concentrations
AvaI	C^PyCGPuG	A,B	8	Can function at 45°C
BamHI	G^GATCC	A,B,C,D	5	Compatible ends; BclI, BglII, MboI, Sau3A
BclI	T^GATCA	A,B,D,E	8	Incubate 50°C, compatible ends; same as BamHI
BglI	$GCC(N)_4$^NGGC	E	29	Can cleave at Sfi sites
BglII	A^GATCT	A,B,D,E	6	Compatible ends; same as BamHI
BssHII	G^CGCGC	A,B,D,E	6	Compatible ends; MluI, most active at 50°
BstEII	G^GTNACC	A,B	13	Incubate at 60° for full activity
ClaI	AT^CGAT	A,B,D,E	15	Bsp 106I is an available isoschizomer
DdeI	C^TNAG	B,E	104	Can slowly cleave single-strand DNA
DpnI	GA^m^TG	A,B,D,E	116	Only cleaves when A is methylated
DraI	TTT^AAA	A,B,C,D	13	Sensitive to dilute reaction volumes
EagI	C^GGCCG	B	2	Compatible ends; Not I, XmaIII
EcoRI	G^AATTC	A,B,E	5	Star activity in low salt, high glycerol
EcoRV	GAT^ATC	B	21	Cleaves in *tet*^{r} gene of pBR322
HaeIII	GG^CC	C,D	149	Can cleave GGC^mC but not GG^mCC
HhaI	GCG^C	A,B,E	215	Can slowly cleave single-strand DNA
HindIII	A^AGCTT	B,D	7	Does not cleave well at DNA termini
HpaI	GTT^AAC	A	14	Some star activity in 5% glycerol
HpaII	C^CGG	C	328	Used with MspI to study methylation
Kpn	GGTAC^C	C	2	Star activity in high glycerol, high pH
MboI	^GATC	A,B,E	116	Compatible ends; same as BamHI
MluI	A^CGCGT	D	7	Compatible ends; BssHII
NaeI	GCC^GGC	A,C	1	Exhibits measurable site preference
NarI	GG^CGCC	A,B,C	1	Exhibits site preference, many compatible ends

Enzyme name	Recognition site	Reaction buffers[a]	Sites in λ	Comments
NcoI	C^CATGG	E	4	High fidelity of cleavage products
NdeI	CA^TATG	B,E	7	Unstable enzyme, sensitive to DNA impurities
NheI	G^CTAGC	A,C,D	1	Compatible ends; XbaI, AvrII, SpeI
NotI	GC^GGCCGC	E	0	Compatible ends; XmaII
NsiI	ATGCA^T	B,E	14	Compatible ends; PstI, isoschizomer of AvaIII
PacI	TTAAT^TAA	A,C	0	Useful for genomic DNA mapping
PstI	CTGCA^G	E	28	Compatible ends; NsiI, star activity
PvuI	CGAT^CG	E	3	Less active in cleaving plasmid DNA than λ
PvuII	CAG^CTG	D	15	A stable enzyme, some star activity
RsaI	GT^AC	C	113	Slowly cleaves single-strand DNA
SacI	GAGCT^C	A,C	2	Inhibited by high salt, isoschizomer is SstI
SacII	CCGC^GG	A	4	Inhibited by high salt, isoschizomer is SstII
SalI	G^TCGAC	E	2	Compatible ends; XhoI, low activity on plasmids
Sau3AI	^GATC	A,D	116	Compatible ends; same as BamHI
SfiI	$GGCC(N)_4$ ^NGGCC	A,D	0	Incubate at 50° C, Sfi sites are cut by BglI
SmaI	CCC^GGG	A	3	Incubate at 25° C, cleaves at XmaI sites
SphI	GCATG^C	B,D,E	6	Compatible ends; NlaIII
SrfI	GCCC^GGGC	A	0	Useful for genomic DNA mapping
TaqI	T^CGA	B	121	Incubate at 65° C, many compatible ends
XbaI	T^CTAGA	A,B,C,E	1	Compatible ends; same as NheI
XhoI	C^TCGAG	B,E	1	Compatible ends; SalI, PaeR71
XmaI	C^CCGGG	A,C,D	3	Sites are cleaved by SmaI

[a]An enzyme reaction buffer is listed if enzyme activity is >80% in that buffer.

Reaction buffer	pH at 37°C	Salt concentration	Mg^{2+} concentration	Tris concentration	DTT concentration
A	8.0	50 m*M* KOAc	10 m*M*	20 m*M*	0.5 m*M*
B	8.0	100 m*M* NaCl	5 m*M*	10 m*M*	1.0 m*M*
C	7.5	none	10 m*M*	10 m*M*	none
D	7.5	50 m*M* NaCl	10 m*M*	10 m*M*	none
E	7.5	100 m*M* NaCl	10 m*M*	50 m*M*	none

APPENDIX

E

E. COLI GENETIC MARKERS

Genotype	Phenotype
(φ80)	The cell carries the lambdoid prophage φ80. A defective φ80 prophage carrying the *lac* M15 deletion is present in some strains
(Mu)	Mu prophage; Mud means the phage is defective
(P1)	The cell carries a P1 prophage. Such strains express the P1 restriction system
(P2)	The cell carries the P2 prophage. This allows selection against $Red^+Gam^+\lambda$ (Spi^- selection)
ara	Mutation causes inability to use arabinose
dam	Endogenous adenine methylation at GATC sequences abolished. *dam* strains have a high recombination frequency, express DNA repair functions constitutively, and are poorly transformed by Dam-modified plasmids. Used for making DNA susceptible to cleavage by some restriction enzymes (e.g., *Bcl* I)
dcm	Endogenous cytosine methylation at CCWGG sequences abolished. Used for making DNA susceptible to cleavage by some restriction enzymes (e.g. Ava II)
dnaJ	One of several "chaperonins" is inactive. This defect has been shown to stabilize certain mutant proteins expressed in *E. coli*
dut	dUTPase activity abolished. In combination with *ung*, allows incorporation of uracil into DNA. Some procedures for oligonucleotide mutagenesis use this
e14	An excisable prophage-like element, present in K-12 but missing from many derivatives. e14 carries the *mcrA* gene among others, so $e14^-$ strains are $McrA^-$.
endA	Activity of nonspecific endonuclease I abolished, DNA preparations are thought to be of higher quality when prepared from *endA* strains
F	A low-copy number self-transmissible plasmid. F′ factors carry portions of the *E. coli* chromosome, most notably the *lac* operon and *proAB* on F′ $lac\text{-}proA^+B^+$
galK	Inability to utilize galactose
gyrA	DNA gyrase subunit A; resistance or sensitivity to nalidixic acid; presence of the *tyrA96* mutation results in resistance to naladixic acid
hflA	Results in high frequency lysogenization by λ
hsdM	*E. coli* (or EcoK) DNA methylase. Mutation blocks sequence-specific adenine methylation in the sequence A^{N6*}ACNNNNNNGTGC or GC^{N6*}ACNNNNNNGTT (*methylated). DNA isolated from a $hsdM^-$ strain will be restricted by a $HsdR^+$ host

Genotype	Phenotype
hsdR	*E. coli* (or *EcoK*) restriction endonuclease. Absence of this activity permits the introduction of DNA propagated from non-*E. coli* sources
hsdS	Specificity determinant for *hsdM* and *hsdR*. Mutation of this gene eliminates both HsdM and HsdR activity
lacI	Repressor protein of lac operon. $lacI^q$ is a mutant of *lacI* that overproduces the repressor protein. Repression is overcome by addition of IPTG to the cells
lacY	Lactose permease activity abolished
lacZ	This gene codes for β-D-galactosidase, a protein involved in lactose utilization. Cells with *lacZ* mutations produce white colonies in the presence of X-gal; wild type produces blue colonies
lacZΔM15	A specific *N*-terminal deletion which permits the α-complementation segment present on pUC-based plasmids, phagemid, or lambda vectors to make a functional LacZ protein
lon	Activity of a protease responsible for degrading aberrant proteins abolished. Some eukaryotic proteins are stabilized in *lon* strains. *E. coli* B naturally lacks Lon
malA	Inability to utilize maltose
malB	The *malB* region encompasses the genes *malEFG* and *malK lamB malM*. Δ (*malB*) deletes most or all of this region and eliminates expression of maltose binding protein (MalE)
mcrA	*E. coli* restriction system that recognizes methylated DNA of sequence 5′C*CGG (*internal cytosine methylated). Mutation in this gene prevents cleavage of this sequence
mcrCB	*E. coli* restriction system that cleaves methylated DNA of sequence 5″G^5*C,5″G^{5h}*C or 5′G^{N4}*C (*methylated cytosine). Mutations in this gene prevent restriction of these sequences. Absence of McrCB activity is important when cloning genomic DNA or methylated cDNA
mrr	*E. coli* restriction system that recognizes methylated DNA of sequence 5′-G*AC or C*AG (*methylated adenine). Mutation in this gene prevents cleavage of these sequences. Mutation also prevents McrF restriction of methylated cytosine sequences
proAB	Mutants require proline for growth in minimal media
recA	This gene is central to general recombination and DNA repair. Mutations in this gene reduce homologous recombination of DNA propagated in this strain and render the bacteria sensitive to UV light
recB	This gene product is involved in general recombination. Strains containing a mutation in both *recB* and *recJ* confer a *recA* phenotype
recD	Exonuclease activity of ExoV abolished, but recombination activity elevated. Inverted repeat sequences in λ can be propagated in *recD* strains. Plasmid replication is aberrant
recF	Plasmid-by-plasmid homologous recombination abolished
recJ	The RecJ exonuclease is involved in recombination pathways alternate to the *recA* pathways. Mutation in conjuction with *sbcC* reduces Z-DNA rearrangements. Mutations in conjuction with *recB* confer a $recA^-$ phenotype
relA	Relaxed phenotype; permits RNA synthesis in the absence of protein synthesis
rpoH	(Also known as htpR) Lack of this heat-shock transcription factor abolishes expression of some stress-induced protease activities in addition to *lon*. Some cloned proteins are more stable in *rpoH* and *sup*Cts strains at high temperature
rpsL	30S ribosomal subunit protein S12. Mutation makes cells resistant to streptomycin; also written *strA*
sbcB	Exonuclease I activity abolished. Strains carrying *recB, recC,* and *sbcB* are usually also *sbcC*. These quadruple mutant strains are recombination-proficient and propagate inverted repeats in λ and in plasmids
supC(ts)	Strains carry a thermosensitive tyrosine-inserting ochre (UAA) and amber (UAG) suppressor. Nonsense mutations in the same strain are suppressed only at low temperatures
supE	Strains carry a glutamine-inserting amber (UAG) suppressor tRNA; required for growth of some phage vectors

Genotype	Phenotype
supF	Strains carry a tyrosine-inserting amber (UAG) suppressor tRNA; required for lytic growth of S7 orS100 λ phage, such as λgt11
traD	Severely reduces the self-transmissibility of the F factor
traD36	Mutation inactivates conjugal transfer of F′ episome
tsp	Deletion eliminates a periplasmic protease that may degrade secreted or cytoplasmically overexpressed proteins after lysis
ung	Uracil *N*-glycosylase activity abolished. Uracil incorporated into DNA is removed by Ung^{+}, leaving baseless site. See *dut*
uvrC, umuC	These genes are components in *UV* repair and *SOS* repair pathways, respectively
	Mutations in these pathways reduce rearrangement of inverted repeats
xylA	Inability to utilize xylose

APPENDIX F

GEL ELECTROPHORESIS INFORMATION

TABLE F.1 Agarose Gel Percentages for Resolution of Linear DNA[a]

Gel percentage (%)	DNA size range (bp)
0.5	1,000–30,000
0.7	800–12,000
1.0	500–10,000
1.2	400–7,000
1.5	200–3,000
2.0	50–2,000

[a]The agarose concentration does not significantly alter the migration of the tracking dyes, xylene cyanol and bromophenol blue, relative to the molecular weight of DNA fragments. Therefore, over the range of agarose gel percentages shown in this table, xylene cyanol will co-migrate with ~5 kb DNA fragments and bromophenol blue will co-migrate with ~0.5 kb DNA fragments.

TABLE F.2 Polyacrylamide Gel Percentages for Resolution of DNA

Gel percentage (%)	DNA size range (bp)
3.5	100–1,000
5.0	75–500
8.0	50–400
12.0	35–250
15.0	20–150
20.0	5–100

TABLE F.3 Dye Migration in Polyacrylamide Nondenaturing Gels

Gel percentage (%)	Bromophenol blue (bp)	Xylene cyanol (bp)
3.5	100	460
5.0	65	260
8.0	45	160
12.0	20	70
15.0	15	60
20.0	12	45

TABLE F.4 Dye Migration in Polyacrylamide Denaturing Gels

Gel percentage (%)	Bromophenol blue (nucleotides)	Xylene cyanol (nucleotides)
5.0	35	140
6.0	26	106
8.0	19	75
10.0	12	55
20.0	8	28

TABLE F.5 Acrylamide Gel Percentages for Resolution of Proteins

Gel percentage (%)	Protein size range (kDa)
8	40–200
10	21–100
12	10–40

TABLE F.6 Formulas for Commonly Used Electrophoresis Buffers

Resolution of nucleic acids		***Resolution of proteins***	
Buffer	Formula	Buffer	Formula
Tris-acetate (TAE)	40 m*M* Tris-acetate 1 m*M* EDTA	Separating gel buffer	1.5 *M* Tris-HCl, pH 8.8 0.4% SDS
Tris-phosphate (TPE)	90 mM Tris-phophate 2 m*M* EDTA	Stacking gel buffer	0.5 *M* Tris-HCl, pH 6.8 0.4% SDS
Tris-borate (TBE)	45 m*M* Tris-borate 1 m*M* EDTA	Tris-glycine running buffer	25 m*M* Tris-base 192 m*M* glycine 0.1% SDS
Alkaline	50 m*N* NaOH 1 m*M* EDTA	Sample buffer	10% glycerol 2.3% SDS 62 m*M* Tris-HCl, pH 6.8

APPENDIX

G

USEFUL INTERNET RESOURCES

TABLE G.1 National Institutes of Health Web Sites

Internet address (http:/)	Description
/www.nih.gov/	National Institutes of Health home page
/www.ncbi.nlm.nih.gov/	National Center for Biotechnology Information home page
/www.ncbi.nlm.nih.gov/BLAST/	Site to initiate a NCBI BLAST homology search
/www.ncbi.nlm.nih.gov/ncicgap/	NCI Cancer Genome Anatomy Project (CGAP) home page
/www.nhgri.nih.gov/	National Human Genome Research Institute home page

TABLE G.2 Protein and Nucleic Acid Resource Web Sites

Internet address (http:/)	Description
/gdbwww.gdb.org/	The Genome Database home page
/www.genome.ad.jp/	Japan GenomeNet home page
/www.tigr.org/	The Institute for Human Genome Research home page
/www-genome.wi.mit.edu/	Massachusetts Institute of Technology Genome Research page
/shgc.stanford.edu/	Stanford University Human Genome Center page
/genome.wustl.edu/est/	Washington University-Merck Human EST project page
/www.pdb.bnl.gov/	Brookhaven National Labs Protein Database home page
/www.biochem.ucl.ac.uk/bsm/pdbsum/	Integrated protein structure database using PDB file data
/www.mdli.com/chemscape/chime/	Netscape plug-in available for viewing protein structures

TABLE G.3 Miscellaneous Web Sites Related to Applied Molecular Genetics

Internet address (http:/)	Description
/www.bio.net/	BIOSCI bio.net newsgroup home page
/www.elsevier.com/locate/tto	Elsevier Publishing Technical Tips home page
/www.cell.com/	Cell Publishing Company home page
/www.nature.com/	Nature Publishing Company home page
/www-jbc.stanford.edu/jbc/	Jounal of Biological Chemistry home page
/www.pnas.org/	Proceedings of the National Academy of Science home page
/www.sciencemag.org/	AAAS Science Magazine home page
/www.interscience.wiley.com/	John Wiley & Sons electronic journals
/www.bio.org/	Biotechnology Industry Organization home page
/www.biomednet.com/	Internet Community for Biological and Medical Researchers

INDEX